To Vic

It is a pleasure to work w[ith]

Fred

DIGITAL LOGIC
circuits and systems

Fred Hilsenrath
Bill Pierce

Delmar Publishers Inc.®

NOTICE TO THE READER

Publisher does not warrant or guarantee any of the products described herein or perform any independent analysis in connection with any of the product information contained herein. Publisher does not assume, and expressly disclaims, any obligation to obtain and include information other than that provided to it by the manufacturer.

The reader is expressly warned to consider and adopt all safety precautions that might be indicated by the activities described herein and to avoid all potential hazards. By following the instructions contained herein, the reader willingly assumes all risks in connection with such instructions.

The publisher makes no representations or warranties of any kind, including but not limited to, the warranties of fitness for particular purpose or merchantability, nor are any such representations implied with respect to the material set forth herein, and the publisher takes no responsibility with respect to such material. The publisher shall not be liable for any special, consequential or exemplary damages resulting, in whole or in part, from the readers' use of, or reliance upon, this material.

Delmar Staff
Managing Editor: Barbara A. Christie
Production Editor: Christine E. Worden

For information, address Delmar Publishers Inc.
2 Computer Drive West, Box 15–015
Albany, New York 12212

Copyright © 1988 by Delmar Publishers Inc.

All rights reserved. No part of this work covered by the copyright hereon may be reproduced or used in any form or by any means—graphic, electronic, or mechanical, including photocopying, recording, taping, or information storage and retrieval systems—without written permission of the publisher.

Printed in the United States of America
Published simultaneously in Canada
by Nelson, Canada
A division of International Thomson Limited

10 9 8 7 6 5 4 3 2 1

Library of Congress Cataloging-in-Publication Data

Hilsenrath, Fred.
  Digital logic.
  Includes index.
  1. Digital electronics. 2. Switching circuits.
3. Logic circuits. I. Pierce, Bill, 1930– .
II. Title.
TK7868.D5H54 1988   621.3815′3   87-20184
ISBN 0-8273-2475-8
ISBN 0-8273-2476-6 (instructor's guide)

This text is dedicated to Kip Sears, a gentleman and gifted professional editor who initiated this project.

contents

Preface   xiii

1 Numbering Systems or Codes   1

1–1 Introduction   1
1–2 Code Conversion   6
1–3 Code Conversion with Fractional Components   9
1–4 Binary Coded Decimal   15
1–5 The Gray Code   16
1–6 A General Method for Conversion to Any Base   20
1–7 Fractional Conversion   23
1–8 Short Form Conversion from Decimal to Any Base   25

2 Transistor Review and TTL Circuits   29

2–1 Transistor Review   29
2–2 TTL   41
2–3 A Typical TTL Inverter Circuit   43
2–4 The TTL NAND Gate   44
2–5 TTL Inputs   50
2–6 TTL Output Specifications and Fanout   51
2–7 Open Collector Output Circuit   53
2–8 Interpreting Specification Data Sheets   54
2–9 Rise Time, Fall Time, and Propagation Delay   58
2–10 Other TTL Series   60

3 Logic Elements   65

3–1 Introduction   65
3–2 The AND Circuit   65
3–3 The OR Gate   66
3–4 The NOT Function   67
3–5 The NAND Gate   67
3–6 The NOR Gate   68
3–7 The Exclusive OR Gate   68

v

3–8 The Exclusive NOR Gate 69
3–9 Logic Element Usage 70
3–10 Sum of Products from the Truth Table 73
3–11 Boolean Algebra 75
3–12 Boolean Derivations and Examples 77
3–13 De Morgan's Theorems 78
3–14 Algebraic Analysis of Existing Circuits 80
3–15 The Product of Sums 81
3–16 Other Theorems 82
3–17 Circuit Simplification 83
3–18 The Karnaugh Map 83
3–19 Complementing the Karnaugh Map 88
3–20 A Five-Variable Karnaugh Map 90
3–21 Practical Uses for the Karnaugh Map 92
3–22 The Universal NAND Gate 94
3–23 Summary 95

4 Flip Flops or Multivibrators 100

4–1 Introduction 100
4–2 The Bistable Multivibrator 100
4–3 The R/S Integrated Flip Flop 101
4–4 The NAND Gate Bistable 103
4–5 The Clocked R/S Flip Flop 104
4–6 The D Flip Flop 106
4–7 The J/K Flip Flop 107
4–8 The J/K Flip Flop with Preset & Clear 110
4–9 The Race Problem 113
4–10 The Master/Slave J/K Flip Flop 114
4–11 The Monostable Multivibrator 115
4–12 The Astable Multivibrator 118
4–13 The 555 Timer 119
4–14 The 555 as an Astable Multivibrator (A Clock) 123
4–15 The 74121 and 74123 Monostable Multivibrators 125
4–16 The Crystal Oscillator 127
4–17 The Schmitt Trigger 127

5 Counters and Latches 135

5–1 Introduction 135
5–2 The Ripple Counter 135
5–3 A Divide-by-Sixteen Ripple Counter 137
5–4 Feedback Counters for Nonstandard Moduli 139
5–5 The Ring Counter 141
5–6 The Johnson Counter 144
5–7 Illegal States 145

Contents   vii

   5–8   A Decade Shift Counter   148
   5–9   Nonstandard Modulus Obtained by Feedback   149
   5–10 Latches   150
   5–11 The *D* Latch   151
   5–12 Parallel-Loaded Latch   152
   5–13 The 74192/74193 Counter   153

6 ECL and CMOS   164

   6–1   Introduction   164
   6–2   Emitter-Coupled Logic   164
   6–3   Design Considerations for ECL   167
   6–4   TTL to ECL Logic Level Conversion   169
   6–5   The Two Families of ECL   170
   6–6   CMOS   170
   6–7   A CMOS NAND Gate   172
   6–8   CMOS Design Considerations   173
   6–9   CMOS to TTL Interfacing   175
   6–10 TTL to CMOS Interfacing   175
   6–11 ECL to CMOS Interfacing   176

7 Chip Survey and Applications   178

   7–1   Introduction   178
   7–2   Data Sheets and Specifications   179
   7–3   Gates   184
   7–4   Buffers, Drivers, and Transceivers   184
   7–5   Decoders, Encoders, Selectors, and Multiplexors   188
   7–6   Arithmetic Circuits and Processor Elements   190
   7–7   Flip Flops, Registers, and Latches   193
   7–8   Counters   194
   7–9   Shift Registers   195
   7–10 Timing Chips   197
   7–11 Memories   199
   7–12 Hybrid Devices   201
   7–13 Comparators   204
   7–14 Optoelectronic Components   206
   7–15 Logic Arrays   207

8 Arithmetic   214

   8–1   Introduction   214
   8–2   Addition   214
   8–3   The Accumulating Adder   220
   8–4   Subtraction by Complement Addition   221
   8–5   Multiplication   230
   8–6   Division   236

- 8–7 Floating Point Operations 240
- 8–8 Decimal Arithmetic 246
- 8–9 Serial Arithmetic 254
- 8–10 The Digital Comparator 258

9 Memory 265

- 9–1 Introduction 265
- 9–2 Memory Components 266
- 9–3 Memory Addressing and Data Line Interconnections 268
- 9–4 Memory Timing 279
- 9–5 Application Examples 283
- 9–6 Read-Only Memories 289
- 9–7 Dynamic Memories 295
- 9–8 Core Memories 297
- 9–9 Bubble Memory 300
- 9–10 Associative Memories 302
- 9–11 CCD Memory 303
- Chapter 9 Appendix: Data Sheets 310

10 Synchronous Sequential Logic 323

- 10–1 Introduction 323
- 10–2 Flip Flop Excitation Tables 324
- 10–3 Synchronous Counters 326

11 Sequential Logic Continued 350

- 11–1 Introduction 350
- 11–2 State Diagrams 350
- 11–3 Sequential Analysis 353
- 11–4 Control Pulse Generation 357
- 11–5 Flip Flop Conversion 374

12 D/A and A/D Converters 380

- 12–1 Introduction 380
- 12–2 Bits and Weights 380
- 12–3 A Simple D/A Converter 381
- 12–4 The Iterative Binary Ladder 383
- 12–5 Binary Ladders with Switch Isolation 385
- 12–6 Voltage Division 386
- 12–7 A BCD *D/A* Converter 388
- 12–8 A Multiplying *D/A* Converter 389
- 12–9 A Dividing *D/A* Converter 392
- 12–10 Specifications 394
- 12–11 The *A/D* Converter 394
- 12–12 The Dual-Slope Integrating *A/D* Converter 395

Contents ix

12–13 The Counter-Type A/D Converter   396
12–14 The Successive Approximation A/D Converter   398
12–15 Sample and Hold Circuits   399
12–16 Flash Converters   400
12–17 Some Performance Considerations   405

13 An Introduction to Computers   409

13–1 Introduction   409
13–2 Essential Computer Blocks   411
13–3 The Bus   411
13–4 The Open Collector Gate as a Bus Driver   411
13–5 The Tristate Gate   414
13–6 I/O Ports   417
13–7 Standard Bus Structures   418
13–8 The Keyboard   421
13–9 The Single-Board Computer LED Display   423
13–10 A Single-Board Microcomputer   426
13–11 The CPU   427
13–12 Completing the System   429
13–13 Data Format and Instruction Set   430
13–14 A Simple Program   432

14 A Survey of Computer Peripherals   434

14–1 Introduction   434
14–2 The CRT Terminal   436
14–3 The Disk Drive   440
14–4 The Disk   441
14–5 The Head   444
14–6 Access Time   446
14–7 Data Recording   446
14–8 Disk Error Correction and Detection   448
14–9 The Tape Transport   452
14–10 Tape Format   453
14–11 Data Recording   454
14–12 Data Transfer and Tape Speed   456
14–13 Tape Error Detection and Correction   456
14–14 The Modem   457
14–15 Transmission Codes   457
14–16 Printers   458

Appendix A: The Operational Amplifier   463

Appendix B: Cabling   479

Index   486

preface

Feedback from our students in digital electronics and from practitioners in industry—as well as our own industrial experience—led us to two observations, from which evolved the decision to write this book. First, available texts did not quite fit the material required for a viable classroom experience. Second, the electronics working environment has changed in the last few years and, with this change, so has occurred changes in the type and depth of knowledge an electronics designer or technician needs.

For many students, a survey of the field to gain a general background and remove the mystery hiding behind the front panel of a computer is sufficient. For those who are concerned with electronics repair, testing, and calibration, procedures given in manuals, coupled with an aquaintance with the input/output specifications of given integrated chips, are sufficient. By way of contrast, the express purpose of this text is to train students to work independently and to use their mental versatility. Many of the design and troubleshooting functions ordinarily carried out by electronics engineers are now transferred to logic designers or development laboratory technicians. Under such conditions, it becomes necessary to "think" electronics rather than follow cookbook rules.

Most engineering functions require mathematics and physics. Logic design is an exception. The required Boolean algebra can be learned quickly without any other formal background. When the rules of binary logic are understood, the game of increasingly complex functions begins. It is like a chess game. No two sequences of moves are alike; each move depends on the situation and the consequences. It can indeed be captivating. It is this excitement that we want to impart to the student. Real understanding of what goes on in an electronic network leads to the kind of creativity that makes work productive and rewarding. No watered-down version of "electronics simplified" can achieve that. Yet by its very nature, the subject is logical and explainable in simple English.

Chapters 1 through 6 acquaint the student with switching circuits and Boolean algebra. It is a new world of 1's and 0's. AND and OR functions are easily understood. Yet it is NAND and NOR functions that are most often used. This text aims to thoroughly familiarize the student with methods of design and analysis of logic functions.

The key to visualizing and simplifying the task is mapping. It has been said that computers can do this job. This is indeed true. Calculators do arithmetic very well, too, but this does not free us from learning how to add, subtract, multiply, and divide. So must the logic designer learn how to deal with Boolean algebra and Karnaugh mapping.

Once the basic tools of logic design have been covered, it is appropriate to teach applications. Arithmetic is one important subject. Chapter 8 takes the theory beyond the binary full adder to binary and BCD arithmetic, signed numbers, and floating point concepts.

Another important subject is memory. Chapter 9 covers the architecture of integrated memory chips and the design of complete memory systems using commercially available memory chips.

The more complex and challenging task of designing an electronic system is often in timing and control. The very nature of the subject requires the design of control pulses that are sequence dependent. Chapters 10 and 11 on sequential logic cover the needed algorithms that reduce complex problems to routine tasks. Combinational logic is essential, but it is not the whole story. Much of logic design is sequential in nature. At first we thought it to be too complex a subject to introduce to students at an undergraduate or community college level. However, we have found that students absorbed the material readily and were fascinated and pleased to learn it.

Chapter 12 covers the bridge between analog and digital circuits: *A/D* and *D/A* converters. They are used in data acquisition systems, control systems, and many other applications. It is a fun subject to teach and learn.

Thus, the first seven chapters, first-term material, teach the student the inner workings, function, and use of standard logic chips. Chapters 8 through 12, second-term material, expand the subject to logic systems, thereby preparing the student to tackle real problems in industry.

The final two chapters of this text provide a survey of computers and peripherals. These subjects are usually covered in depth in subsequent courses. If times permits, the instructor may choose to use the material presented to tie previous subjects together and motivate the students for more exciting things to come.

— Fred Hilsenrath

Acknowledgements

The authors are indebted to the Space Sciences department of the Lockheed Research Laboratory and the Stanford Linear Accelerator Center for helpful discussions and technical support. This textbook is the result of several in-depth reviews and subsequent revisions. The authors and the publisher are especially grateful to the following reviews.

- James Antonakos, Broome Community College
- Robert Arndt, Belleville Area College
- Russell L. Bonine, San Diego City College
- William Campas, John Tyler Community College

Preface

- Frederick Cody
- Frederick Driscoll, Wentworth Institute of Technology
- Leslie Fenical, Tampa Technical Institute
- Rick Hardman, State University of New York at Alfred
- Samuel Kraemer, Oklahoma State University
- John Newell
- Arthur Seidman
- Robert Silva, Middlesex Community College

Additional Outstanding Titles for Electronics Technology

Operational Amplifiers and Linear Integrated Circuits, Honeycutt
Electronic Communications: Systems and Circuits, Adamson
Technician's Guide to Fiber Optics, Sterling
Electronic Devices and Circuit Analysis, Pallas
Electronic Fabrication, Shimizu
Z80 Microprocessor Technology: Hardware, Software and Interfacing, Bignell & Donovan
Introduction to Digital Data Communications, Stein
Practical Electronics Troubleshooting, Perozzo
Microcomputer Troubleshooting, Perozzo
Technician's Guide to Programmable Controllers, Cox
Electronics Mathematics, Power
Modern Electronics Mathematics, Sullivan
Electronic Drafting and Printed Circuit Board Design, Kirkpatrick
PASCAL for Technicians, Gulledge
BASIC for Technicians, Gulledge
BASIC Programming for Engineers and Technicians, Guldner & Guldner
BASIC for Electronics, Silva

chapter 1

numbering systems or codes

1-1 INTRODUCTION

Chapter 1, verse 1 of most digital systems texts would most likely be: "Thou shalt know thy numbering systems." The most universally accepted system is the decimal numbering system, which has ten unique symbols, each of which describes a certain quantity of items. The names given to the symbols vary from language to language, but the symbols are universal to Western culture.

Most of you will have noticed that ten symbols were mentioned, but only nine are shown in table 1-1. If questioned, quite a few of you would give "ten" as the missing name. *Wrong!* The missing symbol is 0, which describes the absence of any items. The symbol 10 simply means that all the basic symbols have been utilized and you are starting over.

The decimal system is called a *base ten* or modulus ten system. This tells us that ten unique states are defined. The origins of the decimal system can be seen by a quick observation of the ten fingers on your two hands. Some might say that we have only eight fingers and two thumbs. If we look at it in this manner, two more numbering systems are now available. The *base eight,* or *octal,* system utilizes only the fingers and has eight unique names and symbols (0, 1, 2, 3, 4, 5, 6, 7). The base two, or binary, system would use only the thumbs and have only two symbols (0, 1) before it repeats itself.

There can be as many different base numbering systems as you wish to choose. Several other systems will be referred to in this text, and we will explain them if not in this chapter, then when they are introduced. The *hexadecimal,* or *base 16,* system follows the decimal system until the decimal system starts to repeat (at the number ten), at which point the hexadecimal (hex) system switches to the alphabet to obtain the 16 unique symbols required (0, 1, 2, 3, 4, 5, 6, 7, 8, 9, A, B, C, D, E, F).

TABLE 1-1

Quantity	Symbol	Name English	French
×	1	one	un
× ×	2	two	deux
× × ×	3	three	trois
× × × ×	4	four	quatre
× × × × ×	5	five	cinq
× × × × × ×	6	six	six
× × × × × × ×	7	seven	sept
× × × × × × × ×	8	eight	huit
× × × × × × × × ×	9	nine	neuf

EXAMPLE 1-1

Compare the four numbering systems, demonstrating the maximum number of events that can be recorded before the system repeats itself. See table 1-2.

When we have exhausted our unique symbols and still have more items to account for, we simply move one position to the left and indicate the number of times that we have cycled through our basic symbols. ∎

EXAMPLE 1-2

The decimal number 16 tells us that we have gone through the ten unique symbols one complete cycle and have advanced to six on the second round before running out of the items that we were counting.

The first digit is called the *unit digit*. The second digit to the left in the decimal system is called the *tens digit* and represents the number of times that you have counted to ten. A digit to the left of this one would indicate the number of times the tens digit has cycled and represents the number of times you have counted to one hundred. A multiple-digit number thus represents the sum of the individual numbers. ∎

EXAMPLE 1-3

The decimal number 245 indicates that the hundreds digit has been completely cycled twice (2 × 100 = 200), the tens have been cycled 4 times (4 × 10 = 40), and the unit digit has recorded five more items. The total number of items is 200 + 40 + 5 = 245.

The first or unit digit represents the number multiplying the base raised to the zero exponent. The second digit represents the number multiplying the base raised

Numbering Systems or Codes

TABLE 1-2

	Number of Items	Binary	Octal	Decimal	Hexadecimal
0	No events	0	0	0	0
1	×	1	1	1	1
2	× ×	Starts	2	2	2
3	× × ×	repeating	3	3	3
4	× × × ×		4	4	4
5	× × × × ×		5	5	5
6	× × × × × ×		6	6	6
7	× × × × × × ×		7	7	7
8	× × × × × × × ×		Starts	8	8
9	× × × × × × × × ×		repeating	9	9
10	× × × × × × × × × ×			Starts	A
11	× × × × × × × × × × ×			repeating	B
12	× × × × × × × × × × × ×				C
13	× × × × × × × × × × × × ×				D
14	× × × × × × × × × × × × × ×				E
15	× × × × × × × × × × × × × × ×				F
16	× × × × × × × × × × × × × × × ×				Starts repeating

to the one exponent. The nth digit to the left of the decimal points represents the number multiplying the base raised to the $(n - 1)$ exponent. Note that by mathematical truth, any number raised to the zero exponent is equal to 1:

$$\left[1 = \frac{x^1}{x^1} = x^1 \cdot x^{-1} = x^0 = 1 \right]$$

■

EXAMPLE 1-4

Examine the number 245 in the binary, decimal, octal, and hex numbering systems. The octal numbering system is a base 8 numbering system. There is a unique name for each of the eight states that occur in this system. The system utilizes the same names as those used by the decimal system. Since the decimal system utilizes ten unique names for the ten states, there are two unused defined states in the octal system—the numbers eight and nine. The hexadecimal system utilizes sixteen states and makes use of the decimal system for the first ten states and then changes to the alphabet for the remaining six states that are not named in the decimal system. The numbers are represented by coefficients that are an indication of the number of times the previous lesser significant digit has been cycled. The multiplication of the base raised to the proper exponent is taken for granted when the number is written. For example, in the decimal number 245 the 4 indicates that the unit digit has been cycled through four times. The 2 indicates that the tens digit has been cycled two times. This is mathematically represented by the following expression:

$$245 = 2 \times 10^2 + 4 \times 10^1 + 5 \times 10^0$$

Any number representing any base can be represented in the same manner:

$$245(\text{base } Z) = 2 \times Z^2 + 4 \times Z^1 + 5 \times Z^0$$

(a) *Binary* (base 2): the number 245 does not exist in the binary system. There are two symbols (1 and 0) in this system. Any number that contains any other symbol is not defined in this system.

(b) *Decimal* (base 10):

$$\begin{aligned} 245 = 5 \times 10^0 &= 5 \\ 4 \times 10^1 &= 40 \\ 2 \times 10^2 &= \underline{200} \\ &\ 245 \end{aligned}$$

(c) *Octal* (base 8):

$$\begin{aligned} 245 = 5 \times 8^0 &= 5 \\ 4 \times 8^1 &= 32 \\ 2 \times 8^2 &= \underline{128} \\ &\ 165 \end{aligned}$$

245 octal = 165 decimal

(d) *Hex* (base 16):

$$\begin{aligned} 245 = 5 \times 16^0 &= 5 \\ 4 \times 16^1 &= 64 \\ 2 \times 16^2 &= \underline{512} \\ &\ 581 \end{aligned}$$

245_H = 581 decimal ∎

The binary number system has only the two symbols 1 and 0. It follows that any binary number must contain only 1's and 0's.

EXAMPLE 1–5

The decimal number 7 is represented by 111 in the binary system. A complete mathematical representation of this binary expression is

$$111 = 1 \times 2^2 + 1 \times 2^1 + 1 \times 2^0$$

Numbering Systems or Codes

The decimal number 9 is represented in binary as follows:

$$1001 = 1 \times 2^3 + 0 \times 2^2 + 0 \times 2^1 + 1 \times 2^0$$ ∎

Binary numbers do follow the same convention as decimal numbers, that is, the *least significant bit* (LSB) to the right and the *most significant bit* (MSB) to the left. What is termed a digit in the decimal, octal, or hex system is called a *bit* in the binary system. The base of this system is 2, and the exponent to which the base is raised starts at 0 immediately to the left of the decimal point, if it exists, and increases by one for each step to the left and decreases by one for each step to the right. The exponent is the power to which the base is raised to give the proper weight or value of the digit or bit in question.

EXAMPLE 1–6

Evaluate the binary number 10110.

$$\begin{aligned}
\text{LSB} \rightarrow \quad 0 \cdot 2^0 &= 0 \cdot 1 = 0 \\
1 \cdot 2^1 &= 1 \cdot 2 = 2 \\
1 \cdot 2^2 &= 1 \cdot 4 = 4 \\
0 \cdot 2^3 &= 0 \cdot 8 = 0 \\
\text{MSB} \rightarrow \quad 1 \cdot 2^4 &= 1 \cdot 16 = \underline{16} \\
& \qquad\qquad\qquad 22
\end{aligned}$$

10110 binary = 22 decimal ∎

A code is weighted when the bit if present in a number indicates a certain number of events present. In the decimal number 35, the weight of 5 is $5 \times 10^0 = 5$ units; the 3 has greater weight and is $3 \times 10^1 = 30$ units. The first digit is weighted by 1, the second by 10, the third by 100, and so on.

Why don't we settle on one system or code and forget the rest? The answer is that each code has some advantages and disadvantages. The decimal code is the most familiar and the only code that is acceptable for transmitting information to the general public. So we are stuck with it. Its biggest disadvantage is that it requires ten unique states per group of ten, or *decade*. The binary system requires only two states, since it has only two unique symbols (0 and 1). The transistor is the fundamental component of all digital systems. In operation it is either on or off, and we can define one of these states as a 1 and the other as a 0. Thus the binary system is the basis for a system that utilizes transistors. The binary system is cumbersome in that it requires more bits (or digits in the other codes) than the other codes require. It takes four bits to represent the decimal number 9 (1001 = 9) for generating the 1 or 0 states of the binary system.

1-2 CODE CONVERSION

Since we all tend to think in terms of the decimal code, we use the decimal code here to check our results.

Decimal to Binary

Successive Approximation. Take a digital number (as an example, use the number 453). Find the largest exponent that 2 can be raised to before the number that results (2^x = number) exceeds the number under consideration. In our example we proceed as follows:

$$2^0 = 1$$
$$2^1 = 2$$
$$2^2 = 4$$
$$2^3 = 8$$
$$2^4 = 16$$
$$2^5 = 32$$
$$2^6 = 64$$
$$2^7 = 128$$
$$2^8 = 256$$
$$2^9 = 512$$

Notice that our example number, 453, falls between 2^8 and 2^9. It can be stated that the decimal number contains 2^8 but does not contain 2^9 or 2 raised to any higher power. Find the largest exponent of the 2 in a binary number and place a 1 in the slot assigned to that exponent.

$$2^8 \ 2^7 \ 2^6 \ 2^5 \ 2^4 \ 2^3 \ 2^2 \ 2^1 \ 2^0$$
$$1 \ \underline{} \ \underline{} \ \underline{} \ \underline{} \ \underline{} \ \underline{} \ \underline{} \ \underline{}$$

Subtract the $2^8 = 256$ from the original decimal number 453 (453 − 256 = 197). Examine the difference to determine whether it is greater than 2 to the next lower exponent (in our case, $2^7 = 128$). Since 197 is greater than 128, we can say that the decimal number 453 also contains the 2^7 component. In the location of 2^7, insert a 1:

$$2^8 \ 2^7 \ 2^6 \ 2^5 \ 2^4 \ 2^3 \ 2^2 \ 2^1 \ 2^0$$
$$1 \ 1 \ \underline{} \ \underline{} \ \underline{} \ \underline{} \ \underline{} \ \underline{} \ \underline{}$$

If the number 128 had turned out to be greater than the difference, a 0 would have been inserted in the 2^7 slot. The process is one of choosing the most significant bit and then adding the next lower significant bit. If the sum is larger than the number

Numbering Systems or Codes

that is to be converted, then that bit cannot be present in the binary representation for that number. The final result must be exactly equal to the number that is being converted.

Continuing on, we subtract the $2^7 = 128$ from the previous difference (197 − 128 = 69). This result is now compared to $2^6 = 64$ and again found to be larger. Thus the number contains the 2^6 component, and a 1 is inserted in this slot:

$2^8\ 2^7\ 2^6\ 2^5\ 2^4\ 2^3\ 2^2\ 2^1\ 2^0$
 1   1   1   —   —   —   —   —   —

The $2^6 = 64$ is subtracted from the previous difference (69) to give us the remainder 5. Comparing $2^5 = 32$ to 5, we see that, since $2^5 = 32$ is greater than our remainder, it is not a component of our number; so we insert a 0 in the appropriate slot. Since 32 is not a component of the number, we do not subtract from the previous remainder (5):

$2^8\ 2^7\ 2^6\ 2^5\ 2^4\ 2^3\ 2^2\ 2^1\ 2^0$
 1   1   1   0   —   —   —   —   —

Comparing $2^4 = 16$ to our remainder, we see that 16 is greater than the remainder 5. So we insert a 0 in the 2^4 slot. Again, no subtraction is required:

$2^8\ 2^7\ 2^6\ 2^5\ 2^4\ 2^3\ 2^2\ 2^1\ 2^0$
 1   1   1   0   0   —   —   —   —

Comparing $2^3 = 8$ to the remainder 5, we again come up with a 0 in the 2^3 slot, and no subtraction occurs:

$2^8\ 2^7\ 2^6\ 2^5\ 2^4\ 2^3\ 2^2\ 2^1\ 2^0$
 1   1   1   0   0   —   —   —   —

Comparing $2^2 = 4$ to the remainder 5 indicates that we must now insert a 1 in the 2^2 location and subtract $2^2 = 4$ from our previous remainder (5 − 4 = 1):

$2^8\ 2^7\ 2^6\ 2^5\ 2^4\ 2^3\ 2^2\ 2^1\ 2^0$
 1   1   1   0   0   0   1   —   —

Comparing $2^1 = 2$ to our new remainder (1) indicates that 2 is not a component of our number, so we insert a 0 and do not subtract:

$2^8\ 2^7\ 2^6\ 2^5\ 2^4\ 2^3\ 2^2\ 2^1\ 2^0$
 1   1   1   0   0   0   1   0   —

Finally, the remainder 1 is determined to be equal to 2^0, so a 1 is placed in the final slot, and the conversion is complete:

$$2^8 \ 2^7 \ 2^6 \ 2^5 \ 2^4 \ 2^3 \ 2^2 \ 2^1 \ 2^0$$
$$1 \ \ 1 \ \ 1 \ \ 0 \ \ 0 \ \ 0 \ \ 1 \ \ 0 \ \ 1 \ = 453 \text{ decimal}$$

The Double Dabble Method. This method consists of taking the decimal number and dividing by 2 until you reach zero. When dividing by 2, you have either a remainder of 1 (if the original number is odd) or a remainder of 0 (if the original number is even). Again using the number 453, we have

$$
\begin{array}{rl}
0 + 1 & \text{(MSB)} \\
2\overline{)1} + 1 & \\
2\overline{)3} + 1 & \\
2\overline{)7} + 0 & \\
2\overline{)14} + 0 & \\
2\overline{)28} + 0 & \\
2\overline{)56} + 1 & \\
2\overline{)113} + 0 & \\
2\overline{)226} + 1 & \text{(LSB)} \\
2\overline{)453} &
\end{array}
$$

Note that we start at the bottom and divide each quotient by 2. The binary number is the remainder with the MSB occurring at the final (top) division. The number is thus

111000101

Any way that you can develop to accomplish the conversion is satisfactory as long as you get the correct answer.

Decimal to Octal

One can use the successive approximation approach to convert a decimal to an octal number except that now you would use 8 raised to the exponents and then, by a similar technique, determine the coefficient or number multiplying the base 8^x. An easier method is to convert the decimal number to a binary number and, starting at the decimal point, group the binary bits by three. Take the individual value of each set of three bits. Note that with three bits the maximum number of combinations is 8. The combinations that are possible are

000 = 0
001 = 1

Numbering Systems or Codes

$$010 = 2$$
$$011 = 3$$
$$100 = 4$$
$$101 = 5$$
$$110 = 6$$
$$111 = 7$$

So the highest value will be $111 = 7$. Taking the decimal number $453 = 111000101$ binary as an example, we get

$$[111]\ [000]\ [101] \qquad 101 = 5$$
$$\ \ 7 \quad\ \ 0 \quad\ \ 5 \qquad\qquad 000 = 0$$
$$\qquad\qquad\qquad\qquad\ 111 = 7$$

The conversion is now complete:

$$453 \text{ decimal} = 111000101 = 705_8$$

If the grouping farthest to the left does not have three bits, fill in with 0's to the left, which add nothing to the value of the number.

Decimal to Hexadecimal

The technique for converting a decimal number to hexadecimal is similar except that the groupings are done in fours instead of threes ($2^3 = 8 =$ octal, $2^4 = 16 =$ hexadecimal). Again taking the decimal number 453 as an example, we have

$$453 = [0001][1100][0101]$$

$$0101 = 5 \text{ decimal} \ = 5 \text{ hex}$$
$$1100 = 12 \text{ decimal} = \text{C hex}$$
$$0001 = 1 \text{ decimal} \ = 1 \text{ hex}$$

The conversion is complete:

$$453 = 1C5_H = 705_8 = 111000101_{binary}$$

1–3 CODE CONVERSION WITH FRACTIONAL COMPONENTS

The conversion of numbers containing a component to the right of the decimal point, that is, a fractional component, is treated in two steps. The first step is to convert the portion to the left of the decimal in the manner previously described.

With any multiple-digit number the value of the number is the sum of the individual digits multiplied by the base raised to the proper exponent. (The first exponent to the left of the decimal point is 0 and increases with each step to the left. The exponent becomes negative on the right of the decimal point and increases negatively with each step to the right.) The evaluations of right and left sides of the decimal point are done separately. These are then reconstructed as a single number.

Decimal System Review

The decimal point is the dividing line for the exponent of the base 10. To the right the exponent becomes negative and increases by one increment for each step to the right. For example,

$$8.5 = 8 \times 10^0 + 5 \times 10^{-1}$$

From basic algebra we know that $10^{-1} = 1/10$, so $5 \times 10^{-1} = .5 = 0.5$, and

$$8.52 = 8 \times 10^0 + 5 \times 10^{-1} + 2 \times 10^{-2}$$
$$= 8 \times 1 + 5 \times 0.1 + 2 \times 0.01$$

Fractional Binary to Decimal Conversion

Examining a binary number that has components to the right of the decimal point is performed in the same manner that we used in converting our decimal numbers. The only difference is that we now have base 2 instead of base 10. For example,

10.1101

The first bit to the right of the decimal point represents the coefficient or multiplier coefficient or multiplier of the 2^{-1} term. The total solution is

$$1 \times 2^1 = 2$$
$$0 \times 2^0 = 0$$
$$1 \times 2^{-1} = 0.5$$
$$1 \times 2^{-2} = 0.25$$
$$0 \times 2^{-3} = 0$$
$$\underline{1 \times 2^{-4} = 0.0625}$$
$$2.8125 \text{ decimal}$$

Note that when you are evaluating the portion of the number to the right of the decimal point, the sum of all of these components will never be equal to or greater than 1. For the sum to equal 1 you would have to have an infinite number of 1's to the right of the decimal point.

Numbering Systems or Codes

Fractional Decimal to Binary Conversion

Probably the easiest way to accomplish the fractional decimal to binary conversion is to use the double dabble method introduced earlier. The portion of the number to the left of the decimal point is treated as before. We keep dividing the number by 2 and recording the remainder as either a 1 or a 0 until there is nothing left to divide by 2. Our binary number is the remainders that were present after each division by 2.

To use the double dabble method to convert fractional numbers, we must start with the portion of the decimal number that is to the right of the decimal point and multiply it by 2. After each multiplication by 2, the product of the multiplication is examined to determine whether it is equal to or greater than 1. If the product is greater than 1, the 1 is removed and placed in the corresponding location for that particular bit. (In this case it is the first bit to the right of the decimal point.) If the product number is less than 1, a 0 is placed in that bit location. Once the state of the bit is set, the product term is multiplied by 2 and examined in the same manner for the state of the second bit. This procedure is continued until the product becomes 1.000 (in which case you have the exact representation of your decimal number) or you have carried the process to the desired accuracy. The minus sign associated with the b term indicates that it will be to the *right* of the binary point.

---EXAMPLE 1–7

(a)

$$
\begin{array}{ll}
5.625 & 0.625 \\
\quad b_1\ b_2\ b_3 & \\
1\ 0\ 1.\ ?\ ?\ ? & \times 2 \\
1\ 0\ 1.\ 1\ ?\ ? & 1.250 \quad b_{-1} = 1 = b_1 \\
 & 0.250 \\
 & \times 2 \\
1\ 0\ 1.\ 1\ 0\ ? & 0.500 \quad b_{-2} = 0 = b_2 \\
 & \times 2 \\
1\ 0\ 1.\ 1\ 0\ 1 & 1.000 \quad b_{-3} = 1 = b_3 \\
\end{array}
$$

(b)

$$
\begin{array}{l}
3.3\ 3\ 3\ 3\ 3 \qquad .333 \\
1\ 1.0\ ?\ ?\ ?\ ?\ ?\ ? \qquad \times 2 \\
\qquad\qquad\qquad 0.666 = 11.0?\ ?\ ?\ ?\ ? \\
\qquad\qquad\qquad \times 2 \\
\qquad\qquad\qquad 1.332 = 11.01?\ ?\ ?\ ? \\
\end{array}
$$

$$\begin{array}{r}\times 2 \\ \hline 0.664 = 11.010???\\ \times 2 \\ \hline 1.328 = 11.01011??\\ \times 2 \\ \hline 0.656 = 11.010101?\\ \times 2 \\ \hline 1.312 = 11.010101\end{array}$$

As you can see, the progression in Example 1–7(b) would continue indefinitely. In this example the least significant bit is a 1. Since it is in the sixth position, the weight of this bit is $1 \times 10^{-6} = 1^1/2^6 = 1/64 = 0.015625$. Each succeeding bit will contribute one half of the previous bit. The results of Example 1–7(b) can be reconverted to a decimal number to determine the loss of accuracy caused by the limited number of fractional bits analyzed:

$$11.010101 = 1 \times 2^1 = 2.0$$
$$1 \times 2^0 \phantom{{}^{-1}} = 1.0$$
$$0 \times 2^{-1} = 0$$
$$1 \times 2^{-2} = 0.25$$
$$0 \times 2^{-3} = 0$$
$$1 \times 2^{-4} = 0.0625$$
$$0 \times 2^{-5} = 0$$
$$1 \times 2^{-6} = 0.015625 = 3.328125$$

Comparing to the 3.333333 value with which we began the conversion, we find an error of $3.333333 - 3.328125 = .005208$. If we were to have another fractional bit displayed, it would be weighted at

$$1 \times 2^{-7} = \frac{1}{128} = 0.0078$$

Notice that the error is less than half the next significant bit. This will thus give you the accuracy to which you have computed your answer. ∎

Accuracy of a Binary Number

When examining a decimal number for accuracy, we look at the least significant digit; the implied accuracy is plus or minus one half of the next lower digit value.

Numbering Systems or Codes

For example, the actual value of the number 12.3 is between 12.25 and 12.35, but it has been rounded off to 12.3. In the binary code, as we move to the right, from the most significant bit toward the least significant bit, the weight of each bit is one half the weight of the previous bit. When one stops at a particular bit, the maximum possible deviation of the true value from the displayed value would occur if all succeeding bits were to be 1's, in which case they would sum to a value that is equivalent to the last bit shown. An unequivocal statement could be made that you are accurate to within the value or weight of the last bit shown; for example, 10.1 = 2.5 can at worst be

$$10.1(111111) \cdots = 2.99999 = 2.5 + 0.5 \text{ (the LSB)}$$

Another example is

$$11.011 = 3.375 \text{ decimal}$$

In reality the true decimal value lies between 3.375 and 3.500, depending on the weight of the bits to the right of the least significant bit. Rounding off in the binary code consists of first determining the accuracy to be displayed, which will dictate the number of bits that must be displayed. For example, if we want to display a binary number that is accurate to 0.1 decimal, we first determine which bit will be less than this value (B = binary, D = decimal):

$$0.1_B = 0.5_D$$
$$0.01_B = 0.250_D$$
$$0.001_B = 0.125_D$$
$$0.0001_B = 0.0625_D$$

The fourth bit to the right of the decimal point is the first bit whose value is less than our required accuracy of 0.1. If all the bits following the fourth bit were to be 1, they would sum to 0.0625. Our accuracy is thus guaranteed. Rounding off in binary consists of observing the bit that is one step beyond the accuracy or number of bits that you wish to carry and, if it is a 1, increasing the value of the expression by 1 in the least significant bit place that is to be carried; if it is a 0, present the number as shown. In our previous example, 11.011 becomes 11.10 when rounded off to two places.

EXAMPLE 1–8

Suppose we wish to display a binary number to an accuracy of at least 0.1 decimal, and we are given the number as

11.01111010111

We now know that for 0.1 decimal accuracy the four bits to the right of the decimal point are required:

1 1.0 1 1 1 (1 0 1 0 1 1 1)

The portion in parentheses is examined, and we see that the next bit is a 1, so we must add this bit to our 1 1.0 1 1 1 to round off our binary number. Since we have not talked about binary addition, for now it will be sufficient to say that our answer is 1 1.1 0 0 0. ∎

Fractional Octal to Decimal and Hex to Decimal

The method used for the decimal and binary system is universal to all number systems. Simply replace the base ten with the appropriate base.

EXAMPLE 1-9

(a) Octal

$$12.312_8 = 1 \times 8^1 = 1 \times 8 = 8$$
$$2 \times 8^0 = 2 \times 1 = 2$$
$$3 \times 8^{-1} = \frac{3}{8} = 0.375$$
$$1 \times 8^{-2} = \frac{1}{64} = 0.0156$$
$$2 \times 8^{-3} = \frac{2}{512}(1) = 0.0039$$
$$\overline{10.394506_{decimal}} = 10.395$$

(b) Hex

$$1A.21_H = 1 \times 16^1 = 16$$
$$10 \times 16^0 = 10$$
$$2 \times 16^{-1} = \frac{2}{16} = 0.125$$
$$11 \times 16^{-2} = \frac{11}{256} = 0.0042968$$
$$1 \times 16^{-3} = \frac{1}{4096} = 0.00024414$$
$$\overline{26.129541_{decimal}}$$ ∎

Numbering Systems or Codes

Fractional Decimal to Octal/Hex Conversion

The simplest way to convert a decimal fraction to octal or hex is to do a conversion to binary and then perform the grouping.

―――――――――――――――――――――――――――――――――――――― EXAMPLE 1–10

Convert 35.625 to octal and hexadecimal. First, convert to binary (use the double dabble method):

    1 0 0 0 1 1 . 1 0 1

Perform the grouping by groups of three for octal and groups of four for hexadecimal. Zeros can be added to the end bits to get the proper number in the group:

    Octal: 100011 · 101

                2   3 · 5_{octal}

    Hex: 100011 · 101

            00100011 · 1010

             1   2 · A_{hex} ∎

1–4 BINARY CODED DECIMAL

Binary coded decimal (BCD) is a combination of the binary and the decimal codes and is used for information interfacing with computers. As was mentioned earlier, the computer operates with binary numbers. The general public gives and takes information in decimal form. The BCD code is a transition code. This code is also given a set of numbers that denote the weights of the respective bits. For example, the decimal numbers 0–9 are expressed in binary 0–9. The BCD code takes a single digit of base ten decimal and converts that digit to binary code. Since there are ten possible states in a decade, we are required to use four bits of binary per decade to represent these states. As was shown earlier, three bits have a maximum of eight states. To represent the decimal system, ten unique states are required, so four bits are required. There are 16 possible states from four bits. Only ten are used in the BCD code; the other six are termed illegal. It is an inefficient system, since four bits actually allow $2^4 = 16$ states. It follows that six states are not used and as such are undefined, illegal bit patterns. There are many BCD codes; the most common is the 8421 code, which simply weights the bits with the same weight as the binary code (LSB = 1, MSB = 8).

EXAMPLE 1–11 Convert 59 decimal to BCD

Each digit is treated separately and represented by four bits:

$$9 = 1001$$
$$5 = 0101$$
$$59 = 0101\ 1001\ (BCD)$$

Note the spacing between the two bit patterns. This is a characteristic of this code and is required to eliminate confusion with the straight binary code. ∎

Fractional numbers are treated as whole numbers with the decimal point displayed when the fractional number begins. For example, converting 59.62, we get

$$0101\quad 1001\ .\ 0110\quad 0010\ (BCD)$$
$$5\qquad 9\ .\ 6\qquad 2\quad (decimal)$$

There are other BCD codes; in fact, you can create your own, the only requirement being that you represent all ten characters and that each representation be unique. For example, in the 4221 BCD code we would have two possible choices for the numbers 2, 3, 4, 5, 6, and 7. Codes are by definition an agreement between parties on the meaning of expressions. We can solve the duality problem by simply agreeing to outlaw the second bit pattern for each of the six numbers shown. For the 4221 code the numbers 1011 and 1101 would both have the value 7. Only one number would be chosen to represent 7, and the other would be termed illegal and would not be allowed to exist in the code. Don't forget that in the 8421 BCD code the bit patterns that would normally represent 10, 11, 12, 13, 14, and 15 are also called illegal and are undefined.

1–5 THE GRAY CODE

Up until this point, all of our codes have been what we call *weighted codes*. This simply means that the position of a bit determines the *weight* or value of the component that is added when that bit is present. Again, to review the binary code, the weights of the bits start at 1 for the bit immediately to the left of the decimal point and increase by a factor of 2 for each bit position to the left. The decimal weight for each bit is as shown:

$$8\ 4\ 2\ 1\ .\ \frac{1}{2}\ \frac{1}{4}\ \frac{1}{8}$$

This weighting allows us to derive the decimal value of the number and/or the decimal equivalent of the number by simply adding those components of the weighted code that are present.

Numbering Systems or Codes

The Gray code (named in honor of its inventor) is not a weighted code. It is a code that is used to indicate such quantities as the angular position of a turning shaft or the linear position of a moving table.

To illustrate the need for the Gray code, let's look at a problem that could occur if we were using the straight binary code for indicating position in a linear table. For this example we will allow the table to have 16 positions and we will designate them 0–15 decimal, which, of course, would be 0–F for hex and 0000–1111 for binary. We require a binary display for the system. To indicate the 16 positions, we must have four switches. In position 0, all the switches are open, and we have the code

```
0 0 0 0
D C B A
```

As we move to position 1, switch A closes, and we have 0001 as our output code. At position 2, switch A would open and B would close for a 0010 output. The progression would continue through 1111, where all switches would be closed.

Since this is a mechanical system, it is impossible for all switches to operate at exactly the same instant. Let's examine what happens if we are at position 7 (0111) and moving to position 8 (1000). For this binary code, switches A, B, and C are to open, and switch D is to close. Let's stop the action at 7 just as it is about to begin the transition from 7 to 8. Remember two things: (1) the switches will not operate at exactly the same instant and (2) the mechanical motion can stop anywhere—including within the transition region.

As we enter the transition region, the 7 position has created the switch conditions of 0111. Let's assume that our switches operate in the sequence A–C–D–B. Of course, the sequence could be any of 24 different ways; we are just assuming this one for an analysis of what goes on.

If we froze the action and then moved it very slowly, we would first see A open, and our output code would read 0110 (position 6), which would of course be in error, since we are between positions 7 and 8, advancing to 8. As we move very slowly toward 8, the next switch to change state would be C, and we would read out 0010 (position 2), again an error. Moving on, D would be the next switch to change state, and we would read out 1010 (position 10), again an error. Next and last to change is B, and our indication would now be correct at 1000 (8). Recognize that, being a mechanical device, the system can and will physically stop very briefly at any and all of these locations during the change from 7 to 8. In going from 7 to 8 we could read out 7–6–2–10–8. This is an unacceptable situation for our readout—we cannot tolerate erroneous readings. The erroneous reading would then be interpreted as an erroneous location of the device. Your device would be useless, since in going from location 7 to location 8 it could and would indicate erroneously that you are in locations 6, 2, 10.

The Gray code has the very special condition that only one bit and one switch will be required to change at each position step. This eliminates the intermediate erroneous readings that were present in stepping from position 7 to position 8 in our example in which all four switches/bits would be changing in the one position step.

To derive the Gray code from the binary code, we must first define some conditions and results. First set up the binary number that is to be converted to Gray code, for example,

010110 binary

The first step is to bring down the most significant 1 and keep it in the same bit location.

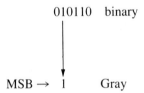

Next add this 1 to the bit that is to the right of this most significant 1. Use the convention

$$1 + 0 = 1$$
$$1 + 1 = 0$$
$$0 + 0 = 0$$
$$0 + 1 = 1$$

Place the results in the bit to the right of the first bit. For the next Gray code number, add the next two bits. Move one bit to the right each time and add the less significant bit, bringing the result down to the Gray code number.

- Step 1. Bring down the MSB 1.
- Step 2. Add the MSB 1 to the adjacent bit 0 and bring down the $(0 + 1) = 1$ Gray code.
- Step 3. Add the second MSB 0 to the third MSB 1 and bring down the $(0 + 1) = 1$ Gray code.
- Step 4. Add the third MSB 1 to the fourth MSB 1 and bring down the $(1 + 1) = 0$ Gray code.
- Step 5. Add the fourth MSB 1 to the LSB 0 and bring down the $(1 + 0) = 1$ Gray code.

Numbering Systems or Codes

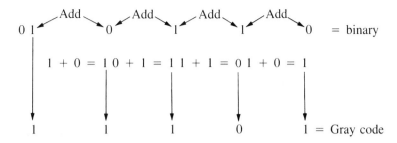

Remember that the Gray code is an unweighted code and as such must be either remembered or derived from the weighted binary code.

Returning from the Gray code to binary utilizes the same addition function. Starting with the Gray code, bring up the most significant 1

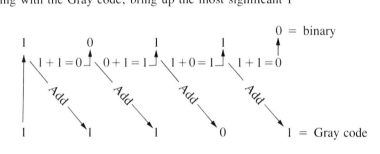

Note that the Gray code is on the bottom.

- Step 1. Bring up the MSB to the binary number.
- Step 2. This is the binary number MSB.
- Step 3. Add the binary MSB to the Gray code second MSB (1 + 1) = 0.
- Step 4. Bring the number (0) up as the binary second MSB.
- Step 5. Add the binary second MSB to the Gray code third MSB (0 + 1) = 1.
- Step 6. Bring this 1 up as the binary third MSB.
- Step 7. Add the binary third MSB to the Gray code fourth MSB (1 + 0) = 1.
- Step 8. Bring up the 1 as the binary fourth MSB.
- Step 9. Add the binary fourth MSB to the Gray code LSB (1 + 1) = 0.
- Step 10. Bring this number up as the binary LSB.

Add this binary bit to the next Gray code bit and place it as the next binary bit. Continue until all bits are accounted for.

EXAMPLE 1–12

Convert 1 0 1 0 1 1 to Gray code and return to binary:

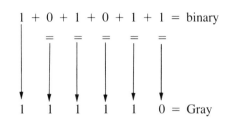

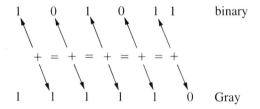

1–6 A GENERAL METHOD FOR CONVERSION TO ANY BASE

Conversion to any other base code should begin with a basic equation that represents the new base. For example, let N = any number in the decimal system and write the equivalent number in the new based system:

(General) $\quad N_{10} = A_x(Base)^x + A_{x-1}(Base)^{x-1} + \cdots + A_0(Base)^0$

(Octal) $\quad N_{10} = B_x(8)^x + B_{x-1}(8)^{x-1} + \cdots + B_0(8)^0$

(Hex) $\quad N_{10} = C_x(16)^x + C_{x-1}(16)^{(x-1)} + \cdots + C_0(16)^0$

(Binary) $\quad N_{10} = D_x(2)^x + D_{x-1}(2)^{x-1} + \cdots + D_0(2)^0$

Note that the A, B, C, and D are the coefficients that multiply the base to the proper power. In the case of the binary system the D_x's would be either 1 or 0; the C_x's could be 0, 1, 2, 3, 4, 5, 6, 7, 8, 9, A, B, C, D, E, or F; the B_x's could be 0, 1, 2, 3, 4, 5, 6, or 7; and of course the A's would depend on what base is chosen. The first exponential term x to be used can quickly be determined by raising the new base to an increasing power and comparing the result to the N_{10}. An exponent Q is chosen such that when the base is raised to Q, it exceeds the number N base 10. Q is selected by starting at 0 exponent and increasing until the number N is

Numbering Systems or Codes

exceeded. Then reduce Q by 1, and you get the highest value the base is raised to in our new base determined number.

EXAMPLE 1–13

(a) Convert 76_{10} to octal:

$(8)^0 = 1$
$(8)^1 = 8$
$(8)^2 = 64$
$(8)^3 = 512$ This is greater than 76_{10}
$Q = 3$ $x = Q - 1 = 2$

The equation will be $76_{10} = B_2 8^2 + B_1 8^1 + B_0 8^0$.

(b) Convert 28_{10} to binary:

$2^0 = 1$
$2^1 = 2$
$2^2 = 4$
$2^3 = 8$
$2^4 = 16$
$2^5 = 32$ Indicates that our first term to the right will be $D_4 2^4$

The equation will become

$$28_{10} = D_4(2)^4 + D_3(2)^3 + D_2(2)^2 + D_1(2)^2 + D_0(2)^0$$

Should you, for any reason, choose a first term that is too large, the coefficient will reduce to zero. Note that the coefficients A, B, C, and D must be defined in the code to which they are being converted. In the octal code, B cannot take on values of 8 or 9. In the binary code, D can take on only the value 0 or 1. ∎

The basis for this technique of numbers conversion is the following mathematical truth: Any equations that are equal and contain both integer and fractional components can be separated into two separate equations. Equate integer part to integer part and fractional part to fractional part. For example, if

$$2x + y + \frac{1}{x + y} = 2y + 3 + \frac{1}{8}$$

then

$$2y + 3 = 2x + y \quad \text{and} \quad \frac{1}{x + y} = \frac{1}{8}$$

The conversion procedure is as follows:

1. Set up the equation. In this example we will convert 76_{10} to octal:

$$76 = B_x^x 8^x + B_{x-1} 8^{x-1} + \cdots + B_0 8^0$$

2. Determine x:

$$8^2 = 64 \qquad 8^3 = 256$$

so $x = 2$. Then our equation is

$$76 = B_2 8^2 + B_1 8^1 + B_0 8^0 = B_2 8^2 + B_1 8^1 + B_0$$

3. Divide both sides of the equation by the base:

$$\frac{76}{8} = B_2 8^1 + B_1 8^0 + \frac{B_0}{8}$$

$$\frac{76}{8} = 9\frac{1}{2} = 9 + \frac{4}{8}$$

4. Separate the fraction and integer components of the equation:

$$9 = B_2 8^1 + B_1 8^0 = B_2 8 + B_1$$

$$\frac{4}{8} = \frac{B_0}{8} \qquad B_0 = 4$$

5. Divide the integer portion of Step 3 by the base 8:

$$\frac{9}{8} = B_2 + \frac{B_1}{8} = 1 + \frac{1}{8}$$

6. Separate integer and fractional parts:

$$B_2 = 1 \qquad B_2 = 1$$

$$\frac{B_1}{8} = \frac{1}{8} \qquad B_1 = 1$$

7. The conversion is complete:

$$\begin{array}{c} B_2 \; B_1 \; B_0 \\ \downarrow \; \downarrow \; \downarrow \end{array}$$

$$76_{10} = B_2 8^2 + B_1 8^1 + B_0 = 1 \quad 1 \quad 4_8$$

$$76_{10} = 114_8$$

Numbering Systems or Codes

EXAMPLE 1-14

Convert 28_{10} to binary:

$$2^5 = 32 \quad \text{Therefore } x = 4$$
$$28 = D_4 2^4 + D_3 2^3 + D_2 2^2 + D_1 2^1 + D_0$$
$$\frac{28}{2} = 14 + \frac{0}{2} = D_4 2^3 + D_3 2^2 + D_2 2^1 + D_1 2^0 + \frac{D_0}{2}$$
$$\frac{D_0}{2} = \frac{0}{2} \quad D_0 = 0$$
$$\frac{14}{2} = 7 + \frac{0}{2} = D_4 2^2 + D_3 2^1 + D_2 2^0 + \frac{D_1}{2}$$
$$\frac{D_1}{2} = \frac{0}{2} \quad D_1 = 0$$
$$\frac{7}{2} = 3 + \frac{1}{2} = D_4 2^1 + D_3 2^0 + \frac{D_2}{2}$$
$$\frac{D_2}{2} = \frac{1}{2} \quad D_2 = 1$$
$$\frac{3}{2} = 1 + \frac{1}{2} = D_4 2^0 + \frac{D_3}{2}$$
$$\frac{D_3}{2} = \frac{1}{2} \quad D_3 = 1$$
$$\frac{1}{2} = \frac{D_4}{2} \quad D_4 = 1$$
$$28_{10} = 1\ 1\ 1\ 0\ 0 \qquad \blacksquare$$

Investigation of this example should show that this is exactly the same procedure as the double dabble method of binary conversion.

1-7 FRACTIONAL CONVERSION

Any term containing both integer and fractional components can be separated into two different problems. The integer conversion was presented in the previous section. This section will deal only with the fractional portion of the number.

EXAMPLE 1–15 Decimal to Octal

$$0.7631_{10} = d_1 8^{-1} + d_2 8^{-2} + d_3 8^{-3} + d_4 8^{-4} + d_5 8^{-5} + \cdots$$

Multiply both sides by 8:

$$6.1048 = d_1 + d_2 8^{-1} + d_3 8^{-2} + d_4 8^{-3} + d_5 8^{-4} + \cdots$$

Separating fractional and integer components, we have

$$6 = d_1$$
$$0.1048 = d_2 8^{-1} + d_3 8^{-2} + d_4 8^{-3} + d_5 8^{-4}$$

Multiply by 8:

$$0.8384 = d_2 + d_3 8^{-1} + d_4 8^{-2} + d_5 8^{-3} + \cdots$$

Separating fractional and integer components, we have

$$0 = d_2$$
$$0.8384 = d_3 8^{-1} + d_4 8^{-2} + d_5 8^{-3} + \cdots$$

Multiply by 8:

$$6.7072 = d_3 + d_4 8^{-1} + d_5 8^{-2} + \cdots$$

Separating, we have

$$6 = d_3$$
$$0.7072 = d_4 8^{-1} + d_5 8^{-2} + \cdots$$

Multiply by 8:

$$5.6576 = d_4 + d_5 8^{-1} + \cdots$$

Separating, we have

$$5 = d_4$$
$$0.6576 = d_5 8^{-1} + \cdots$$

Multiply by 8:

$$5.2608 = d_5 + \cdots$$

Numbering Systems or Codes

$$5 = d_5$$
$$0.7631_{10} = 0.60655_8$$

Note that more terms may be obtained by continuing the process:

$$d_6 = \text{integer portion of } 0.2608 \times 8 = 2.086$$
$$2 = d_6$$
$$0.7631_{10} = 0.606552_8$$

The procedure is similar for all other base codes. ∎

1-8 SHORT FORM CONVERSION FROM DECIMAL TO ANY BASE

This technique was introduced earlier in describing the double dabble method of converting decimal to binary. The same technique will work for converting to any base. The technique is to sequentially divide the decimal number by the new base and retain the remainder as a component number of the new base.

―――――――――――――――――――――――――――――――― EXAMPLE 1-16

Convert 389 decimal to base seven:

$$\frac{389}{7} = 55 + \text{a remainder of } 4 \quad \text{(the new LSB)}$$

$$\frac{55}{7} = 7 + \text{a remainder of } 6 \quad \text{(second LSB)}$$

$$\frac{7}{7} = 1 + \text{a remainder of } 0 \quad \text{(third LSB)}$$

$$\frac{1}{7} = 0 + \text{a remainder of } 1 \quad \text{(the new MSB)}$$

The base 7 number is 1064, which is equivalent to 389 base 10. Check your answer:

$$1064_7 = 4 \times 7^0 + 6 \times 7^1 + 0 \times 7^2 + 1 \times 7^3$$
$$= 4 + 42 + 0 + 343 = 389_{10}$$

∎

The fractional or right-hand side component of decimal conversion is done in the same manner as was demonstrated for the double dabble method.

EXAMPLE 1–17

Convert 28.62_{10} to base 5. Treat this as two separate problems, the left-hand side of the decimal point as one problem and the right-hand side as another.

$$\frac{28}{5} = 5 + \text{remainder of } 3 \quad \text{(LSB)}$$

$$\frac{5}{5} = 1 + \text{no remainder} = 0$$

$$\frac{1}{5} = 0 + \text{remainder of } 1 \quad \text{(MSB)}$$

The left-hand side of the base 5 point is 103. To determine the right-hand side, we multiply by the new base:

$0.62 \times 5 = 5.10$ (5 is the first number to the right of the base 5 point)
$0.10 \times 5 = 0.50$ (0 is the next number)
$0.50 \times 5 = 2.50$ (2 is the next number)

From here on, you can see that the number will continue to repeat itself and 2 will continue on to infinity. Your desired accuracy will dictate how many of the 2's you wish to display.

The answer is the combination of the two solutions:

$$28.62_{10} = 103.502222\cdots_5$$ ∎

PROBLEMS

1. Express each of the following *decimal* numbers as the sum of coefficients times the power of 10. *Example:* $56 = 5 \times 10 + 6 \times 10$.
   (a) 24 (b) 362 (c) 5613 (d) 68.5 (e) 17.25

2. Express each of the following *octal* numbers as the sum of exponentials of 8. *Example:* $27 = 2 \times 8 + 7 \times 8$.
   (a) 31 (b) 256 (c) 3217 (d) 52.3 (e) 73.62

3. Express each of the following *hexadecimal* numbers as the sum of exponentials of 16. *Example:* $29 = 2 \times 16 + 9 \times 16$.
   (a) 43 (b) A21 (c) 6C8.B (d) ABC.DE

4. Express each of the following *binary* numbers as the sum of exponentials of 2. *Example:* $10 = 1 \times 2 + 0 \times 2$.
   (a) 101 (b) 11011 (c) 110.1 (d) 101.1011

5. Convert the following binary numbers to their decimal equivalent, octal equivalent, and hex equivalent.
   (a) 110 (b) 110101 (c) 11101.1 (d) 101.1101

Numbering Systems or Codes

6. Convert the following decimal numbers first to binary then to octal, then to hexadecimal, and also to BCD.
   (a) 58 (b) 287 (c) 139.5 (d) 250.75 (e) 22.222

7. We wish to create a new numbering system with a base 3 and given the following symbols and names:

New Name			Decimal Equivalent	New Symbol
No events	=	hit =	0	&
*	=	ya =	1	$
**	=	got =	2	@

   Convert the numbers given in decimal to this new system. Example:

		Name	Symbol
8	=	got got	@ @
(a) 10	=	_____	_____
(b) 20	=	_____	_____
(c) 15	=	_____	_____

8. Convert the following decimal numbers to BCD.
   (a) 265 (b) 1528 (c) 248.9

9. Given that the following is 8421 BCD code, convert to decimal and binary, octal, and hexadecimal.
   (a) 0101 1001 1000 (b) 1000 1001 . 0101

10. What is 1101 in BCD (8421 code) equal to in decimal?

11. Construct a table that converts the decimal numbers 0–15 first to binary and then to Gray code.

12. Convert the following binary numbers to Gray code:
    (a) 1001011 (b) 1111101 (c) 0010110

13. Convert the following Gray codes to binary:
    (a) 101011 (b) 111101 (c) 000100

14. Using the general method outlined in Section 1–6, convert:
    (a) 289_{10} to octal (b) 289_{10} to hexadecimal

15. Using the general method outlined in Sections 1–6 and 1–7, convert:
    (a) 56.8137 to octal (b) 56.8137 to hexadecimal

16. Convert 135_{10} to a base 5 number.

17. Convert the decimal number 567 to base 6. Check your answer.

18. Convert the decimal number 420.747 to base 7. Check your answer.

19. Convert the base 7 number 314.25 to decimal.

20. Why is it not possible to convert the base 6 number 362.2 to decimal?

21. Convert the decimal number 2398 to base 12. Use the hexadecimal symbols where required.
22. What procedure would you use to convert a number that is in base 5 to an equivalent number in base 6?
23. Convert the base 5 number 3421 to the equivalent base 6 number.
24. Given the hexadecimal number FAD, convert to binary, octal, decimal, and base 3.

chapter 2

transistor review and TTL circuits

2–1 TRANSISTOR REVIEW

This is an attempt to provide the student with an understanding of and feel for the operation of semiconductor diodes and transistors. The student who is intimately familiar with this subject can skip to Section 2–2. Understanding digital systems requires understanding the transistor operation in three modes: off, saturated, and double diode.

Elementary chemistry and physics describe matter in terms of atoms. The internal structures of atoms are made up of three components: protons, electrons, and neutrons. Each chemical element has a unique structure and number of these components. The protons and neutrons compose the nucleus of the atom, with the electrons in finite orbits around this nucleus. The simplest of all elements is hydrogen, which has one proton and one neutron in the nucleus and one electron in orbit. The atomic number of hydrogen is 1. The *atomic number* reflects the number of protons in the nucleus of the atom. (See figure 2–1.) As the atomic number of an element increases, the complexity of the atom increases.

The complex structure of the atom develops in an orderly manner. As the atomic number increases from 1 (hydrogen) to 2 (helium), the number of protons in the nucleus increases to two, and the number of orbiting electrons also increases to two. (See figure 2–2.) There are some basic rules that govern the developing structure of atoms:

1. The atom is always electrically neutral. Electrons are defined as having a negative charge. Protons are positively charged. Oppositely charged particles attract one another, and similarly charged particles repel one another. There are exactly the same number of electrons in orbit as there are protons in the nucleus. The net charge is zero.

FIGURE 2–1 Hydrogen atom, atomic number = 1

FIGURE 2–2 Helium atom, atomic number = 2

  2. The electrons orbit in shells. They choose spherical levels in which to orbit. Each electron orbiting in a sphere will be equidistant from the nucleus. The spheres will be at different locations with respect to the nucleus. The closer an electron is to the nucleus, the more energy is required to free it from the shell.

  3. Each orbital shell has a maximum number of electrons. The first shell can have a maximum of two electrons, the second eight, and the number increases for the shells farther out. (See figure 2–3.) However, the last shell always has a maximum of eight. This shell is called the *valence shell* and determines the chemical properties of the atom. If eight electrons are present in this shell, the element is inert or inactive; the valence shell is complete. Should the element have only seven electrons in the outer shell, it is actively seeking an electron to complete the shell. The elements that have seven electrons in the outer shell are the elements chlorine and fluorine, which are active as acids in attacking and reducing other elements, appropriating their electrons. If an element has only one electron in the outer shell, it gives up the electron rather easily. Such an element would be a good conductor of electric current, since it has electrons available for conduction. Electric current is the movement of electrons.

The elements that have their outer shell filled with eight electrons are the *inert* elements such as argon. These elements are nonconductors of electric current. The elements that have one or two free electrons in their outer shell are called *conductors*, since they readily conduct current. Some examples are gold, silver, and copper. The elements that have four electrons in the outer shell are midway between being a conductor and a nonconductor and so are called *semiconductors*.

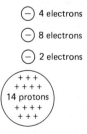

FIGURE 2–3 Silicon atom, atomic number 14

Transistor Review and TTL Circuits

The two most widely used semiconductor elements are silicon and germanium. If we take a bar of silicon and measure its resistance to electrical current flow, we find this resistance to be quite high. In pure silicon, each atom associates closely with four adjacent atoms, sharing one valence electron from each of the adjacent atoms. This sharing of a valence electron results in an apparent valence electron shell of eight electrons. This sharing of electrons and resultant complete valence shell is called a *crystal structure*. The sharing of the electrons and bonding structure is called *covalent bonding*. The resultant material is a very stable material that is a very poor electrical conductor. All valence electrons in silicon are shared by two of the atomic nuclei.

To make the element silicon into a conductor of sorts, one diffuses a material that has five electrons in its outer shell into the bar of pure silicon. Some of the elements that have five electrons in the outer shell are antimony, arsenic, and phosphorus.

Figure 2–4 illustrates the covalent bonding that occurs between silicon atoms. With this type of bonding, one silicon atom would share one of its four valence electrons with each of four adjacent atoms, giving the material the effect of having eight electrons in the outer or valence shell. With the introduction or diffusion of a valence 5 element (antimony, arsenic, or phosphorus) into the silicon, covalent bonding causes atoms of the valence 5 element to be treated as though they were silicon atoms. Since the valence 5 element has five electrons in the outer shell, and only four are accounted for in the covalent bonding, the fifth electron is free to move about; it will change the diffused silicon into a conductor of electrical current.

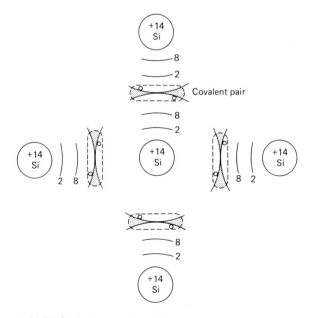

FIGURE 2–4 Covalent bonding

This type of silicon will be called *N type* (N because there are *negatively* charged electrons available for conduction). Note that the element continues to be electrically neutral in that there are exactly the same number of electrons in orbit around each nucleus as there are protons in the nucleus. The free electrons are not totally free and require an energy source to move them. (Application of an external voltage source will cause the electrons to move. They will move toward the more positively charged terminal of the voltage source, since opposite charges attract.)

It is also possible to diffuse a valence 3 element into the silicon. Boron, gallium, and indium are valence 3 elements that are used for this diffusion process. Covalent bonding again occurs; but this time instead of an extra electron being available for conduction, there is a shortage or absence of an electron. The missing electrons are called *holes*. It follows that if there are available electrons, they will be attracted to the hole locations. An electron that leaves one location to move into a hole location will, of course, leave a hole in the location that it has just left. Thus the hole appears to move in the opposite direction from the previously discussed electron movement. This material is called *P type* (P for *positively* charged material).

So electrons are attracted to the positive terminal of an energy source, whereas holes move toward the negative terminal. Electron current through a semiconductor device will flow from the negative terminal to the positive terminal; the hole current will flow from the positive terminal to the negative terminal.

Now that both the N and P type materials exist, let's cause them to be joined together in a *junction*. When this is done, strange things occur at the junction. Since one material has an excess of free electrons and the other has a series of holes that are looking for electrons, some of the electrons will leave the N type material and move toward the P type material. They will cross the junction. Up to this point, everything was electrical neutral. There were always the same number of electrons as protons in both the N and P type materials. However, as the electrons move across the junction, they do not take their charge-balancing proton from the nucleus with them. Consequently, the P type material will accumulate extra electrons and with them an extra negative charge. The more electrons cross the barrier, the higher the negative charge. This increasing negative charge will then create a repulsive force in the P type material against the electrons in the N type material. This will keep all of the electrons from jumping across the junction into the holes of the P type material. (See figure 2–5.)

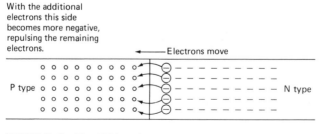

FIGURE 2–5 The PN junction

Transistor Review and TTL Circuits

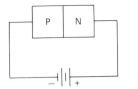

FIGURE 2–6 Reverse bias

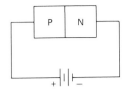

FIGURE 2–7 Forward bias

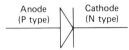

FIGURE 2–8 The diode symbol

Let us put terminals at the opposite ends of the P and N type materials and look at what happens when a voltage is applied in the two different directions. In figure 2–6 the applied voltage is such as to reinforce the negative charge associated with the P type material; any increased flow of electrons is not only discouraged but actually prevented. In figure 2–7 the positive voltage applied by the battery overcomes the negative potential that was preventing electron flow, and electrons move across the junction into the P region and onto the positive terminal of the battery. This electron flow is in the opposite direction from our defined conventional current. The junction is a *PN junction,* commonly called a *diode.* This diode will conduct current if the P terminal is made positive with respect to the N terminal. The voltage required to initiate conduction is commonly 0.7 V for silicon and 0.3 V for germanium. Figure 2–8 is the symbol for this device.

Reviewing the response of this component, we have the following features:

1. The device conducts when the *anode* (P type material) is caused to be more positive than the *cathode* (N type material) by 0.7 V for silicon and 0.3 V for germanium.
2. If the cathode is made more positive than the anode, the device acts as a blocking device for electrical current. All PN junctions respond in this manner.

The Two-Junction Device

Let us now construct a device that has two PN junctions with a common shared P type. (See figure 2–9.) If the P type material is taken to a voltage greater than 0.7 V (we are assuming that the device is silicon), each or both diodes will conduct. (See figure 2–10.) If the left-hand N type material is returned to ground, it will conduct current to the left to ground. If the right-hand side is tied to ground, it will conduct current to the right to ground. If either N side is raised to a voltage greater than ground, that side will cease to conduct. The other side that is connected to ground will continue to conduct to ground. The P terminal is a common source of conventional current, the direction of the current path depending upon which of the

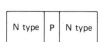

FIGURE 2-9 A two-junction device

FIGURE 2-10 Diode representation of a two-junction device

N type material sections is returned to a voltage that is 0.7 V lower than the P type material. This is not conventional transistor usage, in which one diode junction is always reverse biased. The input transistor for the standard TTL gate circuit is utilized in this manner.

More Normal Operation of the Two-Junction Device

A two-junction device as configured in the previous paragraph would normally be called an *NPN transistor* and would be used in a different manner. The symbol for the device is changed, and the operation of the circuit is totally different from that described in the previous paragraph, but the same device could be used for both types of operation.

Figure 2-11 is the symbol for an NPN transistor, and figure 2-12 is a connection arrangement for the NPN transistor that could cause it to function in a transistor mode as opposed to the dual-diode mode that was described in the previous subsection. The transistor action occurs in the following manner: The 0.7 V between base and emitter is sufficient to cause the base–emitter junction to conduct current in the forward direction, which means that the electrons are moving from the emitter into the base region. Suppose that the base region is made very small in comparison to the emitter and collector regions. Notice that the power supply voltage (V_{cc}) is in a configuration that will cause the collector-to-base junction to be *reverse biased*, that is, it would normally be in the nonconducting mode. With the application of the 0.7 base-to-emitter voltage the electrons in the N type emitter region are given sufficient energy to overcome the junction barrier potential, and they are accelerated

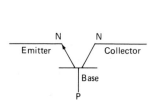

FIGURE 2-11 The NPN transistor symbol

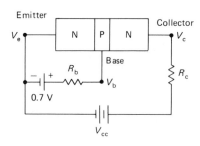

FIGURE 2-12 The way an NPN transistor works

Transistor Review and TTL Circuits

into the base region. If the base region is very small, the electrons that were accelerated into this region to recombine with one of the holes in the P region will continue on and pass into the N type collector region. Once in the collector region, if we have connected a positive voltage to the collector, the electrons will be attracted to this positive voltage and will be collected by the positive voltage V_c. With a resistor R_c in the collector circuit we can measure the voltage drop due to this collector current. A relatively small number of base electrons moving can cause a relatively large change in collector electron flow. This feature allows the transistor to be used as an amplifier. By definition, conventional current flows in the opposite direction from electron flow. Conventional current flows from the positive terminal of the V_{cc} power supply into the collector through the base region into the emitter and from there to the V_{cc} power return. The N type emitter region has available electrons that are accelerated into the small, lightly doped base region where a small percentage ($\sim 1\%$) combine with the holes in the base region and, in so doing, create I_b, the base current. The electrons that move through the base region on into the collector are attracted to the positively biased collector and create the collector current, I_c.

Some parameters that are defined for transistor operation are:

- Alpha (α): ratio of the amount of emitter current that reaches the collector (The value is always less than 1.)
- Beta (β): ratio of the collector current to the current that recombines within the base region (The value will vary from 10 to 1000; a typical value would be 100.)
- V_{ce}: voltage drop from collector to emitter, which is critical in calculating the amount of power dissipated within the transistor
- I_c: collector current
- I_b: base current
- I_e: emitter current ($I_e = I_c + I_b$)
- P_d: power dissipation within the transistor, which is equal to the product of V_{ce} and I_c
- $V_{ce\ sat}$: voltage from collector to emitter, which is the minimum voltage that will appear across the collector to emitter and still permit conduction (In this state, the base current no longer controls the collector current.)

When used in a digital circuit, the transistor is typically used either in the OFF state, in which there is no base current, or the ON state, in which there is sufficient base current to drive the transistor into saturation. For the positive logic state, the OFF state output would be a 1 and the ON state output would be a 0. Figure 2–13 is an example of a transistor generating a 0, and figure 2–14 is a transistor generating a 1.

For the sake of simplicity, the foregoing discussion on diodes and transistors did not take into account diffusion across the junction. Before resuming the dis-

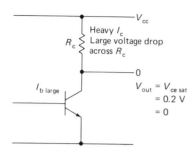

FIGURE 2-13 Zero "0" generating transistor

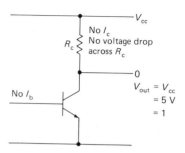

FIGURE 2-14 One "1" generating transistor

cussion on junction behavior a few comments about doped silicon are called for. Doping of pure silicon is accomplished by combining a small percentage of either valence 5 (P type) or valence 3 (N type) material with the pure valence 4 silicon. When P-doped and N-doped silicon are joined to form a diode, they are not glued together or held together with a clamp. Rather, they are fabricated out of a single piece of silicon of which one side is P-doped and the other side is N-doped. The P-doped silicon has many free holes but also a few free electrons. In P type material the holes are the *majority carriers* and electrons are the *minority carriers*. The N-doped silicon has many free electrons but also a few free holes. In N type material the electrons are called the majority carriers and the holes are called minority carriers.

The doping material is diffused into the silicon, and the boundary where the P-doping ends and the N-doping begins is a *transition region.*

It is this transition region that needs examining. What really takes place at the junction is a combination of drift and diffusion. *Drift* is the motion of electrons under the influence of an electric potential. *Diffusion* is the motion of electrons or holes under the influence of a density gradient; that is, particles concentrated in one place tend to spread to a larger area where the particular particle density is less or zero.

Visualize a man smoking a pipe. The smoke emanates from the pipe and gradually spreads through the entire room. The smoke diffuses from the source where the smoke particle density is high to areas where the smoke particle density is zero or low. (See figure 2–15.) Once the smoke particle density has equalized throughout the room, the diffusion stops.

A similar process takes place at the diode junction. Holes diffuse into the N region because the free hole density is greater in the P region than in the N region. Conversely, electrons diffuse into the P region because the free electron density is greater in the N region than in the P region. As holes migrate into the N region and electrons into the P region, they cause an *electric potential* to appear at the junction. This electric potential is positive at the N side of the junction and negative at the P side of the junction. (See figure 2–16.)

The resulting *counterpotential* tends to push the holes back into the P region and the electrons back into the N region. What really happens is that an equilibrium

Transistor Review and TTL Circuits

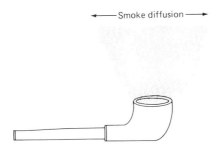

FIGURE 2-15 Pipe smoke diffusion

is eventually reached between the diffusion pressure causing migration in one direction and the counter electric potential causing drift in the opposite direction. When equilibrium is reached, the diffusion stops. In other words, the diffusion mechanism has its own built-in limits. This counterpotential cannot be measured with a voltmeter because the entire diode is still electrically neutral.

Now let us do some mental experiments. We place a battery as in figure 2-17, with a series resistor around the diode such that the positive battery terminal is connected to the P side of the diode. Notice that the battery opposes the counterpotential, thus eliminating the force that stopped the initial diffusion. Diffusion across the junction can again resume, and current will flow as long as the battery depresses the counterpotential at the junction and removes the carriers that crossed the junction.

The current across the junction is *diffusion current*. This is why the diode is called a *diffusion device* (not because the material was doped by a diffusion process). The current is mostly limited by the external resistor.

We can change the experiment by reversing the battery in figure 2-18 and examine what will happen. This time the counterpotential is reinforced and tends to push most diffused holes back into the P region and most diffused electrons back into the N region until the junction is depleted of free carriers, at which point a new equilibrium condition is reached with zero current. There is no current in this mode, since the junction is void of current carriers. Actually, there are some thermally generated carriers in the transition region at the junction that do cause some leakage current to flow, but for all practical purposes we call the current zero.

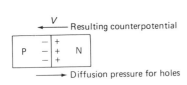

FIGURE 2-16 Diode model

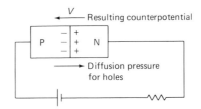

FIGURE 2-17 A diode with forward bias

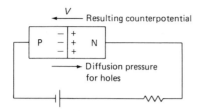

FIGURE 2–18 A diode with reverse bias

Notice, however, an important point: When the reverse battery was first connected, there was a current surge in the reverse diode direction that lasted for a brief interval until minority carriers were swept out of the transition region. In other words, minority carriers flow easily across a *reverse-biased junction* as long as they exist.

As far as the diode is concerned, we have explained current flow in the *forward* direction and the lack thereof (except for a brief initial current pulse) in the *reverse* direction.

Transistor Action

Let us try another experiment. We just talked about the brief current surge in a reverse-biased diode, which stops quickly simply because there are no carriers to allow continued current flow. Suppose we had some means to pour carriers into the depleted transition region or at least close to the junction. (See figure 2–19.) Imagine somebody pouring a bucket of free electrons close to the junction at the P side. These free electrons, minority carriers in this case (since electrons are minority carriers in the P region), will swiftly flow to the N side, and current will cease to flow when the transition region is depleted. Unless we continue to pour electrons, the current will not continue to flow.

In reality we cannot pour electrons with a bucket. We can, however, supply free carriers with a forward-biased diode on the other side of the P region. A *forward-biased diode* means an added N region as shown in figure 2–20. The

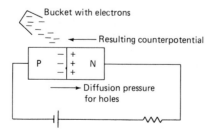

FIGURE 2–19 Reverse-bias diode with added minority carriers

Transistor Review and TTL Circuits

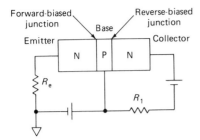

FIGURE 2-20 Properly biased transistor

majority carrier electrons from the N region flow by diffusion into the P region, where they become minority carriers. Minority carriers, as was mentioned before, flow easily across the other reverse-biased junction.

There is a condition to this process. The minority carrier electrons must be injected into the P region close to the reverse-biased junction. As the electrons enter the P region, they will recombine with holes and not constitute free carriers. The longer the path through the P region, the less chance of a free electron making it to the reverse-biased junction. Thus the P region must be very narrow.

- The N region at the forward-biased junction is called the *emitter*.
- The narrow P region is called the *base*.
- The N region at the reverse-biased junction, which receives almost all of the electrons from the emitter, is called the *collector*.

No matter how narrow the base, some of the electrons recombine with holes in the P base and are lost to the collector. This amount is usually less than 1%. Thus 99% (or more) of the emitter current is collected by the collector. The difference constitutes *base current*. To minimize hole current in the forward-biased emitter base junction, the base is less heavily doped than the emitter.

The above is an NPN transistor. Similar reasoning leads to a PNP transistor.

How do we get gain or amplification out of a transistor? The circuit of figure 2-20 is reproduced in figure 2-21 as a more conventional circuit diagram. Notice that the emitter current is solely determined by the emitter circuit, that is, it is primarily controlled by R_e, the external emitter circuit resistance.

The emitter current is almost completely collected by the collector. It is not a function (ideally) of the magnitude of the reverse-biased voltage or of the load resistor R_1. The load resistor can therefore be made arbitrarily large (of course, there are practical circuit limitations), and the voltage across the load resistor is correspondingly large.

The current in the emitter circuit can be controlled by an arbitrarily small resistor, thus producing an ever larger current. (This too has practical limitations.)

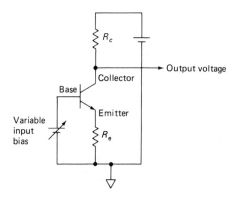

FIGURE 2-21 Transistor circuit with gain or amplification

Thus we have voltage gain.

To repeat: The gain is proportional to the load resistor R_L and inversely proportional to the emitter resistor R_e, or

$$G \sim \frac{R_c}{R_e} \tag{1}$$

Furthermore, the collector current remains constant for a given emitter circuit and is not a function of the load resistor. Thus the collector looks like a current source.

Current relationships in a transistor are simply

Emitter current − Base current = Collector current

or $\quad I_e - I_b = I_c$

The ratio of collector current to emitter current is called alpha (α):

$$\frac{I_c}{I_e} = \alpha \tag{3}$$

Combining equations (2) and (3), we get the ratio of collector current to base current, called beta (β):

$$\frac{I_c}{\alpha} - I_b = I_c$$

or $\quad I_c(1 - \alpha) = \alpha I_b$

$$\frac{I_c}{I_b} = \frac{\alpha}{1 - \alpha} = \beta \tag{4}$$

Transistor Review and TTL Circuits

A Martian Numbering System. Once upon a time, a space ship traveled to Mars. When the astronauts arrived on the surface, they found the ruins of an ancient civilization. Among the ruins was an old high school building. The astronauts entered the building and found themselves in an empty classroom. Papers were still scattered on the teacher's desk, apparently the returns of an algebra quiz. One astronaut picked up a quiz paper marked "correct." The problem stated (of course in English): Given a second-degree equation

$$ax^2 + bx + c = 0$$

where

$$a = 1$$
$$b = 23$$
$$c = 124$$

find the roots x_1 and x_2. The student's answer was

$$x_1 = 12 \quad \text{and} \quad x_2 = 7$$

The astronaut, who was well versed in algebra, did not think this was the right answer. He worked out the problem and found the correct answer to be

$$x_1 = 8.44 \quad \text{and} \quad x_2 = 12.56$$

The student's answer was not correct, yet the teacher had marked it "correct." This led the astronaut to wonder, *"How many fingers did the Martians have?"*

2–2 TTL

The TTL logic family is the oldest of the three (TTL, CMOS, ECL) and the least expensive to purchase. The most common series in this family is the 7400 series. There are many subseries such as the 74LS00, 74S00, and 74F00 series. These systems perform all types of operations, utilizing the binary numbering code, which includes only 1's and 0's. Therefore the system needs to be defined only in the 1 and the 0 states.

The TTL system was designed to work from a 5 V power source, and the 1 and 0 must be defined as voltage levels within the 0–5 V range of the power supply. In this chapter we will work with positive logic, which defines the 1 as the more positive of the two logic levels. In the TTL families a 1 level is required to be >2.0 V at the input to the circuit, and at the output of the circuit it must be >2.4 V. A 0 is required to be <0.8 V at the input and <0.4 V at the output. These voltage level requirements, or definitions, are common to all the TTL families.

The current required at the input varies for the different circuit functions and is different for the different subfamilies. By definition, current that is directed into the input terminal will be designated positive, and current that comes out of the input terminal is designated as negative. The output current polarity is defined as negative when entering the output terminal and positive when leaving the output terminal. To facilitate design and standardize specifications, a standard input has been specified and is called a *unit load* (abbreviated U.L.). A unit load for the 7400 series is as follows:

Input low voltage ≤ 0.8 V

Input low current ≤ -1.6 mA

Input high voltage ≥ 2.0 V

Input high current ≤ 40 μA

Note the difference between the input and output high and low voltage specifications. The input low is ≤ 0.8 V, whereas the output low is required to be ≤ 0.4 V. This guarantees that your output can drive your input and still have a 0.4 V margin for error. This is called the *noise margin*. The high input/output specification also has this same 0.4 V noise margin. As a designer, you would not design your input circuit such that the input low voltage is greater than 0.4 V and your input high is less than 2.4 V. Note that the minus sign indicates that the current is coming out of the input terminal and must be returned to ground. It is your responsibility as the designer to provide a return path to ground for this current. You must not allow this input current, when going through the resistance path to ground, to force the input voltage to be greater than 0.4 V. To allow this to happen would force the input voltage out of the specified low input voltage range. Since we are being very specific in our allowable range of operation, you might wonder what the circuit would do if it were operated in the unspecified input range ($0.8 \leq$ input voltage ≤ 2.0). The answer is simply that the output can be either a high or a low or any voltage in between. The manufacturer specifies only that the output be a certain level when you hold the input to ≤ 0.8 V or ≤ 2.0 V. If the device input is operated between 0.8 V and 2.0 V, the output could be either the high (≥ 2.4 V) or low (≤ 0.4 V) or any place in between. This uncertainty is unacceptable in any digital system. It is quite easy to calculate the largest value of resistance that is allowed for a 1 U.L. input:

Maximum allowed voltage = 0.4 V   (noise factor = 2)

Input low current = -1.6 mA

Ohm's Law gives us 0.4 V/1.6 mA = 250 Ω

From the above calculations, one can conclude that the maximum input drive resistance allowed for a 7400 series TTL circuit is 250 Ω.

2–3 A TYPICAL TTL INVERTER CIRCUIT

The simplest circuit that you will encounter is the simple TTL inverter circuit. This circuit will take whichever TTL input is applied to its input terminals and invert it. If a 1 is applied to the input, the output will be in the 0 state. If the input is in a 0 state, the output will be in a 1 state. The circuit that is used to perform this function is shown in figure 2–22. If the input is at 0, then Q_1 has base current

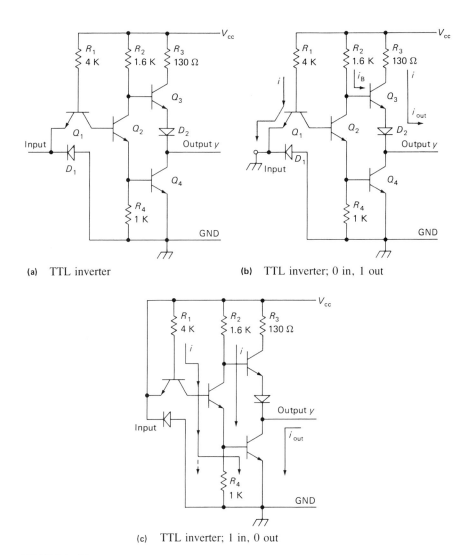

(a) TTL inverter

(b) TTL inverter; 0 in, 1 out

(c) TTL inverter; 1 in, 0 out

FIGURE 2–22

through R_1 and the base–emitter junction of Q_1. There is no path for the Q_1 collector current, so Q_2 will be held off because it has no source of base current. These bias voltages and currents will be discussed in more detail in the next section when we analyze the NAND gate. If Q_2 is held off, the base of Q_3 is returned to V_{cc} (+ 5 V) through a 1.6kΩ resistor. Since the base of Q_3 is returned to V_{cc} through the relatively low value 1.6kΩ resistor R_2, it can be assumed that Q_3 will go into saturation, making Q_3 into a good conductor and virtually connecting the emitter of Q_3 to the collector of Q_3 and through the 130 resistor to V_{cc}. Q_2 was previously determined to be OFF, and so there is no source of base current for Q_4 and Q_4 will be OFF. Since Q_3 is ON and Q_4 is OFF, the output connection is to V_{cc} through the forward conduction diode D_2 and the 130 collector resistor of Q_3. If the input line (emitter of Q_1) is returned to V_{cc}, the base current of Q_1 switches from the emitter and goes through the base collector junction into the base of Q_2, causing Q_2 to turn ON and in the process creating a source of base current for Q_4, which now conducts current and connects the output y to ground through the saturated transistor Q_4. Q_2 was previously determined to be conducting, so the base of Q_3 is no longer tied to V_{cc}. The collector current causes the base voltage to drop below the level required to cause Q_3 to conduct. Figures 2–22(b) and 2–22(c) show the direction of currents when the inverter is in the state that is indicated. The function of the diode D_2 is to force the Q_3 to be in the OFF state when the input is a 1. The diode D_1 is an input-protecting diode that limits the input voltage to a -0.7 V, thus preventing the inputs from being destroyed by large negative voltages on the input. The internal voltages of the NAND gate, which is discussed in the next section, are similar, and the derivation of current values is also pertinent for the next section.

2–4 THE TTL NAND GATE

In this section we look at a typical TTL circuit and attempt to follow the operation of the circuit. Figure 2–27 is a typical TTL NAND gate. A NAND gate is an acronym for a NOT AND gate. An AND gate requires the inputs *A and B and C and . . .* to be high before the output *y* goes high. A NAND gate requires all inputs to be high before the output goes low. To analyze this circuit, we make use of a truth table. A truth table is a complete representation of the circuit response to all possible input conditions. It becomes a table of what the output of the circuit will be for all possible combinations of the input variables. For example, suppose a device exists that has two inputs designated *A* and *B*. The output of our device is

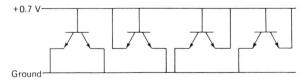

FIGURE 2–23 Four possible TTL input connections

Transistor Review and TTL Circuits

Inputs		Output
B	A	y
0	0	0
0	1	0
1	0	0
1	1	1

FIGURE 2–24 Truth table

designated as y. The device is a digital device; therefore the inputs and outputs can only assume either a 1 or a 0 state. If there are two inputs and each input has two possible states, the total number of combinations that can exist is $2^2 = 4$. If there were four inputs, each capable of assuming two states, then the total number of combinations available would be $2^4 = 16$. Figure 2–23 shows the four possible inputs.

A truth table for a two-input circuit could look like figure 2–24. This truth table would be read in the following manner: If $A = B = 0$, the output y is 0. If $A = 1$ and $B = 0$, the output y is 0. If $A = 0$ and $B = 1$, the output y is 0. If $A = B = 1$, the output y is 1. In summary, the output is 1 only when both A and B are 1. This is a truth table for a two-input AND gate.

To understand the function of each of the circuit elements in the TTL gate of figure 2–27, a short discussion of transistor circuit operation is in order. In the following discussion the NPN transistor will be used. (See figure 2–25.) Operation as a transistor requires that the base–emitter junction be forward biased while the base collector junction is reverse biased. A PN junction is forward biased when a voltage exists on the P terminal that is more positive than the voltage existing on the N terminal. In the normal mode of operation the collector base junction is reverse biased, and the base emitter junction is forward biased.

Most of the integrated circuit semiconductors that are currently available are made of silicon. To facilitate the analysis of the system, we assume that if the junction is conducting in the forward direction, it will require a 0.7 V drop across the junction. When operated in the digital system, the transistor will be operating either in or near saturation. When it is operating in saturation, the transistor has approximately a 0.2 V drop from the collector to the emitter. For circuit analysis we represent the transistor as two PN diodes with their P (anode) terminals connected. (See figure 2–26.)

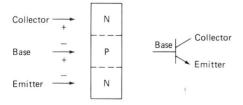

FIGURE 2–25 Transistor symbols

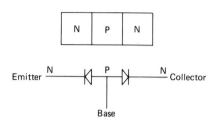

FIGURE 2–26 Transistor or dual diode

Note that if the base is held more positive than 0.7 V above the collector and emitter, the transistor can and will conduct current through both junctions to the lower potential. If the base is held at a positive potential, the current passes through the junction that has the cathode at the lower voltage.

Making use of these observations and approximations, we will now examine the operation of a typical TTL gate. (See figure 2–27.) Note that the inputs are from the left and are the connections to the two emitters of Q_1. There are two inputs, and each input has two possible values. One value is a low, or zero, or it can be connected to any voltage that is less than 0.8 V—notably ground or zero volts. The other possible input value will be a high, sometimes called a 1; in fact it will be a voltage that is >2.0 V. For increased reliability this voltage level is not generally obtained by direct connection to the V_{cc} (+5 V). It is usually connected to V_{cc} through a 1 kΩ resistor that can be shared with up to ten other similar inputs.

To examine the internal operation of the circuit, all possible input combinations must be considered. If we use the truth table, the first combination to be examined will be A = 0, B = 0. Connect both inputs of figure 2–27 to ground. With these input connections it is apparent that Q_1 will conduct through both emitter junctions to ground. The current path will be from the +5 V V_{cc} through R_1, dividing equally at the base with equal values of current passing through each of the two base–emitter junctions. Note that the two emitters are electrically isolated from each other. For this input condition the base voltage will be approximately 0.7 V. The

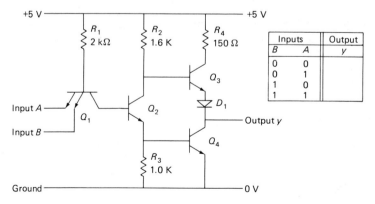

FIGURE 2–27 A TTL NAND gate, inputs open

Transistor Review and TTL Circuits

base of Q_2 is tied to the collector of Q_1. It can be stated that Q_2 is nonconducting because it would require at least 0.7 V on the base of Q_2 before it will conduct. To get 0.7 V on the base of Q_2 requires the base of Q_1 to be at least 0.7 V greater than the emitter. We have already determined that the base of Q_1 is held at 0.7 V. It follows that Q_2 is nonconducting in this state. It also follows that if Q_2 is not conducting, then Q_4 is held OFF, there being no source for the required Q_4 base current. If Q_4 is nonconducting, then we can state that the output y is not connected to ground; it is not a low or 0 state. Since we have determined that Q_2 is OFF, it can be treated as an open circuit. It follows that Q_3 has its base returned to V_{cc} through the 1.6K R_2 resistor. Sufficient base current is available to cause Q_3 to go into saturation, and to a first approximation it can be treated as a short circuit. It follows that the output y is connected to V_{cc} (+5 V) through the 130 R_4 resistor and the diode D_1. It is now apparent that for the input condition $A = B = 0$ the output y is high or 1. (See figure 2–28.) The 1 is entered into the appropriate block in the truth table of figure 2–27.

The next input condition to be examined is the $A = 1, B = 0$ combination. (See figure 2–29.) The only change to the input circuit will be that the A input will be connected to a high or 1 state. V_a will be >2.0 V. The circuit will react in the following manner: The current path is from V_{cc} through R_1 into the base of Q_1. Instead of dividing equally through the two emitters, all the current will pass through junction B to ground. The net change in the circuit voltages will be negligible. The base voltage of Q_1 will remain at 0.7 V. As a result of this voltage holding at 0.7 V, the remainder of the circuit will be in the same state as for the input conditions $A = B = 0$. It follows that the output y will be a high or 1 for this input combination. Enter a 1 in the appropriate block of the truth table of figure 2–29.

The third input combination to be examined is $A = 0, B = 1$. This input combination will result in the base current of Q_1 simply switching from the B emitter to the A emitter with no significant voltage changes within the circuit. It follows that the output y will again be high. Enter this 1 into the appropriate block of the truth table of figure 2–30.

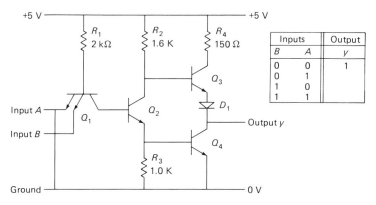

FIGURE 2–28 A TTL NAND gate; input A = ground, input B = open

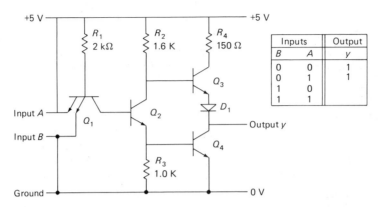

FIGURE 2–29 A TTL NAND gate; input A = high, input B = low

The fourth and last input combination to be considered is $A = B = 1$. (See figure 2–31.) For simplicity's sake we will consider the inputs as being connected to V_{cc}. For this state, both inputs are to +5 V, and the base of Q_1 would drift toward the +5 V supply. At this point, let's look at what it would require for Q_4 to conduct. The base of Q_4 would require positive input current and a voltage of 0.7 V. The Q_4 base current and voltage are supplied from the emitter of Q_2. For Q_2 to supply emitter current and a voltage of 0.7 V, Q_2 must have its base 0.7 V greater than its emitter and receive base current from the collector of Q_1. This will require the base voltage of Q_2 to be 1.4 V. For the Q_1 base–collector junction to conduct current as a forward-biased diode, the Q_1 base must be 0.7 V greater than the collector. It follows that when the base of Q_1 reaches a voltage equivalent to three diode drops (2.1 V), Q_4 will receive sufficient base current to go into saturation. Once Q_4 is in saturation, we can consider it a short circuit from collector to emitter. Under these conditions the output y is shorted to ground. Before we can enter a 0 into the truth table of figure 2–31, it must be shown that Q_3 is OFF. If Q_3 is not OFF, then there is a contest between Q_3 and Q_4 to determine the state of the output

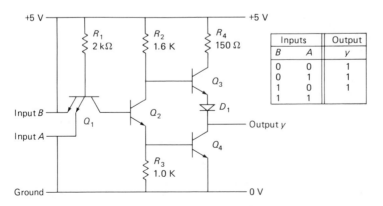

FIGURE 2–30 A TTL NAND gate; input A = low, input B = high

Transistor Review and TTL Circuits

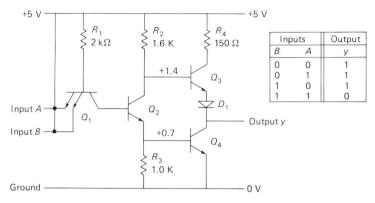

FIGURE 2–31 TTL NAND gate; both inputs high $A = B = 1$

y. We will see that the diode D_1 is required to ensure that Q_3 is held OFF when Q_4 is ON. For this analysis we shall assume that the collector-to-emitter voltage of a saturated transistor is 0.2 V. We assume that the output y is low (<0.4 V).

If Q_3 is to be ON, the base of Q_3 must have a voltage high enough to force the series combination of the diode D_1 and the base–emitter junction of Q_3 to conduct. The total voltage required will be

$$V_{\text{diode}} \ (0.7 \text{ V}) + V_{\text{be}} \text{ of } Q_3 \ (0.7 \text{ V}) + \text{Output voltage } (0.2 \text{ V}) = 1.6 \text{ V}$$

This Q_3 base voltage also happens to be the collector voltage of Q_2. The Q_2 collector voltage is the sum of the Q_2 emitter voltage (0.7 V) and the Q_2 collector-to-emitter voltage (0.2 V) = 0.9 V. This is the actual base voltage of Q_3, which is 0.7 V less than the voltage required to cause Q_3 to conduct. We can now state that Q_3 is OFF and the output y is truly a low or 0 value. The last block of the truth table of figure 2–31 can now be filled in with a 0. This is the truth table for a NAND gate.

This particular type of output circuit is quite popular and is called a *totem pole* output. One advantage to this type of output is that it is relatively fast. A disadvantage is that you cannot connect two outputs directly together. The problem is that both outputs will not always be in the same 1 or 0 state. When they are in opposite states, there will be a conflict between the Q_3 of one and the Q_4 of the other to determine the state of the output y. The result will be that the output will be in no-man's-land until one of the gates fails, causing a breakdown of the complete logic function.

You might wish to look at this in a different manner. Observe Q_1 in figure 2–32, which is an NPN transistor, and treat it as two PN diodes with a common P terminal. The Q_1 base current, which in our circuit will be the common anode current, will flow through the diode whose cathode is returned to the lower voltage.

Our previous calculations indicate that for the right-hand side of Q_1 to conduct, the voltage at the collector terminal of figure 2–32(a), the right cathode of figure 2–32(b), must be at least 1.4 V. The sum of the two base–emitter voltages is Q_2

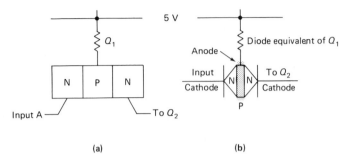

FIGURE 2–32 Double-diode representation of a transistor

and Q_4. It follows that the base voltage of figure 2–32(a) (the P region of figure 2–32(b)) must be at least 2.1 V. In a sense we are controlling the conduction condition of the right-hand side of the circuit by the value of the voltage that we connect to the left-hand side of the circuit, which we are using as our input circuit. It follows that if the input voltage is less than 1.4 V, the left-hand side of our circuit will conduct; if it is greater than 1.4 V, the right-hand side will conduct, eventually turning Q_4 ON. By convention we are limiting our input to a 0 or 1 state. A 0 state is defined as any voltage less than 0.8 V, and the Q_1 base current will be directed through our input circuitry to ground. In this case the input current must return to ground through the input circuitry. When the input is a 1, this voltage is always greater than 2.0 V, and from the previous analysis the Q_1 base current will pass through the right-hand side of figure 2–32(b), which eventually causes Q_4 to conduct and the output y to be in a 0 state.

It is important to note that for the input to be in a 0 state, you must supply a path to ground for the Q_1 base current. As was previously mentioned, this value is specified to be no greater than 1.6 mA. If you do not supply this path to ground for the Q_1 base current, the input will react as if it were connected to V_{cc}. This is why you should consider any open circuit TTL input as being in a 1 or high state. Unused inputs should never be allowed to float. Depending on the function, they should be tied directly to ground or through a resistor to V_{cc}.

2–5 TTL INPUTS

Let us focus our attention on the input circuit of the typical TTL circuit, which was discussed in the previous section. When the inputs are taken to a voltage less than 0.8 V, we are required to *sink*, or return to ground, the base current of Q_1. We have previously defined a unit load (U.L.) as -1.6 mA for $V_{in} < 0.8$ V and $+40$ µA for $V_{in} > 2.0$ V. It follows that the resistor R_1 should have a value determined by the following relationship:

$$V_{cc} - I_x R_1 - 0.7 \text{ V } (V_{be} \text{ of } Q_1) - V_{in} = 0$$

Transistor Review and TTL Circuits

We are given that $I_{max} = 1.6$ mA, and V_{in} can vary over the range 0–0.8 V. It follows that the current will be a maximum when V_{in} is 0 V. Substituting these values in the equation and solving for R_1, we get the following results:

$$5 - 1.6 \times R \text{ (in k}\Omega\text{)} - 0.7 = 0$$
$$R = 2.7 \text{ k}\Omega$$

The current required when the input is in the high or 1 state is the leakage current for the back-biased base–emitter junction of Q_1. This value is specified to be <40 µA. It is important to note that all TTL inputs will not present 1 U.L. as a load to the driving circuit. Some circuits will present a load that is greater than 1 U.L., while others will present fractional U.L. loading.

2–6 TTL OUTPUT SPECIFICATIONS AND FANOUT

The device output specification is given either as the number of unit loads that the chip is capable of driving or as the value of current the chip can source or sink. Quite often, the specification will give different U.L. values for the high and low states.

---------- EXAMPLE 2–1

The specification might read:

$$I_{output} \quad \frac{\text{HIGH/LOW}}{20/10}$$

This will indicate to the user that the I_{oh} (the current supplied or sourced by the output terminal when the output is in the high state) can be as high as 20 U.L. This specifies that the output high current can be as high as 20×40 µA, or you can draw 800 µA from this chip when the output is in the high state. The 10 listed under the low column specifies the I_{ol} (the output current that can be sinked, or returned to ground, by the output terminal when it is in the low state) can be as high as 10 U.L. This translates to a sink current capability of 10×-1.6 mA = -16 mA. ■

Fanout is a term used to describe the ability of a device to drive a similar device with the same series number.

---------- EXAMPLE 2–2

Suppose we are using the 7400 NAND gate and the specification gives the fanout as 10. The output of one of the 7400 NAND gates is capable of driving ten inputs of the 7400 gates tied in parallel. ■

Fanout is the ratio of the drive capability of the device to the input loading effect. This rating determines the capability of a device to drive equivalent devices and a means to calculate drive capacity for nonequivalent devices.

EXAMPLE 2-3

You are using a device that has a fanout rating of 10 and an input load rating of 1 U.L. The device is capable of driving ten units that have the same 1 U.L. rating. To calculate the exact current that the device can supply and the current that it can sink to ground, one simply multiplies the fanout rating times the U.L. value. In this case the positive current that it can supply will be 10×40 µA = 400 µA, and the sink current capacity will be 10×1.6 mA = 16 mA. ∎

It is highly unlikely that one would always be driving devices of the same rating as the device to be driven. If not, there are two ways to calculate the drive capacity.

EXAMPLE 2-4

We wish to use a 7400 to drive the loading inputs ($\overline{PL}$) of as many 74165 latches as possible. The U.L. input rating of the 7400 is 1, and the fanout is 10. The $\overline{PL}$ input of the 74165 has a rating of 2 U.L. The number of latches that could be driven would be 10/2 = 5. ∎

There are situations in which you must reduce the problem to basic current calculations.

EXAMPLE 2-5

The problem is to drive an LED display that requires 32 mA of current through the device to provide sufficient illumination. How could a 7400 NAND gate be connected to drive this LED? The current available from one gate is determined by multiplying fanout times the input rating. The positive current available (out of the output terminal) is

$$40 \text{ µA} \times 10 = 400 \text{ µA} \quad \text{per gate}$$
$$400 \text{ µA} \times 4 = 1.6 \text{ mA} \quad \text{per chip}$$

Obviously, there is no way in which the device can supply sufficient positive or output current from a chip to illuminate this LED. The negative current available (sinked to ground, into the output terminal) is

$$1.6 \text{ mA} \times 10 = 16 \text{ mA} \quad \text{per gate (insufficient)}$$
$$16 \text{ mA} \times 4 = 64 \text{ mA} \quad \text{per chip}$$

Transistor Review and TTL Circuits

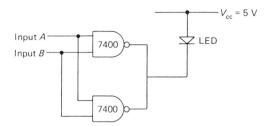

FIGURE 2-33 Undesirable solution for paralleling outputs

Two gates connected in parallel would be sufficient. The circuit is shown in figure 2-33. ∎

The circuit in figure 2-33 should never be used; owing to the nonequal switching times of the gates, there would be finite times when the totem pole outputs would be in the opposite states, and they would be competing for control of the output. This is a very undesirable state and can be avoided by using the open collector output device.

2-7 OPEN COLLECTOR OUTPUT CIRCUIT

The output circuit shown in our typical TTL gate is a totem pole output. This type of circuit has relatively good drive capability, but it is important to note that the outputs cannot be connected directly together (hard wired). If we eliminate Q_3 and D_1 from the TTL output circuit and keep the collector tied to the output lead, the circuit will be called an *open collector output*. The advantage of this output is that it can be hard wired to any number of similar outputs. In this output the transistors are either full ON or OFF. If we treat them as closed switches when they are ON and open switches when they are OFF, it is obvious that there is no conflict between transistors if there are many outputs hard wired together. It takes only one being in the ON state to take the common tie point to ground. The other transistors are quite safely operated in the OFF mode with 0 volts on their collectors. This open collector (O.C.) output works quite well when the output transistor is turned ON and it will respond rapidly with good drive capability because the transistor is a low-impedance device when it is turned ON. When you are driving any device, there is some capacitance associated with the device and/or the connecting lines. Capacitors can be charged more rapidly from low-impedance sources. When the output is switched from a 0 to a 1 state, the output transistor is turned OFF. Any device that is being driven by this circuit will have to depend on external or leakage paths to charge the previously mentioned circuit capacitance.

In using open collector outputs a current path to V_{cc} *must* be provided. An externally connected resistor that provides this current path is called a *pull-up resistor*. To calculate the minimum value of this pull-up resistor, we must know

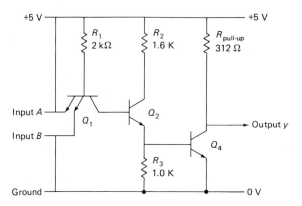

FIGURE 2–34 Open collector circuit with pull-up resistor

the current sinking capability of the output circuit. Suppose we are given an output low spec of 10 U.L. We first convert this to 16 mA. The minimum output voltage that we can anticipate is 0 V. The full 5 V will be across this pull-up resistor, and the value will be

$$R_{\text{pull up}} = \frac{5 \text{ V}}{16 \text{ mA}} = 312 \text{ }\Omega$$

Figure 2–34 is the schematic of an open collector output with the minimum pull-up resistor shown.

2–8 INTERPRETING SPECIFICATION DATA SHEETS

The specification sheets issued by the manufacturers of these devices are intended to make available to you all the information that you would normally need to design the devices into your circuit. Specifications are often not well presented and may require judicial interpretation by the user. Do not be surprised if you sometimes come away from reading a spec with the feeling that the writer is an unemployed political speech writer who is working on a mystery novel. Figure 2–35 is a copy of a reasonable specification for the 7400 NAND gate that we discussed previously.

Starting at the top left-hand side of the specification, we see that this spec represents a total of eight different chips. The 5400, 7400, 54H00, 74H00, 54S00, 74S00, 54LS00, and 74LS00 chips are all specified by this one spec sheet. The 54 prefix indicates that the unit is manufactured to meet the more stringent demands of the military. From the block that carries the commercial specifications and the adjacent block that carries the military spec, the difference between the two specifications can be determined. The military version (54 prefix) has a wider tolerance on the +5 V V_{cc}. The 54 series is guaranteed to operate over a variation of the V_{cc} of 1 V. It can vary from 4.5 to 5.5 V, and the devices must continue to function

Transistor Review and TTL Circuits

54/7400
54H/74H00
54S/74S00
54LS/74LS00
QUAD 2-INPUT NAND GATE

CONNECTION DIAGRAMS
PINOUT A

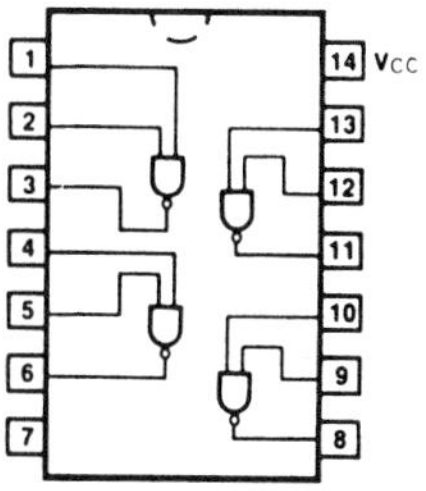

PINOUT B

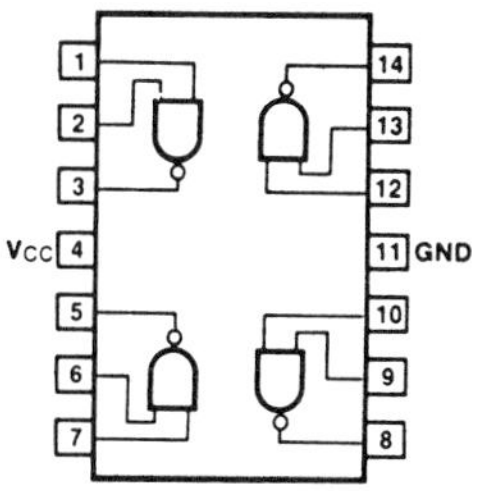

ORDERING CODE: See Section 9

PKGS	PIN OUT	COMMERCIAL GRADE $V_{CC} = +5.0$ V ±5%, $T_A = 0°$C to $+70°$C	MILITARY GRADE $V_{CC} = +5.0$ V ±10%, $T_A = -55°$C to $+125°$C	PKG TYPE
Plastic DIP (P)	A	7400PC, 74H00PC 74LS00PC, 74S00PC		9A
Ceramic DIP (D)	A	7400DC, 74H00DC 74LS00DC, 74S00DC	5400DM, 54H00DM 54LS00DM, 54S00DM	6A
Flatpak (F)	A	74LS00FC, 74S00FC	54LS00FM, 54S00FM	3I
	B	7400FC, 74H00FC	5400FM, 54H00FM	

INPUT LOADING/FAN-OUT: See Section 3 for U.L. definitions

PINS	54/74 (U.L.) HIGH/LOW	54/74H (U.L.) HIGH/LOW	54/74S (U.L.) HIGH/LOW	54/74LS (U.L.) HIGH/LOW
Inputs	1.0/1.0	1.25/1.25	1.25/1.25	0.5/0.25
Outputs	20/10	12.5/12.5	25/12.5	10/5.0 (2.5)

DC AND AC CHARACTERISTICS: See Section 3*

SYMBOL	PARAMETER	54/74 Min Max	54/74H Min Max	54/74S Min Max	54/74LS Min Max	UNITS	CONDITIONS	
I_{CCH}	Power Supply	8.0	16.8	16	1.6	mA	V_{IN} = Gnd	V_{CC} = Max
I_{CCL}	Current	22	40	36	4.4		V_{IN} = Open	
t_{PLH}	Propagation Delay	22	10	2.0 4.5	10	ns	Figs. 3-1, 3-4	
t_{PHL}		15	10	2.0 5.0	10			

*DC limits apply over operating temperature range; AC limits apply at $T_A = +25°$C and $V_{CC} = +5.0$ V.

FIGURE 2–35 Special specification sheet (Courtesy of Fairchild Semiconductor Corporation)

within specifications. The commercial unit (74 series) is guaranteed to operate only within specs for a V_{cc} variation from 4.75 to 5.25 V. The commercial series is guaranteed only to be within specifications over the temperature range of 0°C to 70°C. The military version of the same chip is specified to operate over the temperature range of -55 to $+125$°C. Obviously, the military version of the chip should be more expensive. Generally speaking, a military version can be used in place of a commercial unit, but it would not be acceptable to replace a military version with a commercial version of the same chip.

The upper and middle right-hand side of the specification shows the pin assignments for the two possible packages. The pinout A is the more popular of the two. This assignment is for the dual inline package (abbreviated DIP). Note that the ground pin is assigned to pin 7 and V_{cc} is assigned to pin 14. This is standard for 90% of the 14-pin DIP packages. The pin assignment for the 16-pin DIP package is ground on pin 8 and V_{cc} on pin 16. Again this is for 90% of the available chips. Of course, this requires that you check specifications when using an unfamiliar chip to be sure that the chip is not part of the other 10%. Pinout B is used for the flatpak version of this chip, and only then for the 74H00FC and the 7400FC versions of the device. The flatpak package is a smaller profile that can be designed into the circuit for more efficient space utilization. The DIP package usually plugs into a receptacle or socket and can be easily replaced. The flatpak sets right onto the printed circuit board and can be soldered into place.

Figure 2–35 lists three types of packages. Observe the column listed under PKGS at the left center of the spec. The three packages that are available for this chip are plastic, ceramic, and flatpak. The manufacturer will print the letter P for plastic, D for ceramic, or F for flatpak. The flatpak package is quite different from the DIP package, and you will have no difficulty in recognizing it. The plastic package is less expensive than ceramic and not considered to be as reliable as the ceramic package over the lifetime of the units.

The second column adjacent to the PKGS column is called PINOUT and refers you to the connection diagrams that are located in the upper and center right-hand side of the page.

The next two column headings differentiate between the commercial and military versions of the same chip. The variations allowed in the power supply and ambient temperatures are listed directly under the respective headings. Note that T_A stands for ambient or surrounding temperature as opposed to actual chip temperature, which will be considerably higher than T_A. How much higher it will be will depend on the power dissipated within the chip and on how effectively the heat is conducted away from the device.

As we look at the manufacturer's numbering system, what all the letters represent becomes a little clearer.

EXAMPLE 2–6

- 7400PC: The 7400 indicates that it is a chip that contains four two-input NAND gates. The P indicates a plastic package, and the C tells us that it is the commercial version.

Transistor Review and TTL Circuits

- 54S00DM: The 5400 indicates that it is a chip that contains four two-input NAND gates. The D indicates a ceramic package, and the M indicates a military version. For this chip the 54 indicates military (a 74 prefix would indicate commercial). The S indicates that it is a member of the Schottky subseries of the TTL family. This series will be discussed further in a later section.

The next column encountered is PKG TYPE. Under this column are listed 9a, 6a, and 31. These numbers refer to another section of the spec book where such items as chip height and pin-to-pin dimensions are listed. From this we can deduce that the plastic, ceramic, and flatpak will not have the exact same dimensions. If dimensions are critical to your system, then you must seek these dimensions out and check for system compatibility.

The next section to be examined is the lower midsection. This section is labeled Input Loading/Fanout. The first column is labeled PINS, and the first row under the heading is labeled Inputs. For the case of the units that have the A pinout, this will be pins 1, 2; 4, 5; 9, 10; and 12, 13. This row indicates the relative loading that the particular model will present to the circuit that is driving the input. The loading is dimensioned in unit loads. Reviewing, we note that

$$1 \text{ U.L.} = 40 \text{ }\mu\text{A} \quad \text{when the input is in a 1 state}$$
$$= -1.6 \text{ mA} \quad \text{when the input is in a 0 state}$$

The rows are labeled by series. Underneath the series heading is listed HIGH/LOW, and beneath that are the two numbers 1.0/1.0. The number before the slash is the high loading, and the second number is the low loading. The numbers represent the U.L. multiplier. Looking at the 54/74S column, we see that the input loading is listed as 1.25/1.25. When the input is in the 1 state, the circuit driving the input must be capable of supplying $1.25 \times 40\mu\text{A} = 50 \text{ }\mu\text{A}$. When the input is in the low state, the drive circuit must be capable of sinking $1.25 \times -1.6\text{mA} = -2.0$ mA. The 54/74LS series has the high/low rating of 0.5/0.25 U.L. This circuit will require a high or 1 current at the inputs of $0.5 \times 40\mu\text{A} = 20 \text{ }\mu\text{A}$ and a low or 0 input current of $0.25 \times -1.6\text{mA} = -0.40$ mA.

The second row under pins is titled Outputs and describes the capability of the output circuit to drive other circuits or devices. The outputs are again listed as HIGH/LOW. Again there are four columns that represent the different subfamilies of TTL. The units that are utilized for dimensioning the output drive capability are unit loads. If we observe the high drive capability, it is apparent that the 54/74 series can drive 20 highs (or it can source $20 \times 40\mu\text{A} = 800 \text{ }\mu\text{A}$). The 54/74H series can supply only 12.5 U.L. when in the high mode; it can source $12.5 \times 40\mu\text{A} = 500 \text{ }\mu\text{A}$. The 54/74S series can supply the largest number of high U.L.'s, 25 ($25 \times 40 = 1$ mA). The 54/74LS series has the least drive capacity. It can source only 400 μA when in the high output state. In the low output state the 54/74 series can drive 10 U.L. This means that it can sink a current of $10 \times 1.6\text{mA} = 16$ mA. The 54/74H series can sink $12.5 \times 1.6\text{mA} = 20$ mA. The 54/74S series has the same current sink capacity. The 54/74LS series can sink only $5 \times 1.6\text{mA} = 8$ mA.

A strict interpretation of the definition of fanout would seem to require that we give two values for the Fanout for the 54/74 series. The high output is capable of driving 20 of its own inputs, while the low output is capable of driving only ten equivalent low inputs. Since you will always expect your inputs to be going into both the high and low states, your output must be capable of driving both high and low. It follows that the fanout will be determined by the lower of the two values; in this application it will be listed as 10. Note that the fanout for each of the series shown will be 10, with the exception of the LS series, which is 20 if one uses the value after the slash of 5.0. If you take note of the fact that the author has listed 2.5 in parentheses and speculate that he might be hedging on this value, then you can use only a fanout of 10 for this series.

The last section to be checked out is the power requirement and signal time delays associated with the circuit. This is the section labeled DC AND AC CHARACTERISTICS. This is probably the most revealing of the differences between the series shown. The symbols used are:

- I_{cch} = maximum current that will be drawn from V_{cc} when all four outputs are in the high state.
- I_{ccl} = maximum current that will be drawn from V_{cc} when all outputs are in the low state.
- t_{plh} = time delay to be expected for a pulse passing through the gate when the output is changing from a low to a high state.
- t_{phl} = time delay to be expected when the output is changing from a high to a low state.

Looking at the column under the 54/74 series, we are told that the maximum power supply current that this unit will use when all outputs are in the high state is 8 mA. Next it tells us that the maximum V_{cc} current that will be drawn when the device outputs are all in the low state is 22 mA. Note that the device uses more current and dissipates more power when all outputs are low. The propagation delay when the output is changing from a low to a high state is 22 nsec; this delay is only 15 nsec when the output is changing from a high to a low state. The conditions for these specifications are listed to the extreme right. For the I_{cch} the inputs must be held low. For the I_{ccl} the inputs are held low. This is due to the fact that the circuit that we are examining is a NAND gate that has an inversion when going from input to output. $V_{cc\ max}$ indicates that V_{cc} is at 5.5 V for the 54 (military) version and at 5.25 V for the 74 (commercial) series.

This is a very instructive table to which we will be referring in the next section when we compare the advantages and special uses for the different series. ■

2–9 RISE TIME, FALL TIME, AND PROPAGATION DELAY

In working with any electronic circuitry there is always some capacitance associated with the circuitry. Capacitance is a charge-storing property of all devices that requires energy of a driving source if the source is attempting to change voltages

Transistor Review and TTL Circuits

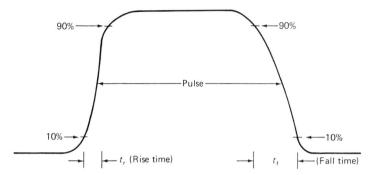

FIGURE 2-36 Rise and fall times

in a circuit. Capacitance is a measure of the energy that must be supplied by the driving source to change voltage on the driven circuitry. It is a physical impossibility to totally eliminate all capacitance. When a circuit has associated capacitance, there are time delays caused by the finite time required to charge this capacitance. A perfect square wave applied to the input of any circuit will be delayed in response owing to this input capacitance. The effect of this delay is expressed by a term that is called *rise time*. Since there is stray capacitance in all circuitry, it is impossible to generate a perfect square wave with zero rise time. Rise time is then used as a measure of the speed of response of a circuit. It is also a quantity that is used to define just how fast the input circuit is being excited. By definition, rise time is the time it takes a waveform to increase from 10% to 90% of the final amplitude of the waveform. *Fall time* is the response of the circuit to the trailing or falling edge of the pulse. Fall time is defined as the time required to fall from the 90% to 10% of the pulse amplitude. See figure 2–36.

Propagation delay (figure 2–37) is the delay that exists between an input change and an output response. If a TTL inverter were to receive an input 0 to 1 step, the propagation delay would be the time that it takes for the output to go to the 0 state

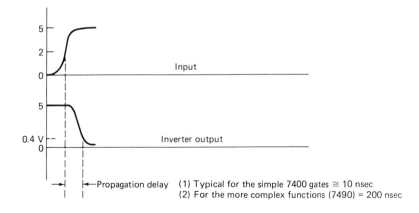

FIGURE 2-37 Propagation delay

referenced to the time when the input 0 to 1 change was initiated. Inherent in this propagation delay is an accumulation of all the rise times experienced within the integrated circuit. The circuits that can be used at the higher frequencies are the ones with the minimum propagation delays.

2–10 OTHER TTL SERIES

The first subseries that we will look at is the S or Schottky series. This series was designed to improve the operating speed of the system. The standard 54/7400 series uses *saturated logic,* which simply means that the transistors that are performing the logic function inside the chip either are working in the OFF state or, if they are ON, are operating in the saturated mode. When a transistor is operated in *saturation,* both the base emitter and base collector are forward-biased and there is a minority current flowing in both junctions. When you have minority current flowing, an increased time period is required to allow the minority carriers to recombine after you have removed the saturation-producing base current and expect the device to turn off. In the case of the S series a Schottky-type diode is placed across the base–collector junction, preventing this junction from going into saturation and thus decreasing the response time of the turn-off function. This phenomenon, along with a discussion of a Schottky diode, is covered elsewhere. Suffice it to say here that the 74S series is designed to be faster than the 7400 series. The cost for the increased speed of response of this series is increased power dissipation. The 54/74LS series is slower than the 54/74S series but has considerably lower power dissipation than either the 54/74S or the 54/7400 series. Unless you absolutely need the S series speed of response, the 54/74LS appears to be the best choice of the three systems for a general design. A comparison of the three systems can be accomplished by observing the AC/DC characteristics given in the specification of the previous section. The power dissipation is in direct proportion to the maximum power supply current, I_{cc}. Comparing the I_{cc} when the output is in the high state, I_{cch}, we get

54/7400   — 8 mA
54/74S00   — 16 mA
54/74LS00 — 1.6 mA

The 54/74S requires twice the current as does the 54/7400 and ten times the current required by the LS series. These devices have the same package size, so the device using the lower current and dissipating the lower power will run at a cooler temperature and will cost less to operate. The I_{ccl} has a different value than the I_{cch}, but the ratio of the currents is very close to that of the I_{cch}. A first comparison of the speed of response of the output transition indicates that the 54/7400 series has a slower response when the output is switching from low to high, $t_{plh} = 22$ nsec as opposed to $t_{phl} = 15$ nsec. The other series have basically the same switching

Transistor Review and TTL Circuits

time for either direction of output toggling. A comparison of the three series that we are discussing shows

t_{plh} = 22 nsec for the 54/7400 series
= 4.5 nsec for the 54/74S00 series
= 10 nsec for the 54/74LS00 series

A summary of these comparisons indicates that the LS series consumes the lowest power and is only half as fast as the S series, but twice as fast as the 54/7400 series. The S series is the fastest of the three but consumes approximately five times as much power as the LS series. Speed of response is very important in that it dictates the speed at which you can make logic decisions and how many decisions you can make in a fixed period of time.

There is occasional use of special unit loads associated with the S and the LS series. They are defined as follows:

High input: S_{ul} = +50 µA LS_{ul} = +20 µA
Low input: S_{ul} = −2 mA LS_{ul} = −0.4 mA

PROBLEMS

1. A 7400 series device is specified to present a load of 2.5 U.L. when the input is high and a load of 4 U.L. when the input is low.
   (a) What are the current requirements for the circuit driving this device?
   (b) What is the maximum value of the output resistance of the driving circuit?

2. A 7400 series device is specified at 2.0 U.L. high and 2.0 U.L. low. Fanout is 10.
   (a) What is the maximum current that can be sourced by this device?
   (b) What is the maximum current that it can sink to ground?

3. The device specified in Problem 2 is to be used to drive or cause an LED to be lit. The LED requires 10 mA to obtain the proper illumination. Show how you would connect the circuit to operate the LED from the output of the specified device. It is your choice whether to light the diode in the output 1 or 0 mode.

4. Fill in a truth table for the circuit shown in figure 2–38 and indicate the ON or OFF status of each of the transistors for all four input combinations. If you are given the following names to choose from, which name would you apply to this circuit?
   (a) AND   (b) NAND   (c) OR   (d) NOR   (e) NOT

5. Repeat Problem 4 for the circuit of figure 2–39.

6. Repeat Problem 4 for the circuit of figure 2–40.

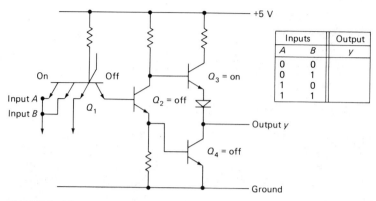

FIGURE 2–38 Circuit for Problem 4

7. Repeat Problem 4 for the circuit of figure 2–41.
8. A 7400 series TTL device has its input returned to ground through a 100 Ω resistor. The voltage measured at the input to ground is 0.4 V. What is the input load measured in unit loads?
9. Upon observing the circuit shown in figure 2–41, what will be the following effective loadings in unit loads when the two inputs are tied in parallel? The specification states the input loading is 1 U.L. high and 1 U.L. low.
   (a) Effective loading high.   (b) Effective loading low.

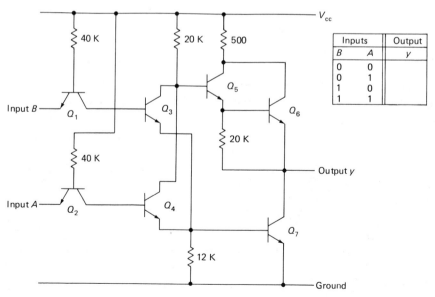

FIGURE 2–39 Circuit for Problem 5

Transistor Review and TTL Circuits

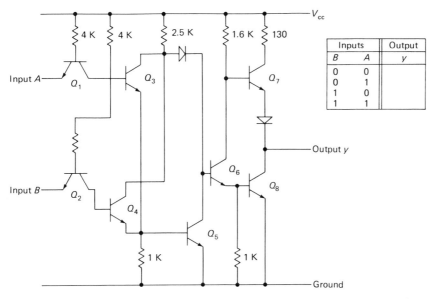

FIGURE 2–40 Circuit for Problem 6

10. A 74S series device is rated 1.5 S_{ul} loading with a fanout of 10.
  (a) What is the maximum current that it can source?
  (b) What is the maximum current that it can sink?

11. A 74LS series device is rated 1.5 LS_{ul} loading with a fanout of 10.
  (a) What is the maximum current that it can source?
  (b) What is the maximum current that it can sink?

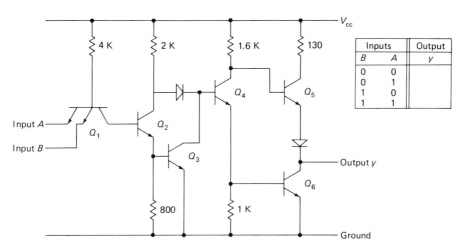

FIGURE 2–41 Circuit for Problem 7

12. A 7400 two-input NAND gate has both inputs tied together. What will be the following current requirements of a circuit that drives this gate?
    (a) I_{high} (b) I_{low}
13. A 7400 NAND gate is used such that all four of the gates are in series. What will be the maximum frequency that can be transmitted through this circuit? (You must check the specifications to obtain the time delay through each gate.)
14. How many 74LS00 inputs can a 7400 output drive?
15. How many 7400 inputs can a 74LS00 output drive?
16. How many 74S00 inputs can a 7400 output drive?
17. How many 74LS00 inputs can a 74S00 output drive?
18. Given the input circuitry in figure 2–42 for a TTL circuit, what is the maximum current the driving circuit can sink to ground?
19. In Problem 18, what would be the maximum current specification if the device had four inputs instead of the two shown in figure 2–42?
20. For the circuit shown in figure 2–42 the current limit is within the specification for a TTL low but would be considered a marginal design. What would be the new value of the resistor if a safety factor of 2 were chosen in place of the marginal design?
21. Given the TTL circuit shown for the NAND gate (figure 2–41), what would happen to the circuitry if the V_{cc} and ground were reversed?
22. Given the same circuit as in Problem 21, what would happen to the device if the V_{cc} terminal were grounded and the ground terminal were returned to -5 V?

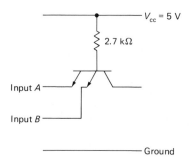

FIGURE 2–42 Circuit for Problems 18 to 20

chapter 3

logic elements

3-1 INTRODUCTION

This chapter will cover the basic digital logic elements that you are most likely to encounter. We shall begin with a presentation of the responses of these basic logic elements—the AND, OR, NAND, NOR, and NOT circuits—and take a look at the XOR and XNOR logic gates. In the discussion to follow, the inputs will be represented by letters A, B, C, and so on, and the output will be labeled y.

3-2 THE AND CIRCUIT

A simple statement of the response of this circuit is that the output y will be a high (a 1 in the 1/0 mode of representation) and this output voltage in a TTL system will be >2.4 V when all the lettered inputs are high. In other words, y is high when A and B and C and . . . (all inputs) are high. It will be necessary to reduce this information to an equation. By convention it is agreed that to communicate the AND function, we simply place a dot, an asterisk, or a $\times$ between the inputs to indicate that they are to be ANDed.

―――――― EXAMPLE 3-1

Suppose you would like a function that will give you a 1 output when your inputs are all high. If the system has two inputs, A and B, the equation will read

$$y = A * B \quad \text{or} \quad y = A \times B$$

If there are three inputs, the equation will read

$$y = A * B * C \quad \text{or} \quad y = A \times B \times C$$

■

Inputs		Output
B	A	y
0	0	0
0	1	0
1	0	0
1	1	1

FIGURE 3–1 Truth table for an AND gate

FIGURE 3–2 The AND gate

The response of the circuit is best presented in tabular form. The table is called a *truth table*. It presents the response of the output y (a 1 or a 0) for all possible combinations of all inputs. A truth table for a two-input AND gate would be as shown in figure 3–1.

Notice that the output y is a 1 only when inputs A and B are both 1. We need to define a graphic representation of this AND circuit that we can use when working with logic diagrams. The AND circuit is represented by the symbol in figure 3–2.

3–3 THE OR GATE

The statement of the OR gate is as follows: The output y will be a 1 when input A *or* input B *or* input C (or any input) is a 1. The equation is written as follows:

Two inputs: $y = A \text{ or } B$
                $y = A + B$

n inputs: $y = A \text{ or } B \text{ or } C \text{ or } \cdots n$
                $y = A + B + C + \cdots n$

The truth table for a two-input OR gate will be as shown in figure 3–3. Notice that the output y is a 1 when A *or* B is a 1. In this case it is also a 1 when A and B are both 1's. The symbol for the OR gate is shown in figure 3–4.

Inputs		Output
B	A	y
0	0	0
0	1	1
1	0	1
1	1	1

FIGURE 3–3 Truth table for an OR gate

FIGURE 3–4 The OR gate

Logic Elements

Input	Output y
0	1
1	0

FIGURE 3–5 Truth table for a NOT gate

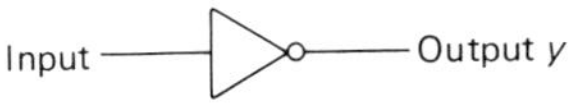

FIGURE 3–6 The NOT gate

3–4 THE NOT FUNCTION

Before continuing to the other common gate functions we must look at an inverting function, which is called the NOT function. This gate simply inverts the input function. If the input is a 0, the output becomes a 1; conversely, if the input is a 1, the output becomes a 0. The simple NOT function is defined for a single input. The truth table for this function is as shown in figure 3–5.

The symbol for this NOT function is shown in figure 3–6. The circle at the output denotes the inverting function. It takes on a second meaning when one moves farther into this field—it indicates that the circuit is active low. In the simplest case we would consider the input to be active when a 1 is received; when that occurs, the output goes active and of course is low. This very simple circuit is important because when it is combined with the previously discussed AND and OR circuits, we get two additional basic gates that add considerable flexibility to the digital design process. The two new gates are formed by passing the outputs of the AND and OR circuits through the NOT gate to give us the NAND and NOR gates, respectively.

3–5 THE NAND GATE

The output of the NAND gate is simply the output of the AND gate inverted. We represent this function in our equation by drawing a bar across the function. For example,

Two-input AND: $y = A * B$

Two-input NAND: $y = \overline{A * B}$

n-input AND: $y = A * B * C * \cdots * n$

n-input NAND: $y = \overline{A * B * C * \cdots * n}$

The truth table for the NAND gate is shown in figure 3–7. The symbol for the NAND gate is the same as the AND gate with a circle added to the output, as is shown in figure 3–8.

Inputs		Output
B	A	y
0	0	1
0	1	1
1	0	1
1	1	0

FIGURE 3–7 Truth table for a NAND gate

FIGURE 3–8 The NAND gate

3–6 THE NOR GATE

The NOR gate is simply the OR gate with its output passed through a NOT circuit. The equation for this circuit becomes

$$\text{Two-input OR gate:} \quad y = A + B$$
$$\text{Two-input NOR gate:} \quad y = \overline{A + B}$$

$$n\text{-input OR gate:} \quad y = A + B + C + \cdots + n$$
$$n\text{-input NOR gate:} \quad y = \overline{A + B + C + \cdots + n}$$

The truth table representation of the NOR gate is shown in figure 3–9. The symbol for the NOR gate is shown in figure 3–10.

3–7 THE EXCLUSIVE OR GATE

The OR gate as shown responds to the AND function in addition to the OR function. The truth table indicates that the output is a 1 when either the A *or* the B input is 1; it is also a 1 when both A *and* B are 1's. In some logic functions this might be an undesirable response. The gate that eliminates this additional AND response is called an Exclusive OR. The truth table for this gate is as follows: The output is true (a 1) when inputs A and B are different. (See figure 3–11.)

The equation that represents the exclusive OR function uses the same plus that is used for the OR function with a circle around the plus sign to denote the exclusivity of the function. The abbreviation for this function is XOR. The equation is

$$y = A \oplus B$$

Inputs		Output
B	A	y
0	0	1
0	1	0
1	0	0
1	1	0

FIGURE 3–9 Truth table for a NOR gate

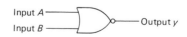

FIGURE 3–10 The NOR gate

Logic Elements

Inputs		Output
B	A	y
0	0	0
0	1	1
1	0	1
1	1	0

FIGURE 3-11 Truth table for an XOR gate

FIGURE 3-12 The XOR gate

For multiple inputs greater than 2, the output is active (1) for the condition of only one input being active or high at a time. If two or more of the inputs are high, the output will be low. The equation for this condition would be written as follows:

$$y = A \oplus B \oplus C \oplus D \oplus \cdots$$

The symbol for the XOR circuit is shown in figure 3-12.

A statement of the XOR function using the two inputs and the NOT function can be made: The output will be high when input A is high and input B is low or when input B is high and input A is low. Since we are working with digital functions, a function must be either a 1 or a 0. A high is a 1, and a low is a 0. The equation for the XOR is

$$y = A \text{ and not } B \text{ or } B \text{ and not } A$$
$$y = A * \overline{B} + B * \overline{A}$$

The internal circuitry of the XOR gate is shown in figure 3-13.

3-8 THE EXCLUSIVE NOR GATE

The Exclusive NOR gate (abbreviated XNOR) has a truth table in which the outputs are the complements of the XOR outputs, as is shown in figure 3-14. The output y will be a 0 when and only when one of the inputs is 1. The output will be a 1

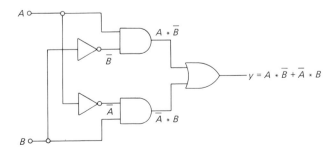

FIGURE 3-13 Internal circuitry of the XOR gate

Inputs		Output
B	A	y
0	0	1
0	1	0
1	0	0
1	1	1

FIGURE 3–14 Truth table for an XNOR gate

FIGURE 3–15 The XNOR gate

for all other input combinations. The output is a 1 when both inputs are equal—either 1's or 0's. The equation for the XNOR gate is

$$y = \overline{A \oplus B}$$

The symbol for this gate is shown in figure 3–15.

3–9 LOGIC ELEMENT USAGE

Now that all the basic elements have been defined, we can explore how and why they are used. A couple of illustrations might be useful at this point.

EXAMPLE 3–2   Use of an AND Gate

Suppose you wish to know when the temperature of a certain room exceeds 70°F when someone is in the room with the lights on, and you wish to have an alarm sound when this happens. You first need a temperature sensor that will give an output level change when 70° is exceeded. Give it the symbol T. In this instance the sensor will change from a 0 to a 1 when the temperature level is exceeded. You also need a sensor to indicate the status of the room lighting. We will label this sensor L and expect that it will change from a 0 to a 1 when the lights are turned on. A simple equation describing the system can now be written:

$$\text{Alarm} = T * L$$

A simple logic circuit that will perform this function is a two-input AND gate, as shown in figure 3–16.   ■

Later in this chapter we will be working with systems that have multiple inputs. The convention that is favored denotes inputs with letters from the beginning of the alphabet; for example, A, B, C, D could be used for a circuit that has four variable inputs. A second convention is to label the output y, or F for function. (F

FIGURE 3–16   Alarm generating circuit

Logic Elements

will be used in the later chapters of this text.) Keep in mind that these lettered inputs represent changing phenomena such as temperature, humidity, vacuum, or switch status or some logic function.

―――――――――――――――――― **EXAMPLE 3–3** Driver Seatbelt Sensor Logic

Suppose you are asked to design a logic system that will perform the following function for an automobile seatbelt system: In a two-seat sport convertible, an alarm is to be sounded if a person is sitting in a seat without the seatbelt fastened and with the driver present.

Implementing the system would require a device to indicate whether a person is sitting on the seat. A pressure-sensitive switch could do the job. Let's label the driver's side sensor S_1 and the passenger's side sensor S_2. The sensor was designed to produce a 1 (>2 V) when someone is sitting in the seat. When no one is sitting in the seat, the sensor will produce a 0 (<0.4 V) at the output. The system also requires a sensor to indicate the status of the seatbelt. Label the driver's seatbelt sensor B_1 and the passenger's seatbelt sensor B_2. These sensors will produce a 1 output when the belts are fastened and a 0 when they are left unfastened. We now make a choice to produce a 1 at the output for any alarm condition. The first thing that we will do is make a truth table for the system, as shown in figure 3–17(a).

The logic statements of each line would be as follows:

- Line 0: Neither passenger nor driver is seated, neither seatbelt is fastened.
- Line 1: Neither passenger nor driver is seated, passenger's seatbelt is fastened.
- Line 2: Passenger is seated but not belted, driver is not seated and driver's seatbelt is not fastened.
- Line 3: Passenger is seated and belted, driver is not seated and driver's seatbelt is not fastened.
- Line 4: Passenger is not seated and not belted, driver is not seated but driver's seatbelt is fastened.
- Line 5: Neither passenger nor driver is seated, both seatbelts are fastened.
- Line 6: Passenger is seated but not belted, driver is not seated but driver's seatbelt is fastened.
- Line 7: Passenger is seated and belted, driver is not seated but driver's seatbelt is fastened.
- Line 8 *(Alarm)*: Driver is seated but not belted, passenger is not seated and passenger's seatbelt is not fastened.
- Line 9 *(Alarm)*: Driver is seated but not belted, passenger is not seated but passenger's seatbelt is fastened.
- Line 10 *(Alarm)*: Driver and passenger are seated, neither seatbelt is fastened.
- Line 11 *(Alarm)*: Driver is seated but not belted, passenger is seated and belted.
- Line 12: Driver is seated and belted, passenger is not seated and passenger's seatbelt is not fastened.

Inputs				Output	
S_1	B_1	S_2	B_2	y	Line Number
0	0	0	0	0	0
0	0	0	1	0	1
0	0	1	0	0	2
0	0	1	1	0	3
0	1	0	0	0	4
0	1	0	1	0	5
0	1	1	0	0	6
0	1	1	1	0	7
* 1	0	0	0	1	* Alarm 8
* 1	0	0	1	1	* Alarm 9
* 1	0	1	0	1	* Alarm 10
* 1	0	1	1	1	* Alarm 11
1	1	0	0	0	12
1	1	0	1	0	13
* 1	1	1	0	1	* Alarm 14
1	1	1	1	0	15

FIGURE 3–17(a) Truth table for seatbelt problem

- Line 13: Driver is seated and belted, passenger is not seated but passenger's seatbelt is fastened.
- Line 14 *(Alarm):* Driver is seated and belted, passenger is seated but not belted.
- Line 15: Driver and passenger are seated and belted.

This is the truth table for the system. Much of this chapter will deal with deriving the equations that represent this table, implementing these equations, and simplifying them as much as possible. For this example we will write the equation that represents the alarm condition and draw the circuit that will implement the simplified equation. The equation is

$$y = \text{Alarm output} = S_1 * \overline{B_1} + S_1 * S_2 * \overline{B_2}$$

Notice that the five alarm terms shown in the truth table are reduced to two in the equation. The reduction techniques will be covered later. This statement can be read as follows: The output will be a 1 when the driver is seated *and* the driver's seatbelt is *not* fastened *or* the driver *and* the passenger are seated *and* the passenger's seatbelt is not fastened. The circuit shown in figure 3–17(b) can be constructed

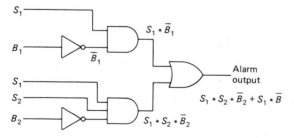

FIGURE 3–17(b) Logic circuit for solution to seatbelt problem

Logic Elements

directly from the statement. The circuit will produce an output of 1 when the alarm conditions specified at the beginning of the example exist. ∎

3-10 SUM OF PRODUCTS FROM THE TRUTH TABLE

It is appropriate to examine the truth table in more detail and analyze why we get what we do from it. First, it is important to recognize that only one of all the input conditions can exist at any point in time. Each row represents a particular unique combination of inputs that is mutually exclusive with respect to the other possible input combinations. This situation requires that when 1's exist in the output column at more than one location, they cannot occur at the same time and if a circuit is to be created to represent this truth table, these output 1's must be combined and represented by the OR circuit. The unique combination of input conditions that specifies the conditions for the 1 to exist must occur simultaneously. Hence this condition is an AND function and must be represented by an AND gate. If we combine the two requirements, it becomes apparent that each of the 1's in the y output column will be an input to an OR gate, and each of these 1's will be created by the proper combination of the input variables. This is an AND function. The number of inputs to the AND gate is determined by the number of input variables. In the truth table of figure 3-17(a) the AND gate would have four inputs. Since there are five 1's or alarm conditions in the output, a total of five of these AND gates are required. The outputs of these five AND gates will be the inputs to the OR gate. The output of the OR gate is then the unsimplified logic function that represents the truth table. This technique will always generate a logic function and a logic circuit that implement the truth table logic. Simplification of this expression and circuitry will be discussed later.

This technique is called the *sum of products* solution to the truth table. The combination of inputs that is required to generate a 1 in the output is created by ANDing to get a product. After the product terms have been created, they must be combined or summed to obtain the final logic function that will represent the truth table. The result is termed a sum of products solution to the truth table. Again, first the ANDing of the inputs creates the products, then the ORing of the products creates the sum condition.

―――――――――――――――――――――――――――――――――――――― EXAMPLE 3-4

If we use the truth table of figure 3-17(a), a sum of products solution will present us with four input variables: S_1 = driver seated, B_1 = driver's seatbelt, S_2 = passenger seated, and B_2 = passenger's seatbelt. Since we have decided to use letters to represent the variables, we will use the following terminology:

$D = S_1$ (driver's seating status)
$C = B_1$ (driver's seatbelt status)

$B = S_2$ (passenger's seating status)

$A = B_2$ (passenger's seatbelt status)

We are to obtain a 1 in the output of our product or AND circuit. To obtain this result from an AND gate, all the inputs must be high (a 1) to get an output high (a 1). The first alarm condition that must be examined is line 8, which is the condition in which the driver is seated but not belted, there is no passenger, and the passenger's seatbelt is not fastened. The truth table will give us $D = 1$ and $A = B = C = 0$. If the inputs A, B, C, and D are direct inputs to an AND gate, the output will be a 1 only when the condition $A = B = C = D = 1$ (line 15) occurs. If we want line 8 to generate a 1, we must take the inputs A, B, and C through inverters to generate the desired output for the line 8 input condition. Once the functions have been routed through an inverter, they will be represented as the NOT of the function. It follows that the inputs to the AND gate that is to generate the 1 for this combination of inputs would be represented by the following function: $D\overline{C}\overline{B}\overline{A}$ = Line 8. Similar logic will give us the remainder of the alarm conditions: Line 9 = $D\overline{C}\overline{B}A$, Line 10 = $D\overline{C}B\overline{A}$, Line 11 = $D\overline{C}BA$, and Line 14 = $DCB\overline{A}$.

The sum of products solution would be the equation

$$y = D\overline{C}\overline{B}\overline{A} + D\overline{C}\overline{B}A + D\overline{C}B\overline{A} + D\overline{C}BA + DCB\overline{A}$$

The circuit is shown in figure 3–18. ■

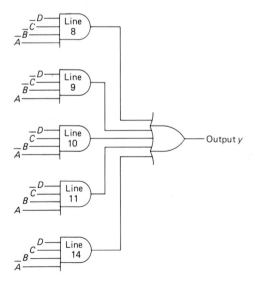

FIGURE 3–18 Sum of products solution for the truth table of figure 3–17

Logic Elements

3-12 BOOLEAN DERIVATIONS AND EXAMPLES

Given the relationship $y = A + AB$, since A is common to both terms, it is acceptable to factor out A:

$$y = A(1 + B)$$

From the previous section it can be stated that any function that is ORed with 1 is reduced to 1. $B + 1 = 1$. It follows that the function now reduces to $y = A(1)$, which further reduces to $y = A$. This is a very useful identity, which will allow you to replace the function $y = A + AB$ with A:

Identity: $A + AB = A$

Given a second relationship $y = A + \overline{A}B$, to simplify the relationship, it must be first expanded. By using the identity in the reverse direction the A function is replaced by $A + AB$:

$$y = (A + AB) + \overline{A}B$$

Remove the parentheses and then enclose the AB terms in parentheses:

$$y = A + (AB + \overline{A}B)$$
$$y = A + B(A + \overline{A})$$

From the previous section we know that $A + \overline{A} = 1$:

$$y = A + B(1) = A + B$$
Identity: $A + \overline{A}B = A + B$

The following identities come from the same derivation. If you have any questions, try the derivations.

Identity: $A + \overline{AB} = A + \overline{B}$
Identity: $\overline{A} + AB = \overline{A} + B$
Identity: $\overline{A} + A\overline{B} = \overline{A} + \overline{B}$

---— EXAMPLE 3-5

Given the expression $y = AB + A\overline{B} + \overline{A}B$, regroup:

$$y = (AB + A\overline{B}) + \overline{A}B$$
$$y = A(B + \overline{B}) + \overline{A}B$$
$$y = A(1) + \overline{A}B$$
$$y = A + \overline{A}B$$
$$y = A + B$$

■

EXAMPLE 3-6

Given the relationship $y = ABC + AB\overline{C} + A\overline{B}C$:

$y = AB(C + \overline{C}) + A\overline{B}C$
$y = AB(1) + A\overline{B}C$
$y = AB + A\overline{B}C$
$y = A(B + \overline{B}C)$
$y = A(B + C)$
$y = AB + AC$ ∎

EXAMPLE 3-7

Given the relationship $y = (ABC)(A\overline{B}) + (AB + A) + (\overline{A} + \overline{A}B)$:

$y = (AAC)(B\overline{B}) + (AB + A) + (\overline{A} + \overline{A}B)$
$y = (AC)(0) + (A) + (\overline{A})$
$y = 0 + (A + \overline{A})$
$y = 0 + 1$
$y = 1$

The $y = 1$ solution to this problem simply indicates that the output is a 1 independent of any and all input variations. As a result, this output or y function would simply be connected to a fixed 1 state. ∎

3-13 DE MORGAN'S THEOREMS

The two theorems that are called De Morgan's Theorems are very simple to present. The power of these theorems will become evident when they are applied to problems of circuit reduction. You will appreciate the power, beauty, and simplicity of the operations after repeated applications of these theorems. Equation representations of these two theorems follow.

De Morgan's Theorem 1: $\overline{A + B} = \overline{A} * \overline{B}$

The complement of an OR function (a NOR function) will have the same response as an AND gate with inverted inputs. The circuit is

Logic Elements

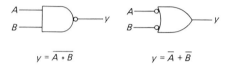

$y = \overline{A + B}$ = $y = \overline{A} * \overline{B}$

De Morgan's Theorem 2: $\overline{A * B} = \overline{A} + \overline{B}$

The complement of an AND function (a NAND function) will have the same response as the output of an OR circuit with inverted inputs. The circuit is

$y = \overline{A * B}$ $y = \overline{A} + \overline{B}$

A one-sentence representation of the two theorems might be "Change the line, change the sign." If the line is continuous across the function and you break it, the function changes at the break in the line. Conversely, if the line is broken and you reconstruct it, the function also changes where the joining of the line takes place.

EXAMPLE 3–8

Given the function

$$y = (\overline{A + B}) * (A * \overline{B})$$
$$= (\overline{A} * \overline{B}) * (A * \overline{B}) \quad \text{(De Morgan's first law)}$$
$$= \overline{A} * A * \overline{B} * \overline{B} \quad \text{(Associative law)}$$
$$= 0 * \overline{B} \quad \text{(Laws from the previous section)}$$
$$= 0$$

∎

EXAMPLE 3–9

Given the function

$$y = (A + B) * \overline{A} * \overline{B}$$
$$y = \overline{(A * \overline{B})} * \overline{A} + \overline{B}$$
$$y = \overline{A} + \overline{\overline{B}} + A + B$$
$$y = \overline{A} * B + A + B$$
$$y = A + \overline{A}B + B$$
$$y = A + B + B$$
$$y = A + B$$

∎

EXAMPLE 3–10

Given the function

$$y = \overline{\overline{A}B} + \overline{A\overline{B}} + \overline{AB}$$
$$y = \overline{\overline{A}B} \cdot \overline{A\overline{B}} \cdot \overline{AB}$$

since $A * \overline{A} = 0$ and $B * 0 = 0$,

$$y = 0$$

EXAMPLE 3–11

Given the function

$$y = \overline{ABC} + \overline{A\overline{B}C} + \overline{A\overline{\overline{C}}}$$

notice that the first terms have a common factor:

$$y = \overline{AC(B + \overline{B})} + \overline{A\overline{\overline{C}}}$$
$$y = \overline{AC} + \overline{A\overline{\overline{C}}}$$
$$y = \overline{AC} * \overline{A\overline{C}}$$
$$y = \overline{AC\overline{\overline{C}}} = A * 0$$
$$y = 0$$

EXAMPLE 3–12

Direct application of De Morgan's Theorems will result in the following conclusions:

1. $y = \overline{A + B} = \overline{A} * \overline{B}$. The NOR gate will perform the same function as an AND gate that has its inputs inverted.
2. $y = \overline{\overline{A} * \overline{B}} = \overline{\overline{A}} + \overline{\overline{B}} = A + B$. An inverted input NAND gate is an OR function.
3. $y = \overline{\overline{A} + \overline{B}} = \overline{\overline{A}} * \overline{\overline{B}} = A * B$. An inverted input NOR gate is an AND function.

3–14 ALGEBRAIC ANALYSIS OF EXISTING CIRCUITS

In this section we will take an existing circuit and derive an expression that will represent that circuit.

EXAMPLE 3–13

Given the circuit in figure 3–21, determine an expression for y.

Logic Elements

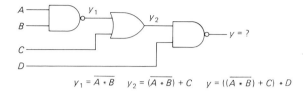

$y_1 = \overline{A * B}$ $y_2 = \overline{(\overline{A * B}) + C}$ $y = \overline{((\overline{A * B}) + C) * D}$

FIGURE 3-21

─── EXAMPLE 3-14

Given the circuit in figure 3-22, derive the expression for y.

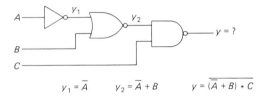

$y_1 = \overline{A}$ $y_2 = \overline{\overline{A} + B}$ $y = \overline{(\overline{A} + B) * C}$

FIGURE 3-22 Logic circuit for Boolean analysis

─── EXAMPLE 3-15

Given the circuit in figure 3-23, determine y.

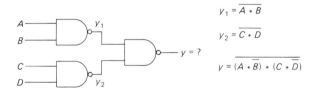

$y_1 = \overline{A * B}$
$y_2 = \overline{C * D}$
$y = \overline{(A * \overline{B}) * (C * \overline{D})}$

From De Morgan's Theorems, $y = A * B + C * D$

FIGURE 3-23

3-15 THE PRODUCT OF SUMS

Just as we can use an AND OR combination to create a solution to a truth table with the sum of products method, a duality exists that allows one to have a product of sums solution to the same truth table. Implementing the solution requires the variables to be ORed and these outputs then to be ANDed. The technique consists of inverting every y output in the truth table and using any technique that you desire to simplify. The answer will be a sum of products solution to the $\overline{y}$ function. De Morgan's Theorems allow the conversion to a sum of products representation.

The solution to the truth table in figure 3–24(b) is as follows:

$$\bar{y} = \overline{C}\overline{B}\overline{A} + C\overline{B}\overline{A}$$
$$\bar{\bar{y}} = y = \overline{\overline{C}\overline{B}\overline{A} + C\overline{B}\overline{A}} = \overline{\overline{C}\overline{B}\overline{A}} * \overline{C\overline{B}\overline{A}}$$
$$y = (\overline{\overline{C}} + \overline{\overline{B}} + \overline{\overline{A}}) * (\overline{C} + \overline{\overline{B}} + \overline{\overline{A}})$$
$$y = (C + B + A) * (\overline{C} + B + A)$$

The circuit to implement this function is shown in figure 3–25.

Inputs			Outputs
C	B	A	y
0	0	0	0
0	0	1	1
0	1	0	1
0	1	1	1
1	0	0	0
1	0	1	1
1	1	0	1
1	1	1	1

FIGURE 3–24(a) Original truth table

Inputs			Outputs
C	B	A	y
0	0	0	1
0	0	1	0
0	1	0	0
0	1	1	0
1	0	0	1
1	0	1	0
1	1	0	0
1	1	1	0

FIGURE 3–24(b) Inverted or complemented truth table

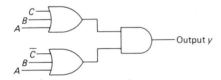

FIGURE 3–25

3–16 OTHER THEOREMS

The following theorems can be derived by using one or more of the previously mentioned theorems.

Theorem	Derivation
$A + AB = A$	$A + AB = A * (1 + B) = A * 1 = A$
$A (A + B) = A$	$A (A + B) = A * A + A * B = A + AB = A$
$(A + B)(A + C) = A + BC$	$= AA + AC + AB + BC = A + AC + AB + BC =$
	$A + AB + BC = A + BC$
$A + \overline{A}B = A + B$	Use $A = A + AB$, then $A + AB + \overline{A}B = A + B(A + \overline{A}) =$
	$A + B * 1 = A + B$
$A(\overline{A} + B) = AB$	$= A\overline{A} + AB = 0 + AB = AB$
$(A + B)(\overline{A} + C) = AC + \overline{A}B$	Derive as an exercise.
$AB + \overline{A}C = (A + C)(\overline{A} + B)$	Derive as an exercise.

Logic Elements

3–17 CIRCUIT SIMPLIFICATION

When presented with a Boolean expression that represents a digital circuit, you will generally try to reduce it to the minimum number of terms possible. Do not forget that each term represents a function, so if you can reduce the number of terms in the expression, you will also be reducing the number of circuit components.

── EXAMPLE 3–16

Suppose we are given a truth table that represents the conditions that our system requires, such as that in figure 3–26. The Boolean expression that can be written directly from the truth table is

$$y = AB\overline{C} + A\overline{B}C + ABC$$

Factor an AB from the first and third terms:

$$y = AB(\overline{C} + C) + A\overline{B}C = AB + A\overline{B}C$$

Factor an A from the remaining two terms:

$$y = A(B + \overline{B}C) = A(B + C)$$

The circuit is reduced from three three-input AND gates feeding one three-input OR gate to a two-input OR gate and a two-input AND gate. ∎

When simplifying circuitry, use as many previously developed relationships as you can as often as you can.

3–18 THE KARNAUGH MAP

The Karnaugh map is a technique to assist you in arriving at the simplest possible circuit in a rather straightforward manner. You rather quickly find out that when you are using the reduction techniques developed in the previous sections, there are generally several different ways to start the reduction, and you are not always

Inputs			Output
C	B	A	y
0	0	0	0
0	0	1	0
0	1	0	0
0	1	1	1
1	0	0	0
1	0	1	1
1	1	0	0
1	1	1	1

FIGURE 3–26

Truth table

Inputs			Output	
C	B	A	y	(block)
0	0	0	#0	
0	0	1	#1	
0	1	0	#2	
0	1	1	#3	
1	0	0	#4	
1	0	1	#5	
1	1	0	#6	
1	1	1	#7	

Karnaugh map

	$\bar{B}\bar{A}$	$\bar{B}A$	BA	$B\bar{A}$
$\bar{C}$	#0	#1	#3	#2
C	#4	#5	#7	#6

FIGURE 3–27

quite sure that you have embarked on the most straightforward way. The Karnaugh map eliminates that doubt. The map is a matrix representation of the system truth table. Each line of the truth table has a particular assigned location within the table. The positioning of the variables from left to right and top to bottom is not in the normal binary sequence. The requirement is that no two variables (A, B, C, D) can change in any adjacent row or column. For this book we chose the arrangement shown in figure 3–27.

The map is constructed by transferring the value of the y output from the truth table to the corresponding box within the map. Note that this value can be either a 1 or a 0. In some cases we will use an X to indicate that for the particular truth table the system does not care whether the value is a 1 or a 0. This will come about when your system does not allow the possibility of that combination of inputs to occur. For example, when you are decoding the BCD code, you need not be concerned about the existence of the codes 1010 through 1111 because they will never occur when this code is used. They are not defined in this code; and since they will never occur, they are represented by an X.

EXAMPLE 3–17

Solve the previous example using the Karnaugh map, as shown in figure 3–28.

Now that the map is constructed, the simplification comes from the fact that if 1's appear in adjacent blocks, they contain terms that can be factored. Look at the two adjacent 1's that appear in the C row. The terms that they represent are

$$y = C\bar{B}A + CBA = CA(B + \bar{B}) = CA$$

The terms located in the AB column are

$$y = \bar{C}BA + CBA = BA(\bar{C} + C) = BA$$

If you can encircle two adjacent 1's, the functions associated with the adjacent 1's will contain both a function and its complement. The simplification process calls for you to drop this term.

Logic Elements

Truth table

Inputs			Output
C	B	A	y
0	0	0	0
0	0	1	0
0	1	0	0
0	1	1	1
1	0	0	0
1	0	1	1
1	1	0	0
1	1	1	1

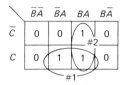

Karnaugh map

FIGURE 3–28 ■

The general rules for simplifying functions using Karnaugh maps are as follows:

1. Find the largest number of 1's that you can circle. You can circle two, four, eight, or sixteen (four input variables).
2. All 1's must be enclosed in a circle.
3. The object is to have the fewest number of circles enclosing the largest number of 1's.
4. Each circle is represented by a term in the final solution.
5. The number of terms that drop from the circled 1's is a function of the number of 1's circled. If you circle two, drop one term. If you circle four, drop two terms. If you circle eight, drop three terms. If you circle sixteen, drop the four variables.
6. Since the assignment of variables to columns and rows is an arbitrary assignment, the outer rows and columns can be treated just as though they were adjacent to one another. This is sometimes called the *rollover effect*.
7. Ones can be circled more than once. If in the process of circling a group of 1's you notice that you are including one or more previously circled 1's, the circle not only can but must be expanded to include them.

The unsimplified equation for the truth table in figure 3–29 is

$$y = \overline{D}\,\overline{C}B\overline{A} + \overline{D}\,\overline{C}BA + \overline{D}C\overline{B}A + \overline{D}CBA + \overline{D}CB\overline{A}$$

Simplification begins by circling the four 1's that are grouped in the upper center of the Karnaugh map (blocks 1, 3, 5, and 7). This leaves the 1 in block 6 alone. Since it must be circled, it is to our advantage to make as large a circle as possible. The best we can do is to enclose it in a circle with the 7 block. (Note that it is acceptable to circle a 1 more than once. It is actually desirable in this case because this will reduce the function associated with this circle by one term.) The expression that is derived from the first circle will contain two of the input variables: $y = \overline{D}A$. The expression that comes from the second circle will lose only one of the input variables and will be $y = \overline{D}CB$. The resultant expression will be the OR function of these two circles: $y = \overline{D}A + \overline{D}CB$.

Truth table

Outputs				Input
D	C	B	A	y
0	0	0	0	0
0	0	0	1	1
0	0	1	0	0
0	0	1	1	1
0	1	0	0	0
0	1	0	1	1
0	1	1	0	1
0	1	1	1	1
1	0	0	0	0
1	0	0	1	0
1	0	1	0	0
1	0	1	1	0
1	1	0	0	0
1	1	0	1	0
1	1	1	0	0
1	1	1	1	0

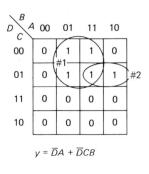

$y = \overline{D}A + \overline{D}CB$

FIGURE 3–29 Using 0's and 1's in place of the *AB* terms shows the direct correlation with the truth table and illustrates the weight of each block in the Karnaugh map.

EXAMPLE 3–18

Suppose we are given the truth table in figure 3–30, which represents the desired system response.

The unsimplified equation for this system is

$$y = \overline{D}\,\overline{C}\,\overline{B}\,\overline{A} + \overline{D}\,\overline{C}B\overline{A} + \overline{D}C\overline{B}A + \overline{D}CBA + D\overline{C}\,\overline{B}\,\overline{A} + D\overline{C}B\overline{A} + DC\overline{B}A + DCBA$$

Truth table

Inputs				Output
D	C	B	A	y
0	0	0	0	1
0	0	0	1	0
0	0	1	0	1
0	0	1	1	0
0	1	0	0	0
0	1	0	1	1
0	1	1	0	0
0	1	1	1	1
1	0	0	0	1
1	0	0	1	0
1	0	1	0	1
1	0	1	1	0
1	1	0	0	0
1	1	0	1	1
1	1	1	0	0
1	1	1	1	1

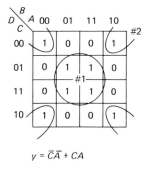

$y = \overline{C}\,\overline{A} + CA$

FIGURE 3–30

Logic Elements

Simplifying, we see that circle 1 encloses the four center 1's and results in the dropping of two terms from the four inputs. Those two remaining variables are $y = AC$. Circle 2 is made up of the four corners. The rollover feature implies that the upper left-hand corner, when rolled horizontally, is adjacent to the upper right-hand corner, and when rolled vertically, these two are adjacent to the lower corners. These four inputs reduce to $y = \overline{CA}$. The simplified equation becomes $y = CA + \overline{CA}$. Note that this is an Exclusive NOR circuit. ∎

EXAMPLE 3–19

Suppose we are given a system that is using BCD code and we wish to decode the number 8. Remember that in BCD code the numbers 10–15 do not exist, and we can therefore choose to treat them in the mapping procedure as either a 1 or a 0. When we set up the map, we insert an X in each of these boxes and then choose the 1 or 0 state to make the circuit reduction most efficient. (See figure 3–31.)

We choose to use the X in the 12 box, directly above the 1 in the 8 box, as a 1. This will eliminate one of the variables. If we look a little further, we see that if the X's located in the opposite column were 1's, we could rollover and eliminate one more input variable. The final Karnaugh map for this system will be as shown in figure 3–32.

The circle will now encompass four 1's and thus will be reduced from four input variables to two:

$$y = D\overline{A}$$

This condition will now guarantee that your output will be a 1 when and only when the input combination is an 8 (1000).

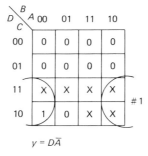

FIGURE 3–31   FIGURE 3–32 ∎

EXAMPLE 3-20

You are given a system that has three inputs, and the statement of the input conditions says that when two adjacent inputs are high, an alarm should occur in the output ($y = 1$). It is further stated that all three inputs cannot exist in the high state together. (See figure 3–33.)

Choose the X to be a 1 state. The solution is $y = B$.

Truth table

Inputs			Output
C	B	A	y
0	0	0	0
0	0	1	0
0	1	0	1
0	1	1	1
1	0	0	0
1	0	1	0
1	1	0	1
1	1	1	X

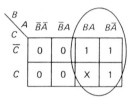

FIGURE 3-33

EXAMPLE 3-21

Four sensors are located on the circumference of a circle. An alarm condition exists when any two nonadjacent sensors are simultaneously high. Since they are on a circle, the A and D sensors are adjacent.

The map can be reduced by making four circles, each circle enclosing four 1's, as shown in figure 3–34:

$$y = CD + AB + AD + BC$$

Truth table

Inputs				Output
D	C	B	A	y
0	0	0	0	0
0	0	0	1	0
0	0	1	0	0
0	0	1	1	1
0	1	0	0	0
0	1	0	1	0
0	1	1	0	1
0	1	1	1	1
1	0	0	0	0
1	0	0	1	1
1	0	1	0	0
1	0	1	1	1
1	1	0	0	1
1	1	0	1	1
1	1	1	0	1
1	1	1	1	1

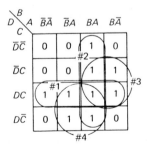

FIGURE 3-34

Logic Elements

3-19 COMPLEMENTING THE KARNAUGH MAP

The solution to the Karnaugh map has been shown to be a sum of products. It is sometimes advantageous to complement the map. To complement the map, each 1 is changed to a 0 and each 0 to a 1.

Using De Morgan's Theorems, let us examine what effect this has on the solution to the map. Choose a typical sum of product answer. Suppose the solution to the noncomplemented map is as follows:

$$y = \overline{A} * B + \overline{B} * C$$

When the map is complemented, a $\overline{y}$ function is created. A typical solution of the map would be a sum of products, but it really represents $\overline{y}$:

$$\overline{y} = \overline{A} * B + \overline{B} * C$$

Using De Morgan's Theorems, we get

$$y = \overline{\overline{A} * B + \overline{B} * C}$$
$$y = (\overline{\overline{A} * B})(\overline{\overline{B} * C})$$
$$= (\overline{\overline{A}} + \overline{B})(\overline{\overline{B}} + \overline{C})$$
$$y = (A + \overline{B})(B + \overline{C})$$
$$= \text{Product of sums solution}$$

---EXAMPLE 3-22

Suppose we are given a truth table for y that produces the Karnaugh map shown in figure 3-35.

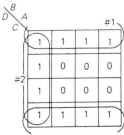

FIGURE 3-35 The solution to this map would be $y = \overline{A} * \overline{B} + \overline{C}$ (sum of products).

Complementing the map produces figure 3-36.

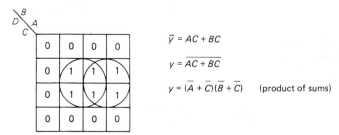

FIGURE 3-36

3-20 A FIVE-VARIABLE KARNAUGH MAP

The basic map is limited to four input variables. A fifth variable can be accommodated in the following manner. A truth table for four input variables will have 32 entries, and each must be accounted for when the map is diagramed. The solution is to make two of the four-variable input maps and use the first one for the condition in which the fifth variable is 0. The second map should then reflect the status of y when the fifth variable is in the 1 state. By using the standard $EDCBA$ as our inputs and allowing the fifth input to be E, the first map would represent the first 16 lines of the truth table, and the second map the last 16 input conditions. Choosing the circles now becomes a three-dimensional problem of selecting spheres. An example should clarify.

EXAMPLE 3-23

Given the truth table for a five-input function shown in figure 3-37, solve for the simplest solution using a Karnaugh map.

To reduce the map, we must look at it as two maps sitting one on top of the other and recognize that adjacent squares in the third-dimensional plane can be treated as adjacent 1's were treated in the four-variable maps. As a consequence, any circle or sphere that includes 1's in both of the maps will result in the E term's being eliminated from the expression that will represent that circle or sphere. For this map, if the circles represent a three-dimensional circle, they will be represented by a double circle. The single circles will be simple planar circles and will be treated in the same manner as was done previously.

- Sphere 1 will enclose positions 0, 1, 4, 5, 16, 17, 20, and 21 and will be represented by the term $\overline{B} * \overline{D}$.
- Sphere 2 will enclose positions 9, 10, 25, and 27 and will be represented by the term $A * \overline{C} * D$.

Logic Elements

#	\multicolumn{5}{c	}{Inputs}	Output
	E	D	C	B	A	y
0	0	0	0	0	0	1
1	0	0	0	0	1	1
2	0	0	0	1	0	0
3	0	0	0	1	1	0
4	0	0	1	0	0	1
5	0	0	1	0	1	1
6	0	0	1	1	0	0
7	0	0	1	1	1	0
8	0	1	0	0	0	0
9	0	1	0	0	1	1
10	0	1	0	1	0	0
11	0	1	0	1	1	1
12	0	1	1	0	0	0
13	0	1	1	0	1	1
14	0	1	1	1	0	0
15	0	1	1	1	1	0
16	1	0	0	0	0	1
17	1	0	0	0	1	1
18	1	0	0	1	0	0
19	1	0	0	1	1	0
20	1	0	1	0	0	1
21	1	0	1	0	1	1
22	1	0	1	1	0	0
23	1	0	1	1	1	0
24	1	1	0	0	0	0
25	1	1	0	0	1	1
26	1	1	0	1	0	0
27	1	1	0	1	1	1
28	1	1	1	0	0	0
29	1	1	1	0	1	0
30	1	1	1	1	0	0
31	1	1	1	1	1	0

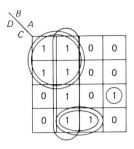

$E = 0$ (Lines 0 through 15)

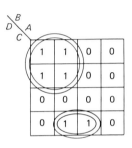

$E = 1$ (Lines 16 through 31)

FIGURE 3–37

- Circle 3 encloses the four positions 1, 5, 9, and 13 on the $\overline{E}$ map and is represented by the term $A * \overline{B} * \overline{E}$.
- Circle 4 encloses only the 14 position of the $\overline{E}$ map and must contain all five of the input variables: $\overline{A} * B * C * D * \overline{E}$.

The equation that represents this truth table is

$$y = \overline{B} * \overline{D} + A * \overline{C} * D + A * \overline{B} * \overline{E} + \overline{A} * B * C * D * \overline{E}$$

3–21 PRACTICAL USES FOR THE KARNAUGH MAP

The Comparator

Quite often it is desirable to compare digital numbers or expressions to determine whether they are equal or whether A is greater or less than B. The simplest possible condition exists when two single bits are compared. When one compares two digital numbers for equality, each of the corresponding bits of each number must be equal for true equality. If $A = 101101$, then for B to equal A, B must equal 101101. To compare A to B for equality, we can simply compare each of the bits; if they are all equal, then the statement can be made that $A = B$. If we look at a very simple Karnaugh map for a single bit, the two functions will be equal only when A and B are either both 1's or both 0's. This corresponds to positions 0 and 3 on the map. If a circuit is required to produce a 1 when the bits are equal, then we can use the

	$\overline{A}$	A
$\overline{B}$	1	0
B	0	1

techniques developed for implementing the solution to this map. The solution to this map is individual circles for each of the 1's. The resulting equation would be

$$y = \overline{A} * \overline{B} + A * B$$

To implement this circuit, we simply recognize this as an XNOR circuit. Should we wish to compare a multibit number, an XNOR circuit would be required for each bit; and since all bits must be equal for the number to be equal, the outputs of the XNOR circuits would then be ANDed. Should we wish to design a circuit that indicates the condition in which A is greater than B, this would be the $A = 1$, $B = 0$ block of our map. The solution to this map would be $y = A * \overline{B}$.

	$\overline{A}$	A
$\overline{B}$	0	1
B	0	0

Decoding for the Seven-Segment Display

As the name implies, this numerical display consists of seven segments, each of which is energized when it is an integral part of the number that is to be displayed. In the following figure the segments are labeled a through g. Note that when the number 8 is to be displayed, all segments will be energized. When the number 3 is to be energized, e and f will not be lit. The segments can be lit in many different

Logic Elements

combinations to form the numbers 0 through 9. The Karnaugh map is a very neat way to determine the optimum circuit for decoding when a particular segment is to be lit. A few of the segments will be analyzed to determine the decoding circuitry.

- a is lit for the following numbers:

    0, 2, 3, 5, 7, 8, 9   (There will be 1's in these map squares.)

- b is lit for the following numbers:

    0, 1, 2, 3, 4, 7, 8, 9

- f is lit for the following numbers:

    0, 4, 5, 6, 8, 9

This display will be driven by the BCD code. Remember that the 10–15 states are not allowed, and so when we do a mapping for this device, these conditions can be treated as "don't care" states. The maps for the segments listed above would be as shown in figure 3–38.

Solving the a segment map, we find that there are four circles and thus four products to be summed:

$$y = D + \overline{A} * \overline{C} + A * C + B * \overline{C}$$

This sum of products circuit would be implemented with logic elements, and the output would be connected to the a segment of the display. The process would be

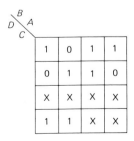
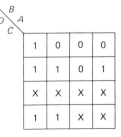

(a) a segment   (b) b segment   (c) f segment

FIGURE 3–38

Decimal	BCD Inputs				Outputs						
Number	D	C	B	A	a	b	c	d	e	f	g
0	0	0	0	0	1	1	1	1	1	1	0
1	0	0	0	1	0	1	1	0	0	0	0
2	0	0	1	0	1	1	0	1	1	0	1
3	0	0	1	1	1	1	1	1	0	0	1
4	0	1	0	0	0	1	1	0	0	1	1
5	0	1	0	1	1	0	1	1	0	1	1
6	0	1	1	0	0	0	1	1	1	1	1
7	0	1	1	1	1	1	1	0	0	0	0
8	1	0	0	0	1	1	1	1	1	1	1
9	1	0	0	1	1	1	1	0	0	1	1
X	1	0	1	0	X	X	X	X	X	X	X
X	1	0	1	1	X	X	X	X	X	X	X
X	1	1	0	0	X	X	X	X	X	X	X
X	1	1	0	1	X	X	X	X	X	X	X
X	1	1	1	0	X	X	X	X	X	X	X
X	1	1	1	1	X	X	X	X	X	X	X

FIGURE 3-39

repeated for each of the segments, and each would require unique circuitry. In general, this decoding is all done either inside a decoding chip or sometimes on the seven-segment display. In the latter case you would be required to connect the BCD code to the display or decoder.

A general truth table that represents the status of all the segments can be constructed, as shown in figure 3–39.

3-22 THE UNIVERSAL NAND GATE

The straightforward implementation of the sum of products solution for the system response will result in a series of multiple-input AND gates feeding into one multiple-input OR gate. With the use of De Morgan's Theorems it can be shown that NAND gates can be used to directly replace both the AND and the OR gates. Since the same equation represents both of the circuits, it is possible to replace one circuit with the other. Generally speaking, the NAND gate is the most prevalent and least expensive of the various gates that are available for purchase. The NAND gate can also be used to generate any of the other standard gate functions, as is shown in figure 3–41. The NOR function will be the same as the OR function with an additional NOT gate added at the output. Figure 3–40(a) is a typical AND/OR circuit implementation. Figure 3–40(b) is the NAND gate equivalent.

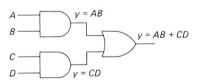

FIGURE 3-40(a) AND-OR; implementation of a straight forward circuit NAND gate equivalent

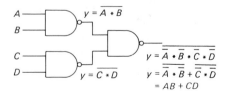

FIGURE 3-40(b) NAND gate equivalent circuit to the AND-OR functions

Logic Elements

Not function:

And function:

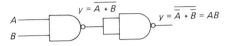

Or function:

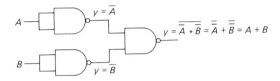

FIGURE 3–41

3–23 SUMMARY

In this chapter you have been exposed to the basic gates with which you must become familiar. You should know the truth tables or be able to derive them for each of the gates that were presented here. The algebra simplification techniques are extremely useful when you are doing design work and an absolute must when you are assigned to do troubleshooting of circuits designed by someone else to their own whims and requirements. The Karnaugh map is an extremely handy tool for simplifying circuits when you are doing the initial design. Quite often, this function will be performed by one of your in-house computers. It is to your advantage to understand how the map is derived and how best to manipulate the map. It is not easy to remember the formulas derived and presented here for algebraic simplification. It is best to have a list of them kept as a ready reference.

PROBLEMS

Using the tools of Boolean algebra developed in this chapter, simplify the expressions in Problems 1 through 10.

1. $y = A + AB$
2. $y = A * B + \bar{A} * B$
3. $y = \bar{A} * \bar{B} + \bar{B} * \bar{C} + \bar{A} * \bar{C}$
4. $y = (A + B + \bar{C}) * \bar{A} * \bar{B} * \bar{C}$
5. $y = (A * B + \bar{C}) + (A + \bar{B} * \bar{C})$

Inputs				Output
D	C	B	A	y
0	0	0	0	0
0	0	0	1	0
0	0	1	0	1
0	0	1	1	0
0	1	0	0	0
0	1	0	1	0
0	1	1	0	0
0	1	1	1	1
1	0	0	0	0
1	0	0	1	0
1	0	1	0	1
1	0	1	1	0
1	1	0	0	0
1	1	0	1	0
1	1	1	0	0
1	1	1	1	1

FIGURE 3–42 Truth table for Problem 11

6. $y = A + \overline{A}B$
7. $y = \overline{A * B} + \overline{A + B}$
8. $y = \overline{A} * \overline{B} * \overline{C} + \overline{A} * \overline{B} * \overline{C} + \overline{A} * \overline{B} * \overline{C}$
9. $y = A * B + \overline{C} * \overline{D} + \overline{A} * \overline{B} * \overline{C} + D$
10. Construct a truth table for the following function, construct a Karnaugh map, and simplify the function:

$$y = C * \overline{B} * A + \overline{C} * B * A + \overline{C} * B * \overline{A}$$

11. Given the truth table in figure 3–42, using a Karnaugh map, simplify the circuit.
12. Given the truth table in figure 3–43, simplify by using a Karnaugh map and draw the circuit that will implement the function.

Inputs				Output
D	C	B	A	y
0	0	0	0	1
0	0	0	1	0
0	0	1	0	1
0	0	1	1	0
0	1	0	0	0
0	1	0	1	0
0	1	1	0	0
0	1	1	1	0
1	0	0	0	1
1	0	0	1	1
1	0	1	0	1
1	0	1	1	1
1	1	0	0	0
1	1	0	1	0
1	1	1	0	0
1	1	1	1	0

FIGURE 3–43 Truth table for Problem 12

Logic Elements

Inputs				Output
D	C	B	A	y
0	0	0	0	0
0	0	0	1	0
0	0	1	0	X
0	0	1	1	X
0	1	0	0	0
0	1	0	1	0
0	1	1	0	0
0	1	1	1	0
1	0	0	0	X
1	0	0	1	X
1	0	1	0	X
1	0	1	1	0
1	1	0	0	X
1	1	0	1	1
1	1	1	0	1
1	1	1	1	0

FIGURE 3–44 Truth table for Problem 13

13. Given the truth table in figure 3–44, simplify by use of the Karnaugh map and draw the resultant circuit.

14. Using NOR gates only, use De Morgan's Theorems to generate the following functions:
    (a) NOT   (b) OR   (c) NAND   (d) AND

15. Show how to use a Karnaugh map to decode the number 9 from a BCD code.

16. Given the truth table in figure 3–45, use the Karnaugh map to simplify.

17. A function has three inputs such that an alarm is desired if any two of the inputs are high. All three inputs *cannot* be high at once. (Treat this condition as a "don't care" condition.)
    (a) Make a truth table for this system.
    (b) Set up and solve a Karnaugh map for the system.
    (c) Implement the simplest possible circuit.

18. A function has three inputs such that an output of 1 is desired if any two of the three inputs are 1 and an output of 0 if all three inputs are high.
    (a) Make a truth table for the system.
    (b) Set up and solve a Karnaugh map for the system.
    (c) Implement the simplest possible circuit.

Inputs			Output
C	B	A	y
0	0	0	0
0	0	1	X
0	1	0	X
0	1	1	1
1	0	0	X
1	0	1	X
1	1	0	1
1	1	1	X

FIGURE 3–45 Truth table for Problem 16

19. A function has four inputs A, B, C, and D, and an output of 1 is desired if any two adjacent inputs are high. Note that A is to be considered as being adjacent to D.
  (a) Make a truth table for the system.
  (b) Solve the Karnaugh map for the system.
  (c) Draw the simplest possible circuit that will implement this function.

20. A function has four inputs A, B, C, and D, and an alarm is desired if any two adjacent inputs are high. Do not consider A as being adjacent to D. Solve the Karnaugh map and implement the function.

21. Show the Karnaugh map for a one-bit comparator that will have a 1 in the output if the A function is less than the B function. Solve the map and draw the circuit that will implement the function.

22. Show the Karnaugh map for a function that will have a 1 in the output when the two inputs are equal. Solve the map and implement the function.

23. Given two three-bit numbers A and B, write an equation such that the output $y = 1$ when the numbers are equal.

24. For the three-bit number in Problem 23, write an equation that will produce a 1 in the output when A is equal to or greater than B.

25. Show that by complementing a Karnaugh map, the equations that represent the function become a product of sums.

26. Given the truth table in figure 3–46, complement the Karnaugh map and solve for a product of sums solution. Using NOR gates, implement a circuit that would produce this function.

27. Given that a function has four inputs, A, B, C, and D, a zero output is to occur any time three adjacent inputs are low. Complement the Karnaugh map and produce a product of sums solution. A is not adjacent to D.

Inputs				Output
D	C	B	A	y
0	0	0	0	1
0	0	0	1	0
0	0	1	0	0
0	0	1	1	1
0	1	0	0	0
0	1	0	1	1
0	1	1	0	1
0	1	1	1	1
1	0	0	0	0
1	0	0	1	1
1	0	1	0	1
1	0	1	1	1
1	1	0	0	1
1	1	0	1	1
1	1	1	0	1
1	1	1	1	1

FIGURE 3–46 Truth table for Problem 26

Logic Elements

FIGURE 3–47 Numerical display for Problems 30–32

28. Given the function in Problem 27, this time consider A and D as being adjacent, and develop a product of sums solution to this problem.

29. Given that a function has four inputs, A, B, C, and D, the input condition of three inputs being low all at the same time is an impossibility. You want a low output when any two nonadjacent inputs are simultaneously high. Use a complemented Karnaugh map to develop a product of sums solution.

30. Given a seven-segment numerical display with the segments shown in figure 3–47, develop an equation for the a segment if the device is driven from an 8421 BCD coded system.

31. If the same system is used as in Problem 30, what is the equation for the f segment?

32. What is the equation for the b segment in figure 3–47?

33. Given the five-input system that is represented by the Karnaugh map, in figure 3–48, where inputs are A, B, C, D, and E, solve for the sum of products solution.

34. Using a Karnaugh map, develop an equation that would be used to compare two two-bit numbers A (A_1, A_0) and B (B_1, B_0) and produce a 1 in the output when A is greater than or equal to B.

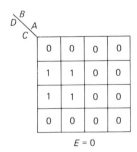 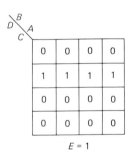

FIGURE 3–48 Karnaugh maps for Problem 33

chapter 4

flip flops or multivibrators

4–1 INTRODUCTION

There are several types of circuits that come under the heading of multivibrators. Monostable, bistable, and astable are three of these types. The circuits are called *flip flops*, a term that describes the action of the transistors or vacuum tubes that are used in the design of these circuits. In this chapter we shall look at a basic circuit built from discrete transistors, then show how the same function can be obtained with the use of the logic gates that were discussed in the previous chapter.

4–2 THE BISTABLE MULTIVIBRATOR

A transistorized version of a very basic bistable multivibrator is shown in figure 4–1. The action of the circuit is as follows: When the power is turned on, one of the transistors will turn on before the other. For analytical purposes, let us assume that Q_1 comes on first and draws base current through the collector resistor of Q_2 and the Q_1 series base resistor. If transistor Q_1 is to go into saturation, we must be able to draw sufficient base current to cause saturation. With sufficient base drive current the collector current is determined by V_{cc} and R_1. The base current is determined by V_{cc} and R_2. The relationship that relates collector and base current is beta (β):

$$\beta = \frac{I_c}{I_b}$$

$$I_b = \frac{V_{cc}}{R_2}$$

Flip Flops or Multivibrators

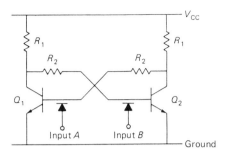

FIGURE 4-1 Discrete bistable multivibrator

$$I_c = \frac{V_{cc}}{R_1}$$

It follows that if the transistor is to go into saturation, the following relationship must hold:

$$\frac{R_2}{R_1} < \beta$$

A good average figure to use for β is 50. This will set the limits on the values for resistors R_1 and R_2.

If we assume that Q_1 comes on when power on occurs, it follows that Q_2 will not have sufficient base current to be ON and hence will be held OFF. The circuit can stay in this state indefinitely. Should an input B current pulse of sufficient magnitude occur to turn the transistor Q_2 on, the results would be regenerative. The decreasing collector voltage of Q_2 is passed on to the base of Q_1, decreasing the base current available to hold Q_1 ON, and Q_1 will *flip* to the OFF state. This is a second stable state that can and will exist until a turn-on pulse is received at the base of Q_1 (input A), at which time the circuit would *flop* back to the first stable state. The circuit has two stable states and is called a *bistable*. In actual practice, one of the collector voltages will be called the Q output and the other will be called $\overline{Q}$. The input that causes the Q output to go high will be termed the S (for *set*) input, and the other input that causes the Q output to go low is termed the R (for *reset*) input. Note that $\overline{Q}$ must always be in the state that is opposite to the state of Q. It is important to note that you cannot apply signals to both inputs at the same time and expect to predict the output of the circuit. The truth table in figure 4-2 represents the responses of this circuit.

4-3 THE R/S INTEGRATED FLIP FLOP

The same circuit response can be obtained with the use of logic gates. Using the NOR gate as shown in figure 4-3 will generate an output response that is identical to the discrete device of figure 4-1. When the circuit is powered on, the circuit

Inputs		Outputs	
R	S	Q	$\bar{Q}$
0	0	No change	
0	1	1	0
1	0	0	1
1	1	Not allowed	

FIGURE 4–2 Truth table for figure 4–1

can go to either of the two stable states. For analysis, let's assume that Q is low and $\bar{Q}$ is high. The cross-connected input to G_1 will be high. Since these are NOR gates, the G_1 output will be forced low. The cross-connected input to G_2 is then low.

Under these conditions the inputs to the NOR gates are as follows:

- NOR gate G_1: Reset and a 1
- NOR gate G_2: Set and a 0

The NOR gate output is a zero if either or both inputs are high. Gate G_1 is inhibited or blocked from responding to the Reset input because the other cross-connected input is already high. On the other hand, gate G_2 has Set and 0 at the inputs. The gate is in a state such that the G_2 output will transition from a 1 to a 0 with the application of a 1 to the set input. This transition will be reflected at the cross-connected input of G_1. As in the previous circuit we will exclude the possibility of both the Set and Reset being high at the same time. It now follows that the output of gate G_1 will toggle from a 0 to a 1. The G_1 output, which we have called Q, has been set to a 1 state. In effect we have *set* the device. By the same logic we can now see that the Set input is effectively blocked and the Reset input is now enabled. Upon receipt of a signal at the Reset input the device will *reset*: $Q = 0$ and $\bar{Q} = 1$.

This circuit is called an *R/S flip flop*. The points to remember are that the Set and Reset refer to the state of the Q output and the Set and Reset inputs are not allowed to be high at the same time. With $R = S = 1$ inputs the circuit is being told to cause the output Q to simultaneously take on the values of 1 and 0. If this

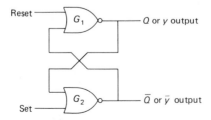

FIGURE 4–3 NOR gate flip flop

Flip Flops or Multivibrators

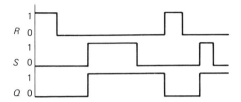

FIGURE 4-4 *R/S* NOR gate response

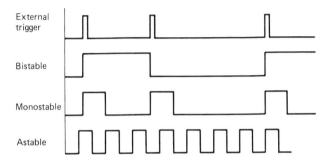

FIGURE 4-5 Response curves to nonsynchronous input excitation

state of $R = S = 1$ is applied to the input, the device will not fail, but the output is undetermined, an unacceptable condition for proper logic design.

The truth table for this circuit is a repeat of the table shown in figure 4-2 for the device made from discrete transistors. Figures 4-4 and 4-5 are response diagrams that demonstrate the responses of the devices to various changes within the circuit. Figure 4-5 shows the general responses of the monostable, bistable, and astable to the required external trigger. The bistable generally changes states upon application of an external trigger. The monostable goes to the unstable state for a specific repeatable time period, then returns to the zero stable state. The astable requires no trigger; it operates at a frequency determined by circuit components.

4-4 THE NAND GATE BISTABLE

A similar bistable latch, or flip flop, can be constructed by using two NAND gates in the same configuration as shown in figure 4-3. The difference in the responses of the two circuits is to the level of the inputs. The Set and Reset inputs will now be active low. The responses are shown in figure 4-6. The low level is the active state, and the $R = S = 0$ input condition is the not-allowed state.

Inputs		Output
R	S	Q
0	0	Not allowed
0	1	0 (Reset)
1	0	1 (Set)
1	1	No change

FIGURE 4-6 R/S flip flop response table

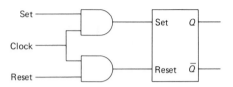

FIGURE 4-7 Clocked R/S flip flop

4-5 THE CLOCKED R/S FLIP FLOP

The device described in the previous section is a level-sensitive circuit that will respond any time an input level is changed at the Set/Reset inputs. In some cases it is highly desirable to cause all the flipping and flopping within a system to be *synchronous*, that is, to occur at the same point in time. If the switching of the devices is to be done synchronously, a master clock will be required to signal when the response is to occur. We use the same device that we described in the previous section; the symbol for this device is a box with S, R, Q, and $\overline{Q}$ indicated in their respective locations. Adding two AND gates at the input will permit us to accomplish the clocking function. (See figure 4-7.)

This circuit will not respond unless the level of the clock is high. When the clock level is high, the device will respond in the same fashion as the previously described R/S flip flop. The truth table that describes this device is shown in figure 4-8.

A further refinement might be desirable. In some cases we might want the output to change only when the clock is changing. This response is called *edge triggering*. This can be accomplished by differentiating (detecting the slope of) the clock pulse. A circuit that can be used to perform this function is a simple *RC* circuit. The time constant of this circuit must be not greater than 1/10 of the width of 1/2 the clock pulse period. (See figure 4-9.) If this circuit is added in front of the clock input to the AND gates, the circuit will respond at the positive-going input of the clock pulse, when the positive-going waveform enables the AND gate

Clock	Inputs		Outputs	
	R	S	Q	$\overline{Q}$
0	X	X	No change	
1	0	0	No change	
1	0	1	1	0
1	1	0	0	1
1	1	1	Not allowed	

X = Does not care

FIGURE 4-8 Response table for a clocked R/S flip flop

Flip Flops or Multivibrators

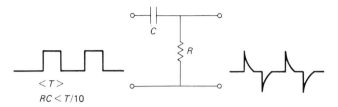

FIGURE 4-9 Discrete differentiator

and allows the Set/Reset function to be active. The symbol for this device (figure 4–10) has a triangle at the clock input to indicate that the device is triggering on the leading edge of the clock pulse. Note that the actual circuitry performing this differentiating function will be located on the chip and might not resemble this RC circuit at all. The response, however, is the same.

Figure 4–11 is a timing diagram for the clocked flip flop of figure 4–10. Figure 4–12 is a timing diagram for an R/S clocked flip flop that is not edge triggered but is level sensitive. A table representation of the response of this circuit is shown in figure 4–13. The device can be made to trigger on the trailing edge of the clock by inserting an inverter at the clock input of the differentiating circuit, as is shown in figure 4–14. The response table is similar to the previous table with the clock slopes changing in the opposite direction.

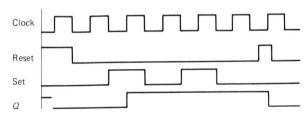

FIGURE 4-10 Symbol for a clocked R/S flip flop

FIGURE 4-11 Timing diagram of a positive edge–triggered flip flop

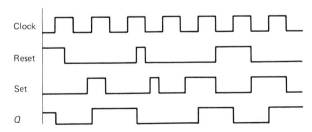

FIGURE 4-12 Response of an R/S flip flop with a level-sensitive clock

As in the discussion of the TTL circuitry in Chapter 2, the time delay experienced by a signal when propagating through a gate is a function of the type of gate. Assume that we have an inverter function that has a time delay of 10 nsec. The same effect that was obtained by the differentiating RC circuit can be obtained by the logic gates arranged as shown in figure 4–15.

The output of the inverter is high when the clock is low. This will enable the input to the AND gate; and upon receipt of the positive-going edge of the clock pulse, the AND output will go high. It will be forced low when the inverter responds to the positive clock by going negative and returning the AND output to the low state. (See figure 4–16.) The output of the AND gate will be high for the length of time that it takes for the pulse to propagate through the inverter (10 nsec). The problem of edge triggering can also be solved by the use of D flip flops, which is covered at the end of Section 4–6.

4–6 THE D FLIP FLOP

One use for flip flops is to store the status of a bit, to latch or hold the 1 or 0 status of a particular bit. A commonly used device is the D latch or flip flop. The device used to create this function is the previously mentioned R/S flip flop, with the Set input connected to the Reset input through an inverter. All the other information remains pertinent. If the R/S is level sensitive, the D flip flop will be level sensitive. If the R/S is edge triggered, the D latch will be edge triggered. In short, all the triggering combinations that are available for the R/S are also available for the D latch. It is impossible to get in to the not-allowed state $(R = S = 1)$ because of the presence of the inverter. A table of responses for the trailing edge–triggered D flip flop is presented in figure 4–17. As an exercise, design a circuit for leading edge triggering of a D flip flop.

The flip flop output of figure 4–18 will assume the level of D but will change state only with the positive edge of the clock. We will examine the circuit behavior of figure 4–18 by examining the logic response for the following cases.

- Case 1: Clock $= 0$. The output of gates 2 and 3 must be 1; therefore Q and $\overline{Q}$ cannot change states. Gates 4 and 1 will toggle if D changes state, but a change in D will have no effect on gates 2 and 3.
- Case 2: Clock changes from 0 to 1. $D = 0$ at that time. The output of gate 4 must be 1. Both inputs to gate 1 are 1; thus the output of gate 1 is 0, and the output of gate 2 is 1. Thus Q will not change.
- Case 3: Clock changes from 0 to 1 with $D = 1$. Both inputs to gate 4 are 1; the output of gate 4 is 0. The output of gate 4 must be 1. Both inputs to gate 2 are 1, the output of gate 2 is 0, and Q changes to a 1.
- Case 4: Clock $= 1$, $Q = 0$, $\overline{Q} = 1$. The output of gate 3 is 0; thus input D cannot get through gate 4 and so will not affect Q.

Flip Flops or Multivibrators

Clock	R	S	Q
⤴	0	0	No change
⤴	0	1	Goes to or stays at the 1 state
⤴	1	0	Goes to or stays at the 0 state
⤴	1	1	Condition not allowed
⤵	X	X	No change

FIGURE 4–13 Clocked R/S flip flop response

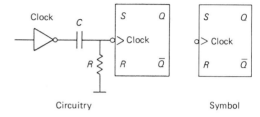

FIGURE 4–14 Trailing edge–triggered flip flop

- Case 5: Clock = 1, Q = 1, and $\overline{Q}$ = 0. The output of gate 2 is 0. The input D can get through gate 4 but not through gate 1; thus Q will be unaffected.

We conclude therefore that D can change any time within the clock cycle but will affect the output Q only when the clock changes from 0 to 1. Thus the circuit can be used to accomplish edge triggering.

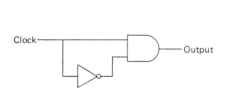

FIGURE 4–15 Edge triggering using chip delay

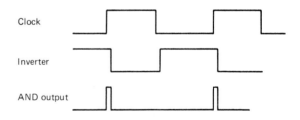

FIGURE 4–16 Chip delay edge-triggering waveforms

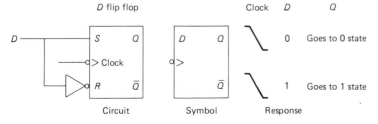

FIGURE 4–17 D flip flop

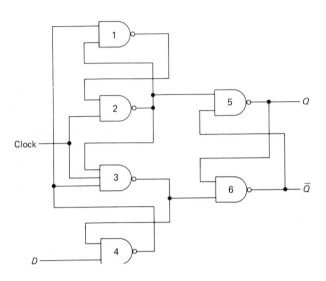

FIGURE 4–18   2N7474 positive edge-triggered flip flop

4–7 THE J/K FLIP FLOP

The most versatile of the flip flops is the *J/K* flip flop, so called because of two inputs that are labeled *J* and *K*. From this flip flop, all the other types can be created. The basic *J/K* flip flop starts with the basic *R/S* flip flop and has added two AND gates with the connections shown in figure 4–19.

Operation of the circuit requires that for the AND gate to activate either the Set or Reset input, all inputs must be high. Q and $\overline{Q}$ are always in opposite or complementary states. Gates G_1 and G_2 will both receive the clock input at the same instant. Note that it will be a differentiated input. If both the *J* and *K* inputs are held low, gates G_1 and G_2 will be held low, and the *R/S* flip flop will receive no forcing inputs; hence the outputs will remain in the no-change state.

For gate G_1 to go high at the output we must have *J* high, the clock must go high, and $\overline{Q}$ must be high, which means that the flip flop is in a Reset state when the clock pulse arrives. If these conditions are met, gate G_1 will go high, setting the Q output high and of course the $\overline{Q}$ output low. In short, we have set the device by programming the *J* input to be high before the clock pulse arrives. Should the flip flop be in the set state at the time of the arrival of the clock pulse, gate G_1 would not respond; since *K* is programmed to the 0 state, G_2 would also fail to respond. The response of the *R/S* to a 0/0 input condition is a no-change situation; the flip flop would remain in the set state. In summary, the *J/K* flip flop will respond to a $J = 1$, $K = 0$ input by causing the Q output to set (go to a high state) or

Flip Flops or Multivibrators

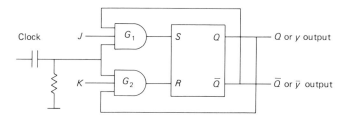

FIGURE 4–19 Positive edge-triggered J/K flip flop

remain set (stay in the high state) *at the time of the arrival of the leading edge of the clock pulse*.

The analysis can be repeated with the condition that $J = 0$ and $K = 1$. For this particular state, G_1 is held off and G_2 is activated only when the output Q is in the set condition. When this condition exists, the output Q will be reset *upon receipt of the next leading edge of the clock pulse after K is taken to a 1 and J is at 0*.

A new and very interesting characteristic of this flip flop occurs for the $J = K = 1$ state. For the previous R/S flip flops this was a not-allowed state. For this device the $J = K = 1$ state results in a condition that will have one of the gates being activated at every clock pulse. The flip flop will thus respond with the Q output changing states with the arrival of the clock pulse. The flip flop is described as being in the toggling mode. When used this way, it can be called a *T* flip flop. A summary of the responses of this J/K is shown in figure 4–20.

The normal operation of the J/K flip flop requires that the J and K inputs be programmed to their desired values a finite time before the arrival of the clock pulse. This will be specified as the setup time. This is the minimum amount of time that the J and K inputs must be held at a stable voltage before the arrival of the clock pulse. It is important to remember that the flip flop will not respond to changes at the J/K inputs until the arrival of the clock pulse.

Clock	J	K	Q
↗	0	0	No change
↗	0	1	If at 0, stays; if at 1, goes to 0
↗	1	0	If at 0, goes to 1; if at 1, stays
↗	1	1	Q Output changes state (toggles)

FIGURE 4–20 J/K response table

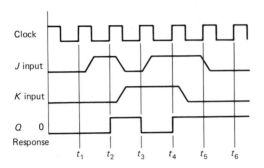

FIGURE 4–21 *J/K* flip flop timing diagram

EXAMPLE 4–1

It should prove helpful to present a possible timing diagram for this particular *J/K* flip flop, as shown in figure 4–21. Note that in this example you are not to concern yourself with how or why the *J* and *K* inputs are programmed. These are determined by someone who wants the circuit to respond in a particular manner. We are concerned only with the *Q* response to the programmed *J/K* inputs.

In this example we must be told that the *Q* output is in the 0 state before the arrival of clock pulse t_1. The responses to clock pulses are as follows:

1. Upon arrival of clock pulse t_1 the *J* and *K* inputs are both in the 0 state. Our response table tells us that for this *J/K* condition the *Q* output will not change; hence *Q* will remain in the 0 state between clock pulses t_1 and t_2.
2. Upon arrival of clock pulse t_2 the *J* input is a 1 and the *K* input is a 0. The response table tells us that the *Q* output must go to a 1 at this time and remain that way until the arrival of the next clock pulse.
3. Upon arrival of clock pulse t_3 the *J* input is a 0 and the *K* input is a 1. From the response table we find that the *Q* output must go to a 0 and remain that way until the arrival of t_4.
4. Upon arrival of clock pulse t_4 the *J* and *K* inputs are both 1. The response to $J = K = 1$ is for the output to toggle. In this case the output will toggle from the 0 to the 1 state and remain that way until the arrival of t_5.
5. Upon arrival of clock pulse t_5 the *J* is found in the 1 state and the *K* is in the 0 state. The response is for the output to be a 1 after the arrival of the clock pulse. Since the *Q* output was a 1 before the arrival of t_5, it will remain in the 1 state until the arrival of t_6.
6. Upon arrival of clock pulse t_6 the *J* and *K* inputs are both in the 0 state, so the *Q* output is programmed for a no-change state. It will remain in the 1 state. ■

4–8 THE *J/K* FLIP FLOP WITH PRESET AND CLEAR

In the previous section we examined how the flip flop responded to changes at the *J* and *K* inputs. It was determined that the flip flop responds to changes at the *J/K* inputs only at the arrival of the next clock pulse. We will examine the response of

Flip Flops or Multivibrators

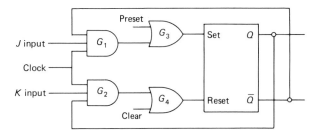

FIGURE 4–22 *J/K* with Preset and Clear

the flip flop to changes at two new inputs, the Preset and Clear inputs. These inputs will be found to be independent of the clock and have priority over all other inputs. There is a limitation on their usage, which is the same as that placed on the *R/S* inputs to the *R/S* flip flop. You cannot call for the output to be preset and cleared at the same time; you would be requesting the *Q* output to be both a 1 and a 0 at the same time. The circuitry that can implement the Preset and Clear functions requires the addition of an OR gate between the output of the AND gate and the *R* or *S* input to the flip flop of figure 4–22.

The Preset and Clear are ORed with the outputs of the G_1 and G_2 AND gates. If you take either the Clear or Preset to the 1 state, the OR gate is activated, and the resultant function—Set or Reset—is activated. Gates G_3 and G_4 follow the AND gates and take priority over the three ANDed functions. The circuit as shown would have the Preset and Clear functions active when they are taken to a high or 1 state. It is quite simple to add an inverter at the input to the OR gates and cause the Preset and Clear functions to become active when in the low or 0 state. The symbols for both gates are shown in figure 4–23.

---- EXAMPLE 4–2

A timing diagram for this circuit (see figure 4–24) will show the immediate response of Preset and Clear inputs. Remember that in such diagrams the *J*, *K*, Preset, and Clear inputs are preprogrammed and are arbitrary. You are to determine only the reaction of the circuit to these inputs. The dependent function is the *Q* output response. You should be able to draw the *Q* response for the given inputs. The

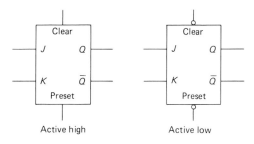

FIGURE 4–23 Symbol for *J/K* with Preset and Clear

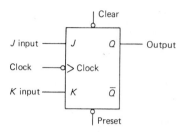

FIGURE 4–24 *J/K* flip flop used for timing diagram

circuit that will be used for this exercise is the active high *J/K* that responds to the falling or trailing edge of the clock pulse, with Preset and Clear active low.

The responses shown in figure 4–25 are as follows:

1. Prior to t_1 the Clear input is low and active, forcing the Q output to zero.
2. At the arrival of clock pulse t_1 the Clear input is still active, overriding everything, and the Q output will stay low.
3. Upon the arrival of clock pulse t_2 the Clear input has been released, but the *J* and *K* inputs are both at 0. This is a no-change condition; thus the Q output remains in the low or 0 state.
4. Upon the arrival of clock pulse t_3, $J = 1$ and $K = 0$. This forces Q to a high or 1 state, so Q will toggle to the high state at the arrival of clock pulse t_3.
5. Upon the arrival of clock pulse t_4 the *J* and *K* inputs are both high. This is a toggle condition, and the Q output will go to the low or 0 state at this clock pulse.
6. Upon the arrival of clock pulse t_5, $J = 0$ and $K = 1$, which forces the Q output to stay in the low or 0 state.

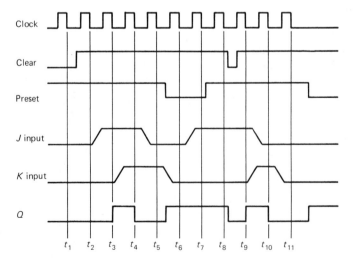

FIGURE 4–25

Flip Flops or Multivibrators

7. Sometime after clock pulse t_5 and before clock pulse t_6, the Preset goes low and becomes active, forcing the Q output to set (go high). It stays high as long as the Preset is low and will remain there until the next clock pulse. In our example that will be clock pulse t_8.
8. Upon the arrival of clock pulse t_8 the $J = 1$, $K = 0$ condition exists, and the flip flop will hold in the high state.
9. Between clock pulses t_8 and t_9 the Clear input goes low and forces the Q output to respond by going to the low state immediately. This Clear input is a narrow pulse and does not last the width of a clock cycle. It forces the Q output to stay low until the arrival of the next clock pulse, t_9.
10. Upon arrival, clock pulse t_9 finds the condition $J = 1$ and $K = 0$. The output Q responds by going high at the clock pulse t_9.
11. Upon arrival, clock pulse t_{10} finds the state $J = 0$, $K = 1$ and responds by taking Q to the low state.
12. Upon the arrival of the last clock pulse, t_{11}, J and K are in the low state, dictating a no-change condition, thus holding the Q output low.
13. After the last clock pulse has been experienced, the Preset goes low. This forces the Q output to go high, and it will stay there indefinitely. (See figure 4–25.) ■

Several problems at the end of the chapter will test your ability to complete a timing diagram. It is sometimes easier to first chart the response of the Clear and Preset inputs. They have a higher priority and do override the J/K inputs. Plot these Clear/Preset responses first, then fill in the J/K response at the next clock pulse.

4–9 THE RACE PROBLEM

A very serious problem that can arise in using a J/K flip flop is the race problem. The race is between the speed of response of the circuit and the clock pulse. If the speed of response of the chip is faster than the clock pulse, the device will give you erroneous results. To understand the problem, examine the J/K flip flop and remember the programmed response. Forget the Clear and Preset inputs for the moment. They do not contribute to the problem directly.

In figure 4–26 we have an elementary J/K flip flop. One of the conditions that makes the J/K flip flop more universal than other types of flip flops is the input condition that calls for $J = K = 1$. In the previous section we noted that this condition causes the device to toggle. This state will be used in Chapter 5 to develop counters. The counters that will be examined will count input clock pulses.

To obtain an accurate count of the clock pulses, it is imperative that the device does not toggle more than once for each clock pulse. Suppose that we toggle the clock pulse high and have the condition set up for $J = K = 1$. If we have one clock pulse, we can accept only one toggling of the Q output for each clock pulse. Note that in this circuit the clock pulse is not differentiated. Assume that before the arrival of the clock pulse, Q is in the low state. It really does not matter which

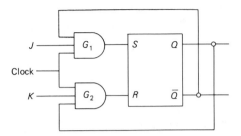

FIGURE 4–26 *J/K* flip flop from an *R/S* flip flop

state we choose; the results will be the same. With $Q = 0$, G_2 is held OFF and $\overline{Q}$ being high will enable G_1. From Chapter 2 we know that these TTL devices will respond (toggle) in 10–20 nsec. Immediately after Q has changed state from a 1 to a 0, gate G_1 will be held OFF and G_2 will be enabled. If the clock remains high, the *J/K* will toggle again, back to the 0 state. It will continue to toggle between these two states for as long as the clock is present on the clock input. If the clock lasts for 1 μsec and the speed of response of the circuit is 20 nsec, the device will toggle $1000/20 = 50$ times. This is an unacceptable, erroneous condition.

In this particular example there is no race at all. The device is simply not usable. If the differentiating circuit is used to reduce the time that the clock pulse is high, then we do have a race. If the clock pulse gates the AND gate ON and then disappears before a second toggle occurs, then you have won the race and the circuit would do the job for you. If the clock remains at the clock input longer than the circuit response time, the race is lost and your system has failed. Unfortunately, you might build a breadboard of the circuit that works, but when you go into production and install a different chip of the same type, you might find that the circuit will not function anymore. This is a marginal design, to be studiously avoided. The problem can be easily avoided by using a master/slave *J/K* flip flop.

4–10 THE MASTER/SLAVE J/K FLIP FLOP

The master/slave *J/K* flip flop is constructed by using two *J/K* flip flops. The output of the first one is used to set the *J/K* inputs of the second. The first is called the master, and the second is called the slave. The master has its *J/K* terminals available for programming. The master responds on the leading edge of the clock pulse, and the slave responds on the trailing edge of the clock pulse. (See figure 4–27.)

The slave is never placed into the $J = K = 1$ mode; it is in either a Set or a Reset mode. The slave follows the lead of the master. There is no race condition in the master because the AND gate that sets the status of the master flip flop is controlled from the output of the slave; and since the slave output will not change until the trailing edge of the clock, the race condition for the master is eliminated. The master responds to the leading edge and the slave to the trailing edge of the clock pulse. The race problem is thus eliminated.

Flip Flops or Multivibrators

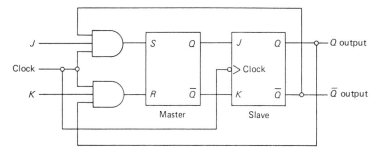

FIGURE 4-27 Master/slave J/K flip flop

Observe what happens during a typical clock pulse (see figure 4–28):

1. The input levels for the master *J/K* inputs must be established before the arrival of the leading edge of clock pulse t_1.
2. At time t_1 this master *J/K* status is transferred to the output of the master *J/K*, which are the *J/K* inputs to the slave.
3. Upon the arrival of the trailing edge of the clock pulse, the slave is caused to respond to the *J/K* status programmed for it by the master. After the trailing edge has gone by, the device has completed a response cycle.
4. During the time interval t_2 to t_3 the *J/K* inputs to the master can be reprogrammed to obtain the desired response at the next clock pulse.
5. Upon arrival of clock pulse t_3 the master will respond to the *J/K* inputs by transferring this information to the *J/K* inputs of the slave.
6. Upon arrival of clock pulse t_4 the slave output will have responded to the *J/K* programming that occurred prior to t_3.

4-11 THE MONOSTABLE MULTIVIBRATOR

The previous sections were devoted to the bistable, which has the two stable states. In this section we will discuss the *monostable*. This device has only one stable state, and if we are talking about the *Q* output, the stable state will be 0. The output can be forced into a 1 state, but this will be an unstable state, and after some time the device will return to the 0 state. These devices are sometimes called *one shots*. They are used to generate time delays, fixed or variable periods of time that can be used to enable circuits, or as it is sometimes described as opening a gate. A

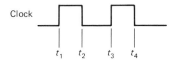

FIGURE 4-28

basic circuit constructed from discrete devices is shown in figure 4–29. The timing diagrams and waveforms are shown in figure 4–30.

The operation of the circuit is as follows: Transistor Q_2 is normally in the saturated ON state. It is held in saturation because the base of Q_2 is connected to V_{cc} through resistor R_5. This resistor must be sufficiently low to allow enough base current for Q_2 to go into saturation. Transistor Q_1 is held off because of the feedback coupling between the emitters of Q_1 and Q_2. The actual voltage on the emitter can be calculated:

$$V_{emitter} = V_{cc} \times \frac{R_4}{R_4 + R_6}$$

If the base voltage of Q_1 is not 0.7 V greater than the emitter, then Q_1 will be held in the OFF state. Since Q_1 is OFF, there is no collector current, and the collector voltage on Q_1 is at V_{cc}. The base voltage of Q_2 is one diode drop (0.7 V) above the previously calculated common emitter voltage. The voltage across the capacitor is V_c (Q_1 collector) minus the voltage on the base of Q_2:

$$\text{Base voltage} = V_{cc} \times \frac{R_4}{R_4 + R_6} + 0.7V$$

$$V_{capacitor} = V_{cc} - (V_{cc} \times \frac{R_4}{R_4 + R_6} + 0.7 \text{ V})$$

These are the stable state voltages for the monostable during the time when it is in the OFF mode. When the voltage on the base of Q_1 is raised to a value that is sufficiently high to turn on transistor Q_1, two immediate responses will occur. First, the collector voltage of Q_1 will drop because Q_1 is now conducting collector current. Second, the common emitter voltage will tend to increase, further decreasing the drive to the base of Q_2. This is positive feedback, and the circuit is said to be *regenerative*. The collector of Q_1 is connected to the base of Q_2 via the capacitor

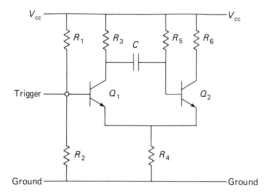

FIGURE 4–29 Discrete monostable

Flip Flops or Multivibrators

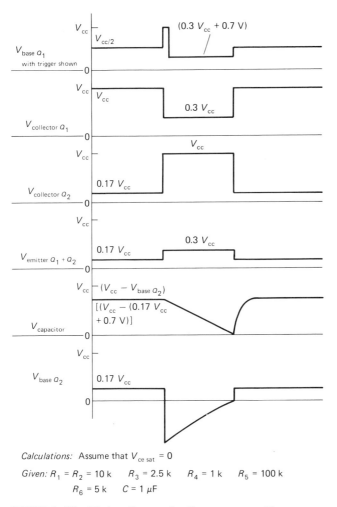

Calculations: Assume that $V_{ce\,sat} = 0$
Given: $R_1 = R_2 = 10\,k$ $R_3 = 2.5\,k$ $R_4 = 1\,k$ $R_5 = 100\,k$
$R_6 = 5\,k$ $C = 1\,\mu F$

FIGURE 4-30 Timing diagram for discrete monostable

C. An important consideration is that the voltage across a capacitor cannot change instantaneously. The signal that turns on Q_1 will cause a quick drop in Q_1 collector voltage. This signal is coupled through the capacitor C to the base of Q_2. This forces the base voltage of Q_2 to drop the same amount as the collector voltage of Q_1. The immediate effect on Q_2 is twofold. The base voltage is lowered, and the emitter voltage is raised. Q_2 will be rapidly cut off. Once Q_2 has been cut off, there will be a charging current into the right side of the capacitor C. This side of the capacitor will be charging toward V_{cc} with a time constant that is determined by R_5 and C. The larger the value of this product, the more slowly the capacitor will charge. When the circuit has gone into this the ON state, the common emitter voltage will now be determined by the R_1 and R_2 combination. The emitter voltage will be one diode drop less than the base voltage of Q_1. This value will be

$$V_{\text{emitter}} = V_{cc} \times \frac{R_2}{R_1 + R_2} - 0.7 \text{ V}$$

Since the right side of the capacitor C is climbing toward V_{cc}, it follows that it must go through a voltage at which Q_2 will again turn on. At this point, Q_1 will be forced OFF, and the device will have returned to the stable state. The device will remain in this stable state until another trigger is applied to force Q_1 into the ON state. Q_1 will stay on only long enough for the capacitor to charge to the voltage that is sufficient to return Q_2 to the ON state. The time required for the capacitor to charge to the voltage that returns Q_2 ON is the time that this one shot is ON. It is a function of R and C, and it can be made relatively constant and accurate. The device is called a one shot because it gives you one output pulse for each input trigger. There are many different ways to trigger the device into the ON mode, but the reset action is basically the same. Chips are available that can perform this function for you. The old reliable 555 chip remains quite popular for this function. A 74122 is also available. The 555 will be discussed further in Section 4–13.

4–12 THE ASTABLE MULTIVIBRATOR

We have looked at the bistable multivibrator, which has two stable states, and the monostable multivibrator, which has only one stable state. We will now focus on the *astable multivibrator,* which has no stable state. It continually switches from the ON to the OFF state. This is, of course, the description of a free-running oscillator, which we will be able to use as a clock for our circuitry.

The discrete circuitry that performs this function is shown in figure 4–31. The two sections are identical—that is, $R_1 = R_3$, $R_2 = R_4$, $C_1 = C_2$—and the transistors Q_1 and Q_2 are identical. Upon power-on condition, one of the transistors will turn on first. Let's assume that it is Q_1 for this analysis. Q_1 turning on draws collector current through R_2, which results in a drop in the collector voltage of Q_1. This voltage drop is transmitted via C_2 to the base of Q_2 as a negative pulse, which forces the transistor Q_2 to be held OFF. The right side of capacitor C_2 is now charging toward V_{cc}. It will pass through the voltage that is necessary to turn Q_2

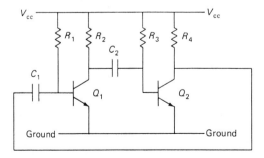

FIGURE 4–31 Discrete astable multivibrator

Flip Flops or Multivibrators

on. In this case the voltage will be 0.7 V; both emitters are returned to ground. When the right side of capacitor C_2 reaches the value 0.7 V, Q_2 will turn on, causing Q_1 to turn off via the coupling capacitor C_1.

The device will continue to oscillate between these two states. It is a *free-running oscillator*. It could be used as a synchronizing clock for your digital circuitry. These discrete devices are rarely used. The functions can be completely contained within a chip. If you are required to design a discrete device, consult just about any transistor text for the design equations. Many chips are available that will perform these functions and are rather inexpensive.

4–13 THE 555 TIMER

A rather old but reliable multipurpose chip that is available for generating both the one shot and the free-running oscillator is the 555. (The 556 is a dual 555.) A sketch of what is in the chip and a functional description follow. The triangle will be used as a symbol for a voltage comparator. In linear circuit analysis the triangle is used to represent an operational amplifier (op amp). (See figure 4–32.) The useful characteristic of the op amp that we wish to use is the very high gain ($\sim$100,000). We must have a differential input to the device such that the input comparing levels can be chosen to be some value other than zero.

For analytical purposes, assume that the gain A is 100,000. Since the input is differential, the output voltage is 100,000 times the difference between input A and input B. If the maximum output voltage variation is limited by power supply considerations to 10 V, then the input change that can cause this output voltage change is 10 V/100,000, or 100 μV. This input voltage swing will cause the output to respond with a full 10 V swing ($+5$ V to -5 V). This analysis is for the typical op amp comparator—not those used in the 555, which uses only a positive voltage $+V_{cc}$.

Assume that input A is set to 1.00000 V. If the input B voltage starts at 0 V and increases toward $+5$ V, the output will start at $+5$ V and remain there until input B is within 100 μV of input A. The output voltage will begin to change. In our analysis we shall assume that the output will start changing when the input B voltage is 50 μV less than the voltage on input A and complete the change when this B voltage is 50 μV greater than the input A voltage. For this example we are using input A at 1.00000 V, but recognize that it can be set to any voltage within the limits of your power supplies minus a small offset voltage that will be specified by the manufacturer.

In this example the output will change the full 10 V (from $+5$ V to -5 V)

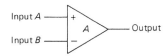

FIGURE 4–32 Operational amplifier symbol

when the input varies from 0.99995 V to 1.00005 V. It follows that we can use either of the inputs as a reference, the other input being our sense input. When the variable input passes within ±50 μV of the reference input voltage, the output voltage should begin toggling. In this analysis, input A is our reference, and the output toggles from +5 V to −5 V when the input B voltage goes through the toggle region. If we had chosen input B as our reference and caused input A to increase through the toggle region, the output would have transitioned from −5 V to +5 V. Be aware of the considerable degree of flexibility that is available form this simple circuit. The 555 chip uses comparators to increase level sensitivity.

Operation of this circuit when it is being used as a monostable or one-shot will require the following connections: The trigger input is held at a dc voltage that is greater than 1/3 V_{cc} (the voltage that is on the other input to comparator C_2). Comparator C_2 will be caused to toggle when this trigger input voltage is caused to go less than $V_{cc}/3$. For this circuit the R/S flip flop will normally be in the Set mode, which means that the output will normally be low. (See figure 4–33.)

An external timing circuit must be supplied to cause the device to act as a monostable. A resistor R_1 will be connected between the discharge terminal and V_{cc}. A capacitor must then be connected between the discharge terminal and ground. This is the timing circuit that determines the width of the pulse during which time the output is caused to go high, the unstable period for this monostable. Note that during the stable period of operation the Q output from the R/S flip flop will be high and the discrete transistor T_1 will be held in saturation, clamping transistor T_1 to ground. The current that is flowing through R_1 will be shunted to ground through transistor T_1. The capacitor will be held at $V_{ce\ sat}$ (~0.1 V) above ground. Upon arrival of the trigger pulse the R/S flip flop will be reset such that Q goes low and $\overline{Q}$ goes high. This action will remove the base current from the discrete transistor T_1, in effect removing it from the circuit.

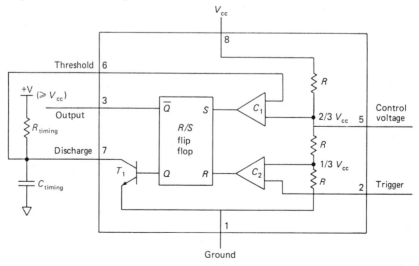

FIGURE 4–33 The 555 timer

Flip Flops or Multivibrators

With transistor T_1 now out of the circuit, the timing capacitor will receive the current that was previously shunted to ground through T_1. C will charge toward V_{cc}. If no other connections were made, C_1 would charge to V_{cc} and the circuit would stay in the ON state indefinitely. This is not our definition of a monostable, so some other connections are required to ensure proper operation. By connecting the threshold input to the capacitor (discharge terminal) the device will sense when the capacitor voltage reaches 2/3 V_{cc}, and comparator C_1 will cause the R/S flip flop to be set. This will cause the discrete transistor T_1 to be forced into saturation and will result in the very low saturation resistance of T_1 to be placed across the capacitor C, in effect causing it to be discharged in a rather short period of time.

The previously mentioned connections are the only ones that are really required to make the device function as a monostable multivibrator. Normally, the control voltage output is bypassed to ground with a .01 µF capacitor. Any variations in this threshold voltage would be reflected as a change in the terminating voltage of the charging capacitor and a subsequent variation in the width of the output pulse. The timing of the device is determined by the R/C charging circuit and by the fact that the device actually charges from 0 V to 2/3 V_{cc} before it is automatically discharged. From your transient analysis course you might remember that the equation that relates charging voltage to time is as follows:

$$\text{Capacitor voltage as a function of time} = v_c(t)$$
$$v_c(t) = V_{cc}(1 - e^{-t/RC})$$

We know the following:

1. Voltage on capacitor starts at 0 V.
2. We are interested in how long it takes the capacitor to charge to the voltage 2/3 V_{cc}. This will be the time that the monostable is ON and will be the width of our output pulse. This time we will call T.

Substituting, we get

$$\frac{2}{3}V_{cc} = V_{cc}(1 - e^{-T/RC})$$

V_{cc} is multiplying both sides of the equation and so can be dropped from the equation completely. We can say that the pulse width is independent of V_{cc}.

$$\frac{2}{3} = (1 - e^{-T/RC})$$

$$-\frac{1}{3} = -e^{-T/RC}$$

$$\frac{1}{3} = e^{-T/RC}$$

Take the natural log of both sides of the equation.

$$\ln \frac{1}{3} = -T/RC(\ln e)$$

$$-1.098 = -T/RC$$

Solving for the pulse width T, we have

$$T = 1.1\, RC$$

From this equation we determine that the width of the pulse is determined only by the R/C combination. If we choose good components that do not vary greatly with time or temperature, we can expect a reasonably accurate, repeatable pulse width.

EXAMPLE 4–3

Design a monostable that will have an output pulse width of 1.0 ms.

$$T = 1.1\, RC = 0.001 \text{ seconds}$$
$$RC = 0.000909$$

This is one equation with two unknowns, which cannot be solved until we have chosen a value for one of the unknowns. You may choose a resistor or a capacitor. Normally, there are fewer common values of capacitance to choose from, so in most cases you would choose a capacitor and use the calculated resistance from the myriad of values of resistors that are available. Once the capacitor has been chosen and the resistor calculated, check to see that the value of resistance that you have calculated is practical. You must be sure that this value is not too low. If it is too low, there will be excessive current drain through the ON transistor when the device is in the OFF state.

    Choose a resistor value to be at least ten times greater than the estimated value of the ON resistance of the transistor. The selection process becomes a two-step procedure: Choose a capacitor value, then calculate the resistor. If the resistor is too small or too large, then readjust your capacitor up or down by as many decades as you desire. In choosing a very large resistor you must be aware that extremely high-value resistors are very expensive and can be less stable than desired, subject to variations in resistance as a function of humidity and even of dust accumulation on the resistor. For an average design it is good design practice to keep the resistor value between 5 kΩ and 10 MΩ. Obviously, you can go outside these values and the circuit can be made to function; but if you are going into production, the resultant variation in the output pulse width might be greater than you can tolerate. For this example, choose $C = 1\, \mu F$. Then

$$R = \frac{0.000909}{0.000001} = 909\, \Omega$$

Flip Flops or Multivibrators

This value of resistance is too low, so we will decrease the chosen C value by two decades to 0.01 μF. This allows us to increase R by the same two decades to 90,900 Ω. Note that the accuracy of your pulse width cannot be any better than the accuracy of your chosen components. If you choose 10% resistors and capacitors, then the output pulse width cannot be predicted to better than this 10% figure. Be very careful with capacitors; the larger values are sometimes not guaranteed to better than $-10, +100\%$. ∎

The trigger circuit must have a pulse width of less than the output pulse width. Otherwise you would have to differentiate this waveform. We have not discussed the Reset input; this input is normally held at V_{cc} unless you either wish to terminate the pulse before it would normally time itself out or wish to inhibit the pulse completely. In either case this function is performed by taking the Reset input to the low or 0 voltage state. A manual triggering circuit can be fashioned from a push-button switch, two resistors, and a capacitor. (See figure 4–34.)

The operation is as follows: With S_1 normally open, the trigger input will stay at V_{cc}, a nontriggering condition. Resistor R_2 will keep capacitor C discharged. When S_1 is depressed, the voltage at the trigger input will be instantly grounded. The device will then be triggered. The trigger input will charge back to V_{cc} through the $R_1 \times C$ time constant. Note that this time must be much less than the desired width of the output pulse. Upon release of S_1, resistor R_2 will again discharge C, priming the circuit for the next manual trigger pulse. (See figure 4–35.)

4–14 THE 555 AS AN ASTABLE MULTIVIBRATOR (A CLOCK)

Reviewing the operation of the 555 chip, we see that the device will trigger when the trigger input is taken from a voltage greater than $V_{cc}/3$, through the $V_{cc}/3$ level, and on down toward ground. When this occurs, the voltage on the timing capacitor will begin to charge toward V_{cc}. When it passes through the 2/3 V_{cc} point, the threshold sensing will dictate that the discharge circuit will initiate the discharge

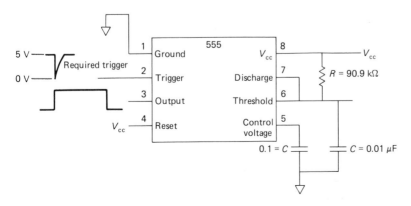

FIGURE 4–34 Pulse-triggered monostable

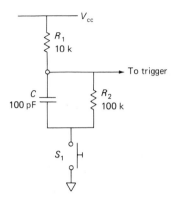

FIGURE 4–35 Manual triggering of a 555 timer

of the timing capacitor C to ground. If left to its own devices, the capacitor would discharge to ground. To make this a free-running or astable multivibrator, we need only tie the trigger input to the discharge terminal of the timing capacitor C. When they are tied together, the capacitor will start to discharge toward 0 V. When it passes through the 1/3 V_{cc} level, the trigger input will be initiated, and the timing capacitor will begin charging toward V_{cc} again. When the timing capacitor reaches the 2/3 V_{cc} level, the threshold sense will initiate discharge again, and the capacitor will again be headed toward 0 V. When it passes through the 1/3 V_{cc} level, the trigger will be reinitiated. The device will continue in this mode of operation indefinitely.

For the previous example, if we were simply to tie the trigger input to the discharge terminal, the ON time would be altered because we are not charging from 0 V to 2/3 V_{cc}. The capacitor will charge from 1/3 V_{cc} to 2/3 V_{cc}. In our previous example,

$$T_{on} = 0.693RC = 0.693 \times 0.00000001 \times 90909 = 0.63 \text{ ms}$$

The discharge time, or OFF time, is the time required to discharge the capacitor from 2/3 V_{cc} to 1/3 V_{cc}. The discharging resistor is not the same as the charging resistor. The discharge resistor is the ON resistance of the discrete discharge transistor Q_1. This value should be somewhere in the region of 100 Ω, and

$$T_{off} = 0.693 R_{transistor} \times 0.01 \text{ μF}$$
$$= 0.693 \times 100 \times 0.00000001 = 0.693 \text{ ms}$$

Figure 4–36 is the output pulse waveform with these external connections.

To increase the discharge time and widen the OFF pulse, we must find a method of inserting some resistance in the discharge path. There are several possibilities. The first thought that comes to mind is to insert the same value of resistance for discharge that you have for charging, and you should end up with a symmetrical waveform, with equal ON and OFF times.

Flip Flops or Multivibrators

FIGURE 4–36 Monostable output pulse

There is a serious problem with that concept. Review this circuitry in figure 4–37. Notice that if the charging resistor equals the discharge resistor and if we assume an ON resistance of only 100 Ω for the transistor, the voltage on the capacitor will be discharged only to $V_{cc}/2$, and the trigger will never reach the required 1/3 V_{cc} for it to restart the charging cycle.

If we are not concerned about obtaining a symmetrical waveform, it is functionally correct to insert a resistor as shown in figure 4–38. One must be certain to limit the value of this resistance to a value that will allow the discharge point and hence the trigger input to drop below the 1/3 V_{cc} level. This will require that the $R_{discharge}$ when added to the ON resistance of the transistor must be less than half the value of the charging resistor. A second connection that will produce a nonsymmetrical waveform is shown in figure 4–39. The capacitor charges through both resistors R_1 and R_2 but discharges through only R_2 (plus the ON resistance of the discharge transistor).

A symmetrical ON/OFF time can be obtained by making $R_1 = R_2$ in figure 4–39 and then bypassing R_2 in the charging mode with a diode. In this mode of operation the capacitor charges through R_1 and the diode; then when discharging, the diode is reverse biased, and resistor R_2 is the discharge resistor. See figure 4–40 for equal ON/OFF time circuitry.

4–15 THE 74121 AND 74123 MONOSTABLE MULTIVIBRATORS

These two chips are dedicated to monostable operation. The manufacturer's specifications supply one with the necessary design equations for designing a monostable. Each chip has two independent monostable circuits. The inputs are Schmitt triggers, which will be discussed later in this chapter. They sharpen any slowly changing input pulses. These devices are operational over six decades of timing capacitors

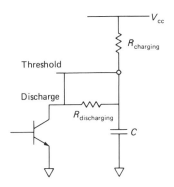

FIGURE 4–37 Charge/discharge circuit

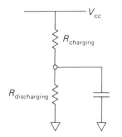

If $R_{charging} = R_{discharging}$, the lowest voltage that this point will go to is $V_{cc}/2$.

FIGURE 4–38 Discharging circuit

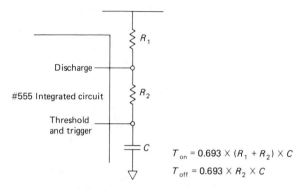

FIGURE 4–39 Charging circuit with nonequal ON/OFF times

(10 pF to 10 μF) and a full decade of resistance (2 kΩ to 40 kΩ). The device has a built-in resistance of 2 kΩ and a minimum internal capacitance such that if no external components are used to generate a timing pulse, the minimum pulse width obtainable is typically 35 nsec.

The timing equation for the 74121 is

$$t_w = 0.7\, R_{ext} \times C_{ext}$$

The equation for the 74123 is

$$t_w = \sim 0.28\, R_{ext} \times C_{ext}$$

When designing with these two devices, be careful to return the capacitor to the C_{ext} terminal; the normally used return to ground for the capacitor will not work with these two devices.

The difference between the two devices is that the 74123 monostable is retriggerable. A monostable is retriggerable if the timing cycle can be restarted before

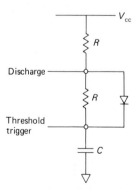

FIGURE 4–40 Charging circuit with equal ON/OFF times

Flip Flops or Multivibrators

a previous one has been completed. The 74121 is nonretriggerable. An example of a use for a retriggerable monostable would be a missing pulse detector. If you monitor a waveform with a particular frequency f_0 and make the t_w (ON) time of the 74123 slightly longer than one period of the f_0, but less than twice the period of f_0, the monostable would return to its normal OFF state only if a pulse is missing in the f_0 pulse string.

4–16 THE CRYSTAL OSCILLATOR

Most clocks encountered in logic and microprocessor systems are crystal controlled. These clocks are extremely accurate and stable with time and temperature. Most digital systems requiring precise clock timing will use either a crystal or the 60 Hz power line as a reference. The crystal will be cut to oscillate at a specific frequency. Many of the systems that are currently in use utilize crystals cut to resonate around 1–2 MHz. Chips are available that require only a connection to the crystal, all other circuitry being supplied. Any and all clocks that are used in a crystal-controlled system will derive the clocks by counting down from the higher crystal frequency.

EXAMPLE 4–4

Suppose you have a 2 MHz crystal-controlled oscillator and your system requires a signal of one pulse per second. A frequency divider or counter circuit would be required to reduce the two million pulses per second to one pulse per second. You would require a circuit to divide by two million. If one flip flop divides by 2, how many flip flops would be required? Solve for n: $2,000,000 = 2^n$. For this case, n would be 21. Twenty-one stages of flip flops would be required ($2^{21} = 2,097,152$). ■

4–17 THE SCHMITT TRIGGER

The Schmitt trigger is a bistable multivibrator that has some very special applications. It can be used as a voltage level–sensing switch. It can be used for improving the rise times of pulse trains. If your data or clock edges are slow in comparison to the rest of your circuit, quite often you can improve circuit performance by using a Schmitt trigger on these inputs to sharpen the edges of your pulses.

Assume for a start that the variable input voltage is set to $-V_{cc}$, which is shown as -5 V in figure 4–41. This condition will force transistor Q_1 to be OFF. The required 0.7 V drop from base to emitter is not present. Since Q_1 is not conducting, it can be treated as an open circuit. The resultant condition for transistor Q_2 has the base tied to a voltage that is determined by the voltage divider R_1, R_4, and R_5. It can be assumed that Q_2 will be ON and in saturation. (We will assume an ideal transistor with 0 V drop from collector to emitter when in saturation.) There is now a current path through transistor Q_2 in series with R_2 and R_3. The sum of R_2 and R_3 will determine the value of this current. In this example the current will be 10 V/(4 + 1)k = 2 mA. The voltage on the common emitters can

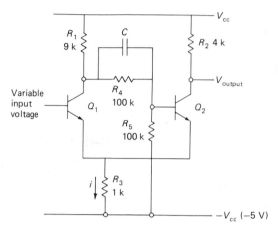

FIGURE 4-41 Discrete Schmitt trigger

be calculated to be $-5\,V + i \times R_3 = -3\,V$. If the input voltage is increased from $-5\,V$ toward $+5\,V$, it will pass through a level that will cause Q_1 to turn ON. This level is the voltage on the common emitters plus the required 0.7 V base-to-emitter voltage drop. It follows that when the input voltage exceeds this $-2.3\,V$ level, transistor Q_1 will turn on.

The action is regenerative. It will feed on itself, that is, it will *bootstrap*. When Q_1 turns ON, it forces Q_2 to go OFF. This bootstrap action is achieved through the common emitter resistor R_3 and through the Q_2 base connection to the collector of Q_1. As Q_1 begins to draw current, it will force the common emitter voltage to increase. This will cause Q_2 to draw less current. The increased current through Q_1 will result in a drop in the collector voltage of Q_1 and a drop in the base voltage of Q_2. The emitter voltage of Q_2 is increasing, and the base voltage of Q_2 is decreasing. Both actions will result in a decrease in Q_2 current. This will decrease the base–emitter voltage of Q_2, which will force more collector current through Q_1, and the device will quickly switch to a second state. In this state there will be a current through Q_1 that is determined by the combination of R_1 and R_3. In this example it will be $10\,V/(9+1)k = 1\,mA$. The common emitter voltage will now be at $-5\,V + 1\,mA \times 1\,k = -4\,V$.

If the input voltage were to be decreased toward the value $-5\,V$, the device would not switch back to the initial state at the same voltage at which it switched when going in the positive direction. This feature is called *circuit hysteresis*. (See figure 4–42.) For this example there is a difference of 1 V between the two trip levels; consequently, the hysteresis has a value of 1 V. A little investigation into the circuit reveals that the value of this hysteresis is set by the relationship of the two resistors R_1 and R_2. If these two resistors are equal, there is no hysteresis.

Your next question should be, Why do we want this feature? Recalling the reasons for using this circuit, you are trying to get one transition of the output when the input exceeds this preset voltage level. Suppose that you have no hysteresis built into the circuit and your input signal has some high-frequency sinusoidal noise

Flip Flops or Multivibrators

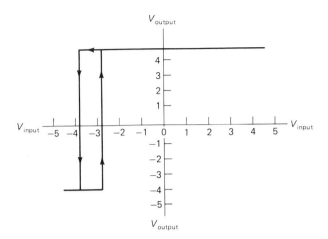

FIGURE 4–42 Hysteresis effect

superimposed. As your input is slowly increasing and approaches the trip level of the circuit, the high-frequency noise signal could take the composite input signal above, then below, then back above the trip level many times during the normal transition time for one single input transition. The output would indicate multiple transitions of this preset level, whereas you would really expect only a single transition.

Note that the voltages chosen for plus and minus V_{cc} are arbitrary choices. There is no requirement that there even be a negative voltage present. If you wish to transition at or below the zero voltage level, then a negative voltage on the $-V_{cc}$ is required. If you are using TTL circuitry and wish to use the standard TTL lows and high voltages, then a $+5$ V source would be adequate. Many Schmitt triggers are available in chip form; consult the digital design manuals for specific chip numbers and specifications. The capacitor C acts as a speed-up capacitor. The function of this capacitor is to cause the regeneration to occur in less time than if C were left out of the circuit. A small 100 pF capacitor is usually adequate for this component.

PROBLEMS

1. Using two NAND gates instead of the NOR gates used in the illustrative example for the integrated R/S flip flop, make a truth table for the circuit and draw the equivalent symbol, using 0 to indicate active low inputs.

2. For the D flip flop in figure 4–43, construct a truth table and plot the Q response.

3. If the waveform in figure 4–44 is the input to the three circuits shown, plot the anticipated output waveforms for each of the three circuits.

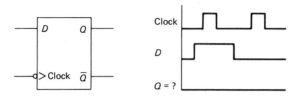

FIGURE 4–43 D flip flop for Problem 2

4. Given the circuit that would replace the R/C network for differentiating the clock shown in figure 4–45, if the response time, or t_d, of each circuit is 10 nsec, what is the width of the internal clock pulse?

5. For the bistable flip flops shown in figure 4–46, draw the appropriate symbols and truth tables for each and compare.

6. Sketch in the Q response for the NOR gate R/S flip flop shown in figure 4–47. Indicate any illegal states.

7. Sketch the Q response for the conditions of Problem 6 for a NAND gate R/S flip flop.

8. You are given a J/K flip flop that is not a master/slave type. The speed of response of this circuit is 20 nsec. The inputs are programmed with $J = K = 1$. How many times will the output change state if the clock pulse remains active for 100 nsec?

9. You are given the program shown in figure 4–48 for the Preset, Clear, J, and K inputs. Show the Q response for the master/slave J/K flip flop.

10. Sketch the Q response for the master/slave J/K flip flop shown in figure 4–49. The inputs are programmed as shown.

11. Given the master/slave J/K flip flop shown in figure 4–50 and the desired Q response, show the necessary J, K, Clear, and Preset inputs to obtain the desired response.

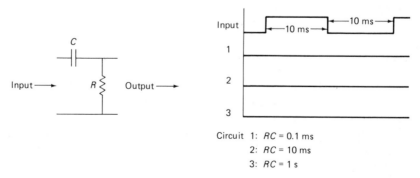

FIGURE 4–44 Circuits for Problem 3

Flip Flops or Multivibrators

FIGURE 4–45 Clock for Problem 4

FIGURE 4–46 Bistable flip flops for Problem 5

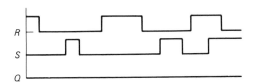

FIGURE 4–47 NOR gate *R/S* flip flop for Problems 6 and 7

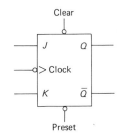

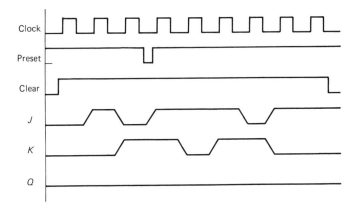

FIGURE 4–48 Master/slave *J/K* flip flop for Problem 9

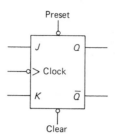

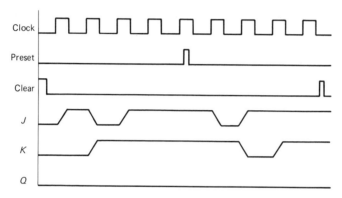

FIGURE 4–49 Master/slave J/K flip flop for Problem 10

12. Repeat Problem 11 for the Q response shown in figure 4–51.
13. Using the standard master/slave J/K flip flop, show the necessary connections to make it function as a D flip flop.
14. Using the standard master/slave J/K flip flop, show the necessary connections to make it function as a T flip flop (toggles every clock pulse).
15. Using the 555 timer chip, design a monostable multivibrator that will have an output pulse width of 10 ms. Design for manual trigger operation.
16. Using the 555 timer chip, design a monostable multivibrator that will have a variable output pulse width of from 1 to 10 ms. Design for manual triggering.
17. Using the 555 timer chip, design a free-running clock that will have an ON time of 20 μs and an OFF time of 10 μs.
18. Using the 555 timer chips, design a free-running clock that will have equal ON and OFF times of 50 μs.
19. Using two 555 timer chips, design a manual-pulsed oscillator that will be ON for 1 s for each manual closing of a switch and will produce a free-running 10 kHz signal when it is in the ON mode.
20. For the circuit shown in figure 4–52, determine the trip levels and plot the hysteresis curve.

Flip Flops or Multivibrators

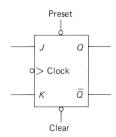

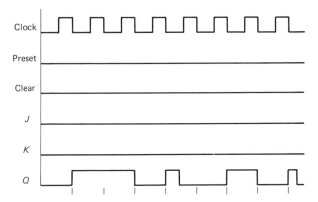

FIGURE 4-50 Master/slave J/K flip flop for Problem 11

21. Devise a scheme to make the upper trip level variable over the range of 1–3 V. What is the hysteresis at the 1 V and 3 V trip points?
22. Using a 74123 retriggerable monostable, design a circuit that will give an output when one pulse of a 1 kHz square wave is missing.
23. How many flip flops are required if the circuit you are working with has a crystal frequency of 1 MHz and a 400 Hz secondary clock is required?
24. Using the discrete astable circuitry shown in figure 4–31, sketch the waveforms for C_1 and C_2 for a set of values for R_1 through R_4 and C_1 and C_2 of your choice.
25. A T flip flop is a toggle flip flop. Show the necessary connections to make a T flip flop from a J/K flip flop.

FIGURE 4-51 Q response for Problem 12

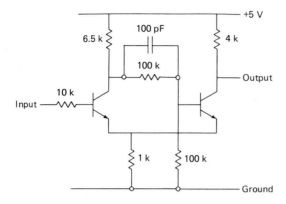

FIGURE 4–52 Circuit for Problems 20 and 21

chapter 5

counters and latches

5-1 INTRODUCTION

In this chapter we will study the use of flip flops in counting and storage circuits. Most computer operations and digital systems use the binary code. Circuits are required to store binary numbers that represent these operations or information. The information storage circuits are latches and registers.

Another common functional circuit is the counter circuit. The most familiar use for this circuit is probably the digital clock that we have come to accept as the standard for displaying the time of day. The *counter circuit* basically counts the number of clock pulses that are directed to the clock input and presents an output that is in some type of previously discussed code. The first counter that we will discuss is the ripple counter. This counter counts clock pulses and presents the sum of the input pulses at the output in binary code. Later on we will analyze some other types of counters such as the ring and Johnson counters.

5-2 THE RIPPLE COUNTER

The *ripple counter* derives its name from the fact that the counting effect is somewhat like the ripple effect that occurs when you drop a rock into a calm body of water. The count moves through the stages of the ripple counter in this manner. This in comparison to the counters that we will discuss later, in which each part of the counter receives its count pulse and responds at the same time. This second type of counter is called synchronous. The ripple counter requires that the first stage of the counter respond before the second stage can respond. This can be easily seen when the circuit is examined. We will examine a very simple two-stage counter first. The circuit requires two *J/K* flip flops. We use the master/slave *J/K* to avoid the race problem that was discussed in the previous chapter. (See figure 5-1.) For

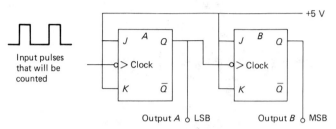

FIGURE 5-1 Two-stage ripple counter. LSB stands for least significant bit; MSB stands for most significant bit.

the initial analysis, do not concern yourself with the Clear and Preset inputs, but recognize that they must be eventually considered and properly connected. The flip flop is connected as a T type ($J = K = 1$). Assume that the flip flops are in the 0 count state at the start of our count sequence. The waveforms that represent the various count states are shown in figure 5-2.

If we use the status of Q_A and Q_B as our outputs, we can see that the circuit will count up in a straight binary sequence, as is shown in figure 5-3. Once the circuit has returned to the 0/0 state, the next clock pulse will simply repeat the previous sequence. Each stage of ripple counting results in a division by 2. Two stages of ripple counters will result in the input frequency being divided by 4. This dividing factor is called the *count modulus*. It is directly related to the number of stages that are present. In the simplest ripple counter without feedback the count modulus is equal to the binary base 2 raised to the exponent that is equal to the number of stages or J/K flip flops that are used. For example a four-stage ripple counter will have a counting modulus of 2 raised to the fourth power (16). An eight-stage counter will have a counting modulus of 2^8, which is 256.

With the addition of count detection circuitry and/or feedback we can create counters with any modulus desired. Since we live in a world that uses the decade decimal system, we will be quite interested in a counter that counts with a modulus of 10. The circuit that we wish to create will have a clock input that is some frequency f. The counter will have a modulus of 10, which means that at some point in the circuit we must have an output pulse that occurs once for every ten input pulses. When we have accomplished this function, we will have created a circuit that divides the clock input by 10. The counter now becomes a frequency

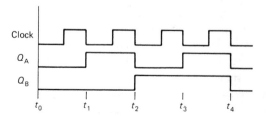

FIGURE 5-2 Ripple counter waveforms

Counters and Latches

Time	Q_B	Q_A	Decimal Equivalent
$t_0 \to t_1$	0	0	0
$t_1 \to t_2$	0	1	1
$t_2 \to t_3$	1	0	2
$t_3 \to t_4$	1	1	3
$> t_4$	0	0	0

FIGURE 5–3 Waveform interpretation

divider. An inviolable rule will be that the ripple (or any other type) counter cannot divide by a greater number than the previously mentioned relationship: $f_{output} = f_{input}/2^n$, where n is the number of stages or flip flops in the ripple counter. To develop a divide-by-ten counter, we must use at least four stages in the ripple counter. With three stages the maximum modulus is eight, and with four it is sixteen. Feedback and any other techniques devised can decrease this modulus, but it cannot be increased.

5–3 A DIVIDE-BY-SIXTEEN RIPPLE COUNTER

The circuit associated with the divide-by-sixteen ripple counter will contain four J/K flip flops. (See figure 5–4.) Recognize that the Clear and Preset inputs must be accounted for, and in most cases the Clear circuitry will be activated at least during power-on periods. For this analysis we will not consider the Clear and Preset inputs. The J and K inputs will both be tied high, causing the flip flops to operate in the T or toggle mode. The Q_A, Q_B, Q_C, and Q_D will be the outputs that record the count sequence in binary form. Q_A will be the least significant bit, and Q_D will be the most significant bit. The J and K inputs are all tied to a TTL high or 1 state. The timing diagram is shown in figure 5–5. From this point on, the count sequence will repeat. From figure 5–5, one can determine that the count modulus (or base) is 16.

To repeat, if a divide-by-sixteen circuit is required, the output must be taken to be Q_D. The output Q_C is a division by 8, Q_B by 4, and Q_A by 2. The counter is called a *count-up* device, since the count sequence is from 0 to 15, then returns

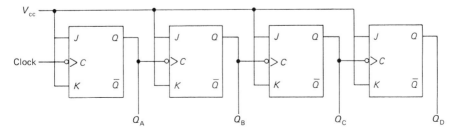

FIGURE 5–4 Divide-by-sixteen ripple counter

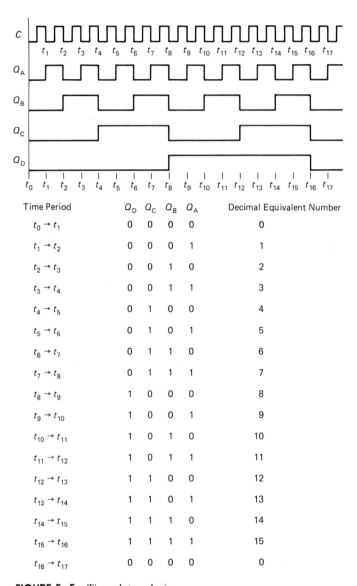

FIGURE 5-5 Time slot analysis

to zero. As an exercise, try using J/K flip flops that are positive edge–triggered. Connect the outputs in the same manner to the clock inputs of the succeeding stage. The counter now responds to the leading edge of the clock pulse, and the count sequence is a *count-down* sequence. It counts from 15 to 0, then jumps back to the count 15. There is another method of obtaining this count-down sequence: Connect the trailing edge clock of each flip flop of figure 5–4 to the $\overline{Q}$ of the previous stage. This connection allows the designer to switch between the two sequences by controlling which of the outputs are connected to the clock input. Figure 5–6 shows

Counters and Latches

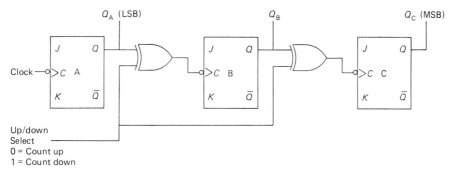

FIGURE 5-6 Selectable up/down counter

a diagram of a three-stage counter (divide by 8) that has the choice of counting up or counting down. The circuit will use exclusive OR logic gates instead of discrete switches for the implementation of this selecting process.

With the Select input low, the exclusive OR circuit acts as a simple buffer circuit, and the Q signal is transmitted directly through to the clock input of the following J/K flip flop. With the Select input high, the Q signal is inverted and consequently converted to the $\overline{Q}$ signal with the counter becoming a down counter.

5-4 FEEDBACK COUNTERS FOR NONSTANDARD MODULI

There are many occasions when you will need a counter that counts to a modulus other than a power of 2. If you want to use the ripple counter, there are useful techniques for obtaining any base that is required. The technique is to skip certain counts that would normally be counted. For instance, if the count modulus is eight, the counts are 0 through 7. If it is not desired to monitor the count output of your counter to display the count in progress, it does not matter which count or counts are skipped. However, if you are detecting and displaying the output count sequence, it is more reasonable to skip the higher counts that you are eliminating from the modulus. For example, if the counter is a three-stage device, the base count for this ripple counter is eight, and the count sequence is 0 through 7. If a modulus of five is required, and the output count sequence is not displayed, then you may skip any three of the eight possible states that would be present in the three-stage counter. When the display is required, it is to your advantage to skip the count states five (101), six (110), and seven (111). There are many ways to accomplish this count-skipping routine. A simple technique for this particular sequence would be to detect the five state (101) and use this detected signal to reset all three J/K flip flops to zero before the arrival of the next clock pulse. The count sequence would then be 0, 1, 2, 3, 4. Upon arrival of the next clock pulse, the circuit would go to the five state. The added circuitry would detect this state in a matter of nanoseconds and reset the counters to the zero count. A circuit that performs this function is shown in figure 5-7.

The NAND gate requires all inputs to be high before the output will go low.

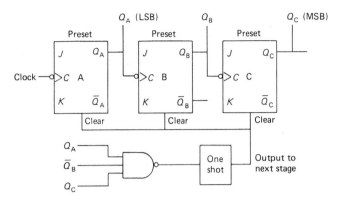

FIGURE 5-7 Truncated counter

For this arrangement, all outputs will be high when the count 101 is present in the counter. This will occur upon arrival of the fifth clock pulse. At this instant the counter will indicate a five count in the output for the period of time required for the NAND gate to sense this five count and reset the J/K flip flops to the zero state. This response time will be in nanoseconds. This circuitry works well in theory; but in actual operating conditions, one of the J/K flip flops can reset before another, removing the NAND gate signal before the other or others respond to the Clear pulse. This condition will result in not clearing to zero, with an erroneous count sequence following. The problem can be avoided by adding a one shot between the NAND gate output and the Clear inputs. (Some prefer a flip flop with a half clock cycle output to the one shot.) The waveforms from the circuit are presented in figure 5-8.

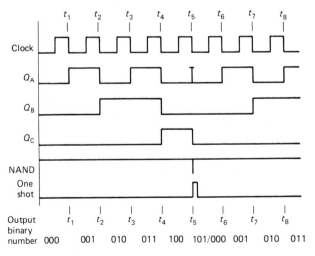

FIGURE 5-8 Truncated counter waveforms

Counters and Latches

The ripple counter is a nonsynchronous counter, simple to construct and relatively easy to understand. *Synchronous counters* by definition are counters in which all the flip flops responding to the clock pulse respond simultaneously.

A disadvantage of the ripple counter is the time delay associated with the transmission of the clock pulses through the flip flops before this Q output becomes the clock pulses for the succeeding counting stages. As the frequency of the clock input increases, it can become faster than the transition time of the flip flops, and at that clock frequency the counter will be generating erroneous output counts. This condition will determine the upper limiting frequency of the ripple counter if the requirement is for a continuous display or monitoring of the instantaneous count that is present in the counting flip flops. (Sometimes called an *on the fly* reading.) If the devices are not to be read on the fly, they will function at a higher frequency. The output of the NAND gate and the output of the one shot are signals that can be fed forward if the counting circuit is used in cascade with other counting circuits. There are any number of combinations that one can use to obtain a special modulus other than the standard 2^n modulus. One is limited only by one's own ingenuity. In addition to driving the Clear or Preset inputs, one can manipulate the J and/or K inputs to achieve a desired result. Another problem that occurs with ripple counters is the existance of erroneous states during transition times.

The ripple counter is an example of an *asynchronous counter*. With the ripple counter the clocks to each of the J/K inputs are not occurring at the same time. The next counter that we will analyze is a synchronous counter in which all changing of the J/K flip flop outputs is performed at the clock time.

5-5 THE RING COUNTER

The name *ring counter* applies to the connection of a series of J/K flip flops in a configuration that can best be described as a ring. The Q outputs of a particular flip flop are connected to the J input of a following flip flop, while the $\overline{Q}$ output of the first flip flop is connected to the K input of the same following flip flop. The outputs of the last flip flop are connected to the J/K inputs of the first flip flop. This configuration is the *ring effect*. Figure 5–9 shows an $n = 4$ ring counter.

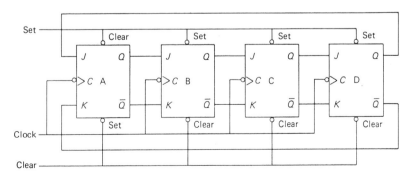

FIGURE 5–9 Four-bit ring counter

Note that the Set and Clear inputs to flip flop A are connected to the opposite inputs of flip flops B, C, and D. This will allow us to set a 1 into flip flop A when we clear the others. If all Sets and Clears are tied together, this counter would never respond or change with any clock pulse input. For the circuit to function, at least one 1, but not all 1's, must be set into the flip flops. If either all 1's or all 0's are set into the flip flops, the Q outputs can never be caused to change upon receipt of a clock pulse. The waveforms associated with this circuit are shown in figure 5–10.

With the application of the Clear pulse the Q_A, Q_B, and Q_C flip flops are cleared to the 0 state, and the Q_A flip flop is set to the 1 state. Upon receiving the first clock pulse the J_A/K_A inputs are programmed 0/1, which will force the Q_A output to the 0 state. The J_B/K_B inputs are programmed 1/0, which will force the Q_B output to go to the 1 state. Both the J_C/K_C and J_D/K_D inputs are programmed 0/1, which will cause them to hold their respective Q outputs in the 0 state. After receipt of the first clock pulse the Q_B output will be the only output that is in the 1 state. The Q_B output is the J_C input, and of course the $\overline{Q_B}$ output is the K_C input. Flip flop C is programmed to go to the 1 state after the second clock pulse. All the other flip flops are programmed $J = 0$, $K = 1$, which will force them to either go to or stay in the $Q = 0$ state. This progression continues with the receipt of the next clock pulse, and the singular high output will cycle around the counter, stepping to the next flip flop at the receipt of a clock pulse. Unlike the ripple counters, in which each stage has a different output frequency, the outputs from all the ring counter flip flops will be at the same frequency. This output frequency will be the clock frequency divided by n, where n is the number of stages, or flip flops, in the ring. The effort required for decoding is minimal. A light or any other type indicator can be connected to the Q or $\overline{Q}$ outputs directly. If the connections shown in figure 5–11 are made, the LED would indicate the count shown.

Since only one Q is high at a time, then only one $\overline{Q}_n$ will be low at one time, and this will be the LED that is activated. When the circuitry is cleared at the start

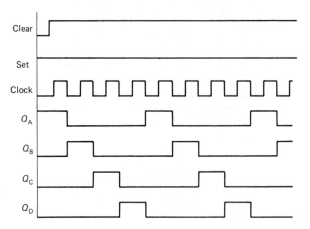

FIGURE 5–10 Four-bit ring counter waveforms with positive pulse

Counters and Latches

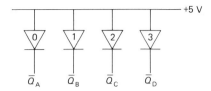

FIGURE 5–11 Binary display

of the count sequence, the 0 LED will be turned on, and all the others will be off. After the receipt of clock pulse 1, LED 1 will light up and LED 0 will go out. If the display is direct, no AND or NAND gates are required for decoding. Good design generally requires that displays such as this be driven from a buffer amplifier, rather than directly from the flip flop outputs. If instead of activating the Clear line, we activated the Set, the waveforms shown in figure 5–12 would be generated.

To implement the same display as in the previous example, the connections will be made to the Q outputs rather than the $\overline{Q}$ outputs. The counting frequency is again determined by the number of flip flops in the ring; in this case the counting modulus is four. Each of the outputs will be at the same frequency, but phase displaced. Should you wish to design a divide-by-ten counter that has direct decimal readout, ten of the J/K flip flops must be connected in the ring configuration. The result will be a divide-by-ten counter with the option of using any of the Q outputs as inputs to any succeeding stages. Should you be required to do a direct decimal decode, the extra decoding logic that is required with the binary counter is not required with the ring counter configuration. The circuit would be as shown in figure 5–13.

For proper operation one of the Set inputs must be connected to the other nine Clear inputs. The decoding is inherent in the counting circuitry. The output waveforms are similar to the previous example, in which only one of the outputs was initially set high. There will be ten output pulses, each one clock pulse wide and all ten phase displaced by one clock pulse. It is easy to generate a series of square

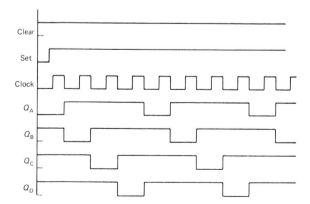

FIGURE 5–12 Four-bit ring counter with negative circulating pulse

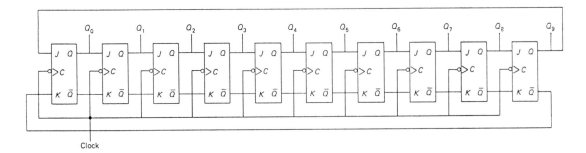

FIGURE 5–13 Ten-bit ring counter

waves with this circuit by connecting five adjacent Clear inputs to the remaining five Clear inputs. Following the logic through on this circuit shows that each J/K will remain on for five clock pulses and then turn off for the next five. All ten Q outputs will be at the same frequency (clock/10). Each adjacent flip flop is phase displaced by one clock pulse width. Some advantages of this type of circuit are that the speed of response is faster than the ripple circuit, the decoding can be inherent to the circuit, and the capacity to get multiple-phase outputs of the same frequency exists. The disadvantage is that the circuit requires more flip flops to achieve the same counting modulus as the ripple counters.

5–6 THE JOHNSON COUNTER

The *Johnson counter* (sometimes called a shift counter) looks very much like a ring counter when one compares the two circuits. The difference in the connections comes with the Q output of the last flip flop being connected to the K input of the first flip flop and the $\bar{Q}$ output of the last flip flop being connected to the J input of the first flip flop. This is a reversal from the ring counter connections. With this reversal the connections to the Set and Clear inputs can all be homogeneous, Sets to Sets and Clears to Clears. The output waveforms take on completely different states than those of the ring counter. A typical three-stage Johnson counter is shown in figure 5–14, and the resultant waveforms are shown in figure 5–15.

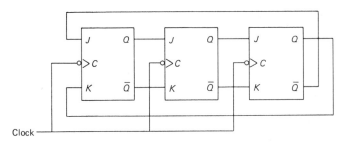

FIGURE 5–14 Three-bit shift register or Johnson counter

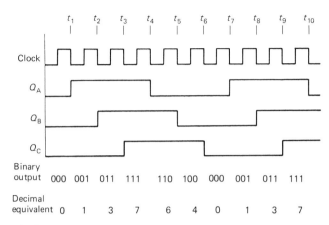

FIGURE 5-15 Three-bit Johnson counter waveforms

Although not shown, all the Set inputs are assumed to be tied together, as are the Clear inputs. To start the circuit in the 000 state for a start count of zero, the Clear inputs are activated. The count sequence is now a nonbinary sequence that is repetitive.

One would obviously not use this type of counter to count events that must have either decimal or binary output displays. The output of this counter is a series of three outputs that are all the same frequency, each phase being displaced by one clock period. The modulus of this counter is six.

We can summarize the results of the three different types of counters. The ripple counter has a base modulus of 2^n, the ring counter has a base modulus of n, and the Johnson counter has a base modulus of $2n$. The base modulus of the Johnson counter can be modified down by the addition of feedback to the circuit. We will discuss this further later in the chapter.

5-7 ILLEGAL STATES

The Johnson counter is a good example of a counter that in the counting process skips certain counts, which we will call *illegal counts*. They are illegal in the sense that if the circuit gets into one of these count states, the count sequence will be permanently altered. This will generally create a different counting modulus and, of course, erroneous results. The counter in the previous section has the legal count sequence of 0, 1, 3, 7, 6, and 4. Since this is a three-bit counter, there are eight possible states, leaving the states 2 and 5 as unused or illegal states. The purpose of this section is to determine what happens to the circuit when these counts accidentally occur and what, if anything, can be done to correct and prevent this condition. The first question should be "How do these illegal states come about?" They can be caused to occur by spurious noise spikes on the signal lines, or possibly the power supply has a dip or spike on it that triggers the three flip flops into the

$Q_C = 0$, $Q_B = 1$, $Q_A = 0$ (the 2 count) state or the $Q_A = 1$, $Q_B = 0$, $Q_C = 1$ (the 5 count) state. Power on can create these states.

Once the counter is in these illegal states, we must examine what happens to it. To do this, it is helpful to create a timing diagram that depicts one of these illegal states and then apply a clock pulse and determine the next count sequence. If the counter were to step to one of the legal states, then the counter would experience a single perturbation and return to a correct counting sequence. The circuit would have corrected itself. Once the counter was back in a legal count, the counting would proceed in the regular sequence.

Suppose something happened to the circuit to cause the device to go into the 2 state. We will examine the circuit response and deduce what action should be taken.

Consider the previous three-stage Johnson counter. If for some reason the counter goes into the 2 state, the circuit will respond as shown in figure 5–16. From these results we can conclude that once in the illegal 2 state, the device will, at the next clock pulse, move to the 5 state. This is another illegal state. Upon receipt of the second clock pulse the device will return to the 2 state. Our analysis need go no further; we see that the device will oscillate between the two illegal states.

Since one must expect noise pulses in any and all systems, it appears highly likely that this condition will occur at some time in the useful life of the circuit. It is the designers' responsibility to ensure that the condition will correct itself should it occur. The solution is straightforward. Detect either of the illegal states and force the counter into a legal state. To accomplish this task, detect the 2 condition and force the circuit into the 0 state.

A simple three-input NAND gate that is connected to the $\overline{Q}_A$, Q_B, and $\overline{Q}_C$ outputs will be used to clear the counter when the illegal state occurs. The output of the NAND gate is used to activate the Clear line, thus clearing all the J/K flip flops and establishing the zero count. The device will now be back in a legal count routine.

A general statement can be made to the effect that in designing counters that do not utilize all possible counts, the counts that are not used and are considered to be illegal counts must be investigated to determine whether they are automatically

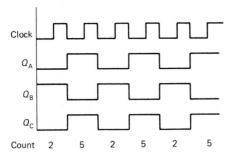

FIGURE 5–16 Illegal count response

Counters and Latches

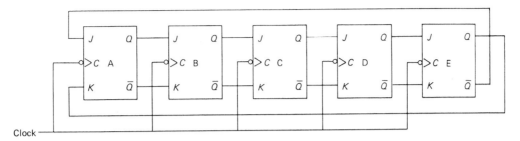

FIGURE 5-17 Divide-by-ten Johnson counter

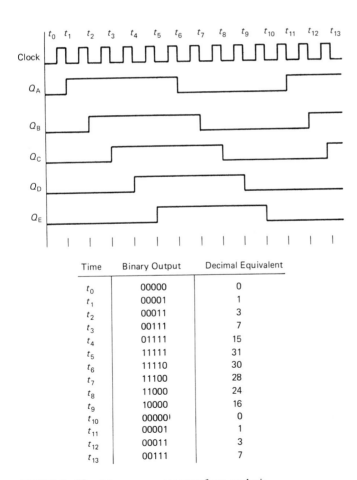

Time	Binary Output	Decimal Equivalent
t_0	00000	0
t_1	00001	1
t_2	00011	3
t_3	00111	7
t_4	01111	15
t_5	11111	31
t_6	11110	30
t_7	11100	28
t_8	11000	24
t_9	10000	16
t_{10}	00000'	0
t_{11}	00001	1
t_{12}	00011	3
t_{13}	00111	7

FIGURE 5-18 Johnson counter waveform analysis

reset to a normal count or whether a special illegal count-detecting circuit, with count reset, is required.

5–8 A DECADE SHIFT COUNTER

A divide-by-ten shift counter will require five flip flops. (See figure 5–17.) The Q output of each flip flop will produce a frequency that is the clock frequency divided by 10. The Q outputs will be phase displaced by one clock pulse. Since there are five flip flops in use, there are $2^5 = 32$ possible output states. We will be using only ten of these states. The circuitry is similar to the previously discussed three flip flop shift counter, with the addition of two more flip flops in the same configuration. A distinguishing feature of this circuit is the criss-crossing of the last Q_E output to the K_A input and the $\overline{Q_E}$ output to the J_A input. The resultant waveforms are shown in figure 5–18.

If display of the count sequence is required, decoding gates must be added to the circuit. The decode circuits would be as shown in figure 5–19. With this particular decoding scheme, the device is a decade counter that includes decade decode.

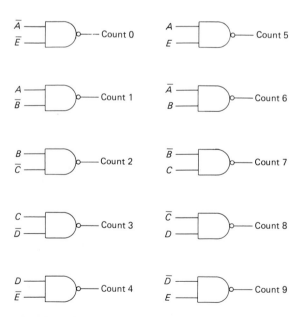

FIGURE 5–19 Decoding circuits

Counters and Latches

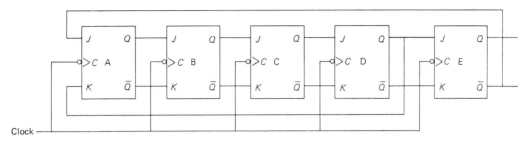

FIGURE 5-20 Truncated Johnson counter

5-9 NONSTANDARD MODULUS OBTAINED BY FEEDBACK

The previous decade counter can be quite easily modified to obtain a counting modulus of nine by feeding back from the Q_D output to the K_A input. The original feedback was from the $\overline{Q_E}$ output to the K_A input.

The circuit would be as shown in figure 5-20. The waveforms associated with this circuit would be as shown in figure 5-21.

The J_A connection that is made to the Q_D output causes the A output to be reset one pulse early with respect to the standard $n = 5$ divide-by-ten Johnson counter. A general rule can be stated: To reduce the count modulus of the Johnson shift counter by one, move the J_A connection from the last Q output to the next-to-last Q output. For the $n = 5$ counter the J_A is moved to the Q_D output. In using an $n = 4$ divide-by-eight counter, to change the modulus to a divide-by-seven, the J_A input would be connected to the Q_C output. Similarly, the $n = 3$ divide-by-six counter would be changed to a divide-by-five by connecting the J_A to the Q_B output.

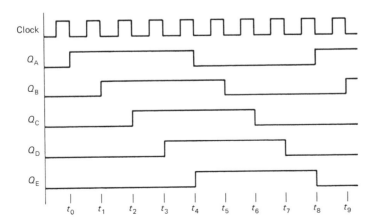

FIGURE 5-21 Waveforms for the Johnson truncated counter

5-10 LATCHES

Latches are used to store binary bits. To store an eight-bit binary word, eight latches are required. The actual circuit that stores the bit can be any of the bistable flip flops that were discussed. The D, R/S, and J/K flip flops are commonly used when the latching function is required. There are two schemes for loading the bits into a multibit latch. Serial loading will be discussed first. For the first example we will store a four-bit word. We require four flip flops to store this word. The circuit would be as shown in figure 5-22.

The circuit will function by programming the J and K inputs sequentially to cause the first or D flip flop to go to the state of the corresponding bit of the four-bit word. For this circuit the D flip flop is to contain the most significant bit when we have finally stored the full four bits.

As an example, suppose we wish to store the binary four-bit word 1011. The least significant bit is a 1, so the J_D input is programmed to be a 1 and the K_D input is programmed to be a 0. Upon receipt of the first clock pulse the D flip flop will go to the 1 state. Before the arrival of the second clock pulse the J and K inputs must be programmed to cause the second least significant bit to be programmed into the D flip flop. The second least significant bit is also a 1, so the J input will stay in the 1 state and the K input will stay in the 0 state. Upon receipt of the second clock pulse the C flip flop will be forced to the 1 state owing to the programming of the C J/K inputs by the outputs of the D flip flop. The D flip flop will respond to the D J/K programmed inputs and in this case will stay in the 1 state. Before the arrival of the third clock pulse the J/K inputs must be programmed for insertion of the third bit—in this case a 0. The J input must be taken to a 0 and the K input to a 1. Upon receipt of the third clock pulse the B flip flop will be responding to the output of the C flip flop; the C flip flop will be responding to the output of the D flip flop; and the D flip flop will be responding to the programmed J/K inputs of the D flip flop. At this point (after receipt of the third clock pulse but before the fourth clock pulse) the outputs will read as follows: $D = 0$, $C = 1$, $B = 1$, $A = $ irrelevant. Before receipt of the fourth and last clock pulse the J/K inputs will be programmed to force the MSB (a 1) into the D flip flop: $J_D = 1$, and $K_D = 0$. Upon receipt of this fourth clock pulse the flip flops will respond with the following outputs: $D = 1$, $C = 0$, $B = 1$, $A = 1$. This is the four-bit word that we wished to store.

It is absolutely necessary that the clock pulses be inhibited or stopped after the

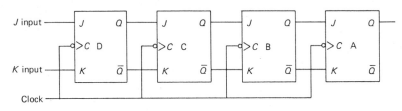

FIGURE 5-22 Serial latch

Counters and Latches

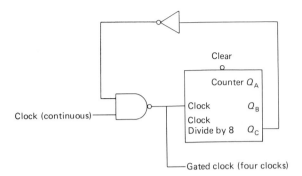

FIGURE 5-23 Gated clock for a serial latch

word is stored. It is not sufficient to take the D J/K inputs to 0 and reason that you are directing the system to a no-change mode. If this is done, the first or D flip flop will remain in the 1 state, and the remaining flip flops will respond in the same manner until all four are in the 1 state. This, of course, is not the word that you are attempting to store. The clock circuit could be handled as shown in figure 5–23.

The four clock pulses will be initiated when the Clear input to the counter is activated. If the desire is to store an eight-bit word, then eight flip flops and eight clock pulses are required. The counter would be a divide-by-sixteen, and the Q_D output would be used instead of the Q_C output. A timing diagram of the system is shown in figure 5–24.

Note that the number 0000 is assumed when the sequence is started. This is not a requirement for the system to function; whatever the previously stored number, after the four clock pulses the latch will contain the newly programmed number. An eight-bit or sixteen-bit serial storage would be accomplished in much the same manner, with a corresponding increase in the number of flip flops and clock pulses that would be required. The J and K inputs would be programmed by some other circuitry that would cause these inputs to be in a stable state before the arrival of the clock pulse. Serial storage operation is obviously slower than the equivalent parallel storage operation. This can be an obvious disadvantage. For serial storage the number of clock cycles required to store a word is equal to the number of bits in that word. The advantages of the serial system are its simplicity and the fewer programming circuits that are required.

5-11 THE D LATCH

A very commonly used latch is the 7475 quad latch. It has four D type flip flops on one chip with an enable function. Programming of the D latch is obviously easier than using an equivalent J/K latch function. With the 7475, when the enable is high, the output will follow the D input. When the enable function goes low, the output is latched to the state of the D input when the enable input went low. For this particular chip there are only two enable inputs, a situation that requires that the latches be used as pairs, storing parts of the same word.

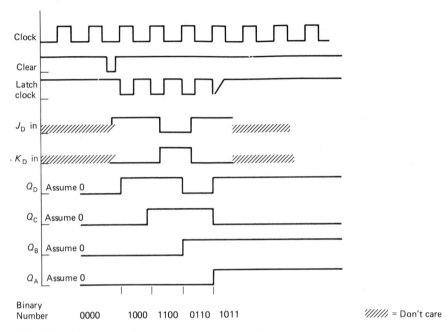

FIGURE 5–24 Timing diagram for a four-bit serial latch

5–12 PARALLEL-LOADED LATCH

The parallel-loaded latch circuit will require the same number of flip flops as the serial-loaded latch. Each of the J/K inputs are required to be programmed individually and simultaneously. The word will be stored with the execution of a single clock pulse. (See figure 5–25.)

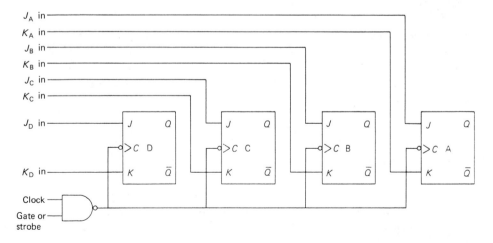

FIGURE 5–25 Parallel storage of a four-bit word

Counters and Latches

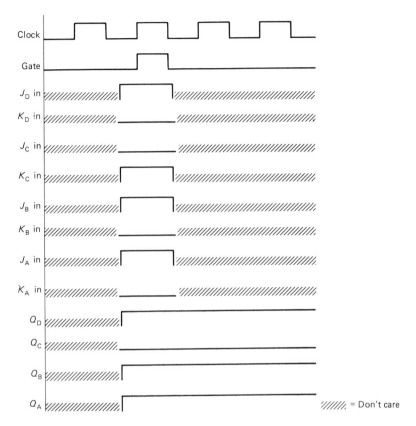

FIGURE 5-26 Number stored is 1011

The waveforms for this circuit to store the word 1011 would be as shown in figure 5-26. Parallel load latches are usually D type.

5-13 THE 74192/74193 COUNTER

The two numbers for the 74192/74193 counter are for two counters that have much in common but are functionally different. In this case the devices have the same pin connections but differ in that the 74192 is a BCD counter and the 74193 is a straight binary counter. The BCD device will expect only the BCD code, while the binary device is a straight divide-by-sixteen four-bit counter.

As a designer using any device, the first item to check will be power dissipation. The specification shown in figure 5-27 verifies that the power connections are standard for TTL. This standard will have V_{cc} (+5 V) on the highest-numbered pin (16) and power return (ground) on the diagonal from the V_{cc} pin (pin 8). The current requirements are typically listed at 15 mA but never more than 26 mA. This parameter is listed under the dc electrical characteristic as I_{cc} (total supply current).

These devices are called *presettable counters*. In the counters that we have studied so far, most started from zero and, in the case of a four-bit counter, would count to 15 and then return to zero. If the counter was a count-down device, it would begin at 15 and count down to zero. With a presettable counter the device can be set to any number, and it will begin counting with that count as your first, or zero clock pulses received, state. With the four-bit counters presented in previous sections the outputs were labeled A, B, C, and D, A being the least significant bit and D the most significant bit. With this device the Q_0 output will correspond to our Q_A, and the Q_3 output will correspond to Q_D (MSB). If we want to start our count sequence at a count other than zero, the inputs labeled D_0, D_1, D_2, and D_3 are inputs to the A, B, C, and D flip flops. This is an asynchronous input that is activated when the P_1 (program load) input is pulsed low.

EXAMPLE 5–1

If you wish to begin your count sequence at 5 and count up to 15, you would program 5 into the D_0–D_3 inputs: $1001 = 5$, so $D_3 = 1$, $D_2 = 0$, $D_1 = 0$, and $D_0 = 1$. The programming may be done by switches, dc voltages, or pulses. These inputs must be in the state shown when the P_1 input is taken low. Immediately upon receipt of this P_1 signal, the Q_3, Q_2, Q_1, and Q_0 outputs will go to the programmed states and will stay there as long as the P_1 input is held low and until the arrival of the first clock pulse following the return of the P_1 input to the high state. ∎

These devices have two clock inputs, pins 4 and 5 for the DIP package. They are labeled C_{pu} for counting up and C_{pd} for counting down. Only one input can be used at any one time; the other input must be tied to an equivalent TTL high state. If you leave the unused clock input floating or unconnected, the counter will not function properly.

Two very useful outputs from these devices are the T_{cu} and T_{cd} terminals (pins 12 and 13, respectively). The equations given by the manufacturer for these functions are as follows.

For 74192:

$$\overline{T_{cu}} = Q_0 * Q_3 * \overline{C_{pu}}$$
$$\overline{T_{cd}} = \overline{Q_0} * \overline{Q_1} * \overline{Q_2} * \overline{Q_3} * \overline{C_{pu}}$$

For 74193:

$$\overline{T_{cu}} = Q_0 * Q_1 * Q_2 * Q_3 * \overline{C_{pu}}$$
$$\overline{T_{cd}} = \overline{Q_0} * \overline{Q_1} * \overline{Q_2} * \overline{Q_3} * \overline{C_{pd}}$$

These equations are the key to understanding the operation of the devices when they are cascaded and/or autoreset to a preset number.

Reviewing the specification of figure 5–27, we note that the device is syn-

Counters and Latches

54/74193
54LS/74LS193

DESCRIPTION

The "193" is an Up/Down Modulo-16 Binary Counter utilizing separate Count Up and Count Down Clocks. The individual circuits operate synchronously in either counting mode changing state on the LOW-to-HIGH transitions on the clock inputs.

To simplify multistage counter designs separate Terminal Count Up and Terminal Count Down outputs are provided which are used as clocks for subsequent stages without extra logic. Individual preset inputs allow the circuit to be used as a programmable counter. Both the Parallel Load ($\overline{PL}$) and the Master Reset (MR) inputs override the clocks and asynchronously load or clear the counter.

FEATURES

- Synchronous reversible 4-bit binary counting
- Asynchronous parallel load
- Asynchronous reset (clear)
- Expandable without external logic

LOGIC SYMBOL

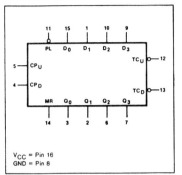

V_{CC} = Pin 16
GND = Pin 8

PIN CONFIGURATION

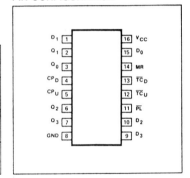

ORDERING CODE (See Section 9 for further Package and Ordering Information)

PACKAGES	COMMERCIAL RANGES V_{CC}=5V ± 5%; T_A=0°C to +70°C		MILITARY RANGES V_{CC}=5V ± 10%; T_A=−55°C to +125°C	
Plastic DIP	N74193N	•	N74LS193N	
Ceramic DIP	N74193F	•	N74LS193F	S54193F • S54LS193F
Flatpak				S54193W • S54LS193W

INPUT AND OUTPUT LOADING AND FAN-OUT TABLE[a]

PINS	DESCRIPTION			54/74	54S/74S	54LS/74LS
CP_U	Count up Clock Pulse input	I_{IH} (μA)	I_{IL} (mA)	40 −1.6		20 −0.4
CP_D	Count down Clock Pulse input	I_{IH} (μA)	I_{IL} (mA)	40 −1.6		20 −0.4
MR	Master Reset (active HIGH) input	I_{IH} (μA)	I_{IL} (mA)	40 −1.6		20 −0.4
$\overline{PL}$	Parallel Load (active LOW) input	I_{IH} (μA)	I_{IL} (mA)	40 −1.6		20 −0.4
D_n	Parallel Data inputs	I_{IH} (μA)	I_{IL} (mA)	40 −1.6		20 −0.4
Q_n	Counter outputs	I_{OH} (μA)	I_{OL} (mA)	−400 16		−400 4/8[a]
$\overline{TC}_U$	Terminal Count up (carry) output	I_{OH} (μA)	I_{OL} (mA)	−400 16		−400 4/8[a]
$\overline{TC}_D$	Terminal Count down (borrow) output	I_{OH} (μA)	I_{OL} (mA)	−400 16		−400 4/8[a]

NOTE
a. The slashed numbers indicate different parametric values for Military/Commercial temperature ranges respectively.

FIGURE 5–27 Data sheet for 74193 (Courtesy of Signetics Corporation)

continues

FUNCTIONAL DESCRIPTION

The "193" is an asynchronously presettable, Up/Down (reversible) 4-Bit Binary Counter. It contains four master/slave flip-flops with internal gating and steering logic to provide asynchronous Master Reset (clear) and Parallel load, and synchronous Count-up and Count-down operations.

Each flip-flop contains JK feedback from slave to master such that a LOW-to-HIGH transition on the clock inputs causes the Q outputs to change state synchronously. A LOW-to-HIGH transition on the Count Down Clock Pulse (CP_D) input will decrease the count by one, while a similar transition on the Count Up Clock Pulse (CP_U) input will advance the count by one. One clock should be held HIGH while counting with the other, because the circuit will either count by twos or not at all, depending on the state of the first flip-flop, which cannot toggle as long as either clock input is LOW. Applications requiring reversible operation must make the reversing decision while the activating clock is HIGH to avoid erroneous counts.

The Terminal Count Up ($\overline{TC_U}$) and Terminal Count Down ($\overline{TC_D}$) outputs are normally HIGH. When the circuit has reached the maximum count state of "15", the next HIGH-to-LOW transition of CP_U will cause $\overline{TC_U}$ to go LOW. $\overline{TC_U}$ will stay LOW until CP_U goes HIGH again, duplicating the Count Up Clock although delayed by two gate delays. Likewise, the $\overline{TC_D}$ output will go LOW when the circuit is in the zero state and the CP_D goes LOW. The $\overline{TC}$ outputs can be used as the clock input signals to the next higher order circuit in a multistage counter, since they duplicate the clock waveforms. Multistage counters will not be fully synchronous, since there is a two gate delay time difference added for each stage that is added.

The counter may be preset by the asynchronous parallel load capability of the circuit. Information present on the parallel Data inputs (D_0-D_3) is loaded into the counter and appears on the outputs regardless of the conditions of the clock inputs when the Parallel Load ($\overline{PL}$) input is LOW. A HIGH level on the Master Reset (MR) input will disable the parallel load gates, override both clock inputs, and set all Q outputs LOW. If one of the clock inputs is LOW during and after a reset or load operation, the next LOW-to-HIGH transition of that Clock will be interpreted as a legitimate signal and will be counted.

STATE DIAGRAM

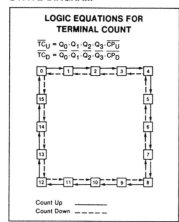

LOGIC EQUATIONS FOR TERMINAL COUNT

$\overline{TC_U} = Q_0 \cdot Q_1 \cdot Q_2 \cdot Q_3 \cdot \overline{CP_U}$
$\overline{TC_D} = \overline{Q_0} \cdot \overline{Q_1} \cdot \overline{Q_2} \cdot \overline{Q_3} \cdot CP_D$

Count Up ———
Count Down - - - - -

MODE SELECT-FUNCTION TABLE

OPERATING MODE	INPUTS					OUTPUTS		
	MR	$\overline{PL}$	CP_U	CP_D	D_0, D_1, D_2, D_3	Q_0, Q_1, Q_2, Q_3	$\overline{TC_U}$	$\overline{TC_D}$
Reset (clear)	H	X	X	L	X X X X	L L L L	H	L
	H	X	X	H	X X X X	L L L L	H	H
Parallel load	L	L	X	L	L L L L	L L L L	H	L
	L	L	X	H	L L L L	L L L L	H	H
	L	L	L	X	H H H H	H H H H	L	H
	L	L	H	X	H H H H	H H H H	H	H
Count up	L	H	↑	H	X X X X	Count up	H(b)	H
Count down	L	H	H	↑	X X X X	Count down	H	H(c)

H = HIGH voltage level
L = LOW voltage level
X = Don't care
↑ = LOW-to-HIGH clock transition

NOTES
b. $\overline{TC_U}$ = CP_U at terminal count up (HHHH)
c. $\overline{TC_D}$ = CP_D at terminal count down (LLLL)

FIGURE 5-27 (Continued)

Counters and Latches

LOGIC DIAGRAM

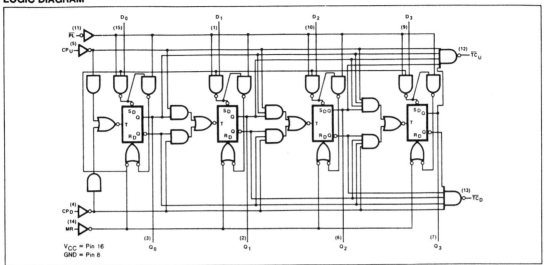

DC CHARACTERISTICS OVER OPERATING TEMPERATURE RANGE[d]

	PARAMETER	TEST CONDITIONS		54/74 Min	54/74 Max	54S/74S Min	54S/74S Max	54LS/74LS Min	54LS/74LS Max	UNIT
I_{OS}	Output short circuit current	V_{CC} = Max	Mil	−20	−65			−15	−100	mA
			Com	−18	−65			−15	−100	mA
I_{CC}	Supply current	V_{CC} = Max	Mil		89				34	mA
			Com		102				34	mA

AC CHARACTERISTICS: T_A=25°C (See Section 4 for Test Circuits and Conditions)

	PARAMETER	TEST CONDITIONS	54/74 C_L = 15pF R_L = 400Ω Min	54/74 C_L = 15pF R_L = 400Ω Max	54S/74S Min	54S/74S Max	54LS/74LS C_L = 15pF R_L = 2kΩ Min	54LS/74LS C_L = 15pF R_L = 2kΩ Max	UNIT
f_{MAX}	Maximum input count frequency	Figure 1	25				25		MHz
t_{PLH} t_{PHL}	Propagation delay CP_U input to $\overline{TC}_U$ output	Figure 2		26 24				26 24	ns ns
t_{PLH} t_{PHL}	Propagation delay CP_D input to $\overline{TC}_D$ output	Figure 2		24 24				24 24	ns ns
t_{PLH} t_{PHL}	Propagation delay CP_U or CP_D to Q_n outputs	Figure 1		38 47				38 47	ns ns
t_{PLH} t_{PHL}	Propagation delay $\overline{PL}$ input to Q_n output	Figure 3		40 40				40 40	ns ns
t_{PHL}	Propagation delay MR to output	Figure 4		35				35	ns

NOTE
d. For family dc characteristics, see inside front cover for 54/74 and 54H/74H, and see inside back cover for 54S/74S and 54LS/74LS specifications.

FIGURE 5-27 (Continued)

continues

AC SETUP REQUIREMENTS: $T_A = 25°C$ (See Section 4 for Test Circuits and Conditions)

	PARAMETER	TEST CONDITIONS	54/74 Min	54/74 Max	54S/74S Min	54S/74S Max	54LS/74LS Min	54LS/74LS Max	UNIT
t_W	CP_U pulse width	Figure 1	26				20		ns
t_W	CP_D pulse width	Figure 1	26				20		ns
t_W	$\overline{PL}$ pulse width	Figure 3	20				20		ns
t_W	MR pulse width	Figure 4	20				20		ns
t_s	Setup time Data to $\overline{PL}$	Figure 5	20				20		ns
t_h	Hold time Data to $\overline{PL}$	Figure 5	0				0		ns
t_{rec}	Recovery time, $\overline{PL}$ to CP	Figure 3	40				40		ns
t_{rec}	Recovery time, MR to CP	Figure 4	40				40		ns

AC WAVEFORMS

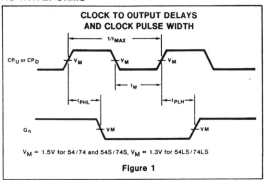

Figure 1

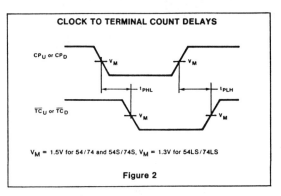

Figure 2

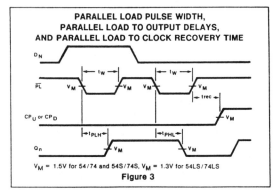

Figure 3

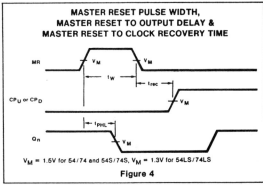

Figure 4

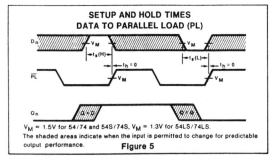

Figure 5

FIGURE 5-27 (Concluded)

Counters and Latches

chronous and responds to the leading edge of the clock pulse. To understand the meaning of the equations, first look at the 74193 specification. The T_{cu} output will be low when the $Q_0 = Q_1 = Q_2 = Q_3 = 1$ (is high) *and* C_{pu} (clock pulse up) is low. The last terminal to discuss is the *MR* (pin 14) terminal, which is the master reset for all the flip flops. This is an asynchronous input that is active high. When this terminal is taken to a high state, the Q outputs are all forced to a low state; the count is returned to zero.

───────────────────────────────────── EXAMPLE 5–2

Design a count-up counter with a modulus of 10, with no count display required.

(a) Connect pin 16 to a +5 V power supply.
(b) Connect pin 8 to power return (usually ground).
(c) Use the internal loading capability by connecting $\overline{T_{cu}}$ (pin 12) to the program load input $\overline{P_1}$ (pin 11). Since we are not required to decode the count, it is simpler to use the T_{cu} pulse to reset our counter and generate a modulus 10 counter. The equation

$$\overline{T_{cu}} = Q_A * Q_B * Q_C * Q_D * \overline{C_{pu}}$$

indicates that when the chip is in the 15 state and the clock pulse goes to the low state, T_{cu} will go low. This signal will be used to load the counter with a preset number that when subtracted from 15 will give the mod 10 counter desired. For this example the number that must be programmed into the D_0, D_1, D_2, and D_3 inputs is 5 (0101). Upon receipt of the first clock pulse after programming has occurred, the count will change to 6. After receipt of the tenth clock pulse, the count will go to 15. It follows that the counts within this modulus would be 5, 6, 7, 8, 9, 10, 11, 12, 13, 14 (15 is a temporary state that is forced to the 5 count before the next clock pulse).
(d) The clock input that will be used for a count-up sequence is C_{pu} (pin 5). Pin 6 (C_{pd}) must be held high when using C_{pu}.
(e) *MR* (master reset) is not used, so it will be tied low to ground. Should you wish to reset this counter to zero, this terminal (pin 14) would have to be pulsed high.

The circuitry is shown in figure 5–28. Waveforms for the circuit are shown in figure 5–29.

This particular chip is useful in creating a counter with a modulus in the range of 2–16. It is easily cascaded with other counters of the same type. Care must be taken if one wishes to maintain a synchronous counting system. Additional circuitry will be required. ■

In summary, to determine which of the available counters will best fit the requirements of the design, it is best to go to a manufacturer's specification book,

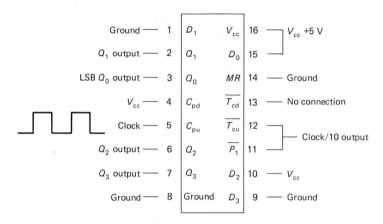

FIGURE 5–28 Divide-by-ten counter

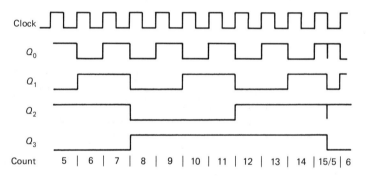

FIGURE 5–29 Waveforms for 74193 modulus 10 count-up counter

which will list for all of the counters that the supplier manufactures. The delineations will be number of bits, preset or no preset, ripple, synchronous, BCD, binary, ring, Johnson, and so on. See table 5–1 for some examples.

PROBLEMS

1. Using the 7473 dual J/K flip flop, design a ripple counter that will divide by 4. Show all waveforms including the clock.
2. How many 7473 chips would be required to make a divide-by-sixty counter?
3. Design a divide-by-seven ripple counter, using the 7473 chip. Show all waveforms.
4. Design a decode circuit that will decode the number 6 from the counter designed in Problem 3.

Counters and Latches

TABLE 5–1 Available counters

Part Number	Counter Description	Family	Number of Pins
7490	Ripple, decade. Moduli available are 2 and 5.	TTL, LS	14
7492	Ripple, div by 12. Mods available are 2 and 6.	TTL, LS	14
7493	Ripple, binary. Mods available are 2 and 8.	TTL, LS	14
74160	Decade, sync, BCD, with master reset	TTL, LS	16
74163	Binary, sync, 4 bit, sync reset	TTL, LS	16
74193	Binary, programmable, 4 bit, up/down counter with separate clocks	TTL, LS	16
74390	Ripple, dual decade—$2 \times 5 \times 2 \times 5$	LS	16
74393	Ripple, 8 bit, mod 16×16	LS	14
74592	Sync, 8 bit, programmable	LS	16

5. If the propagation delay from clock to output for the 7473 is 40 nsec, determine the upper frequency limit of a counter that has a modulus of
   (a) 60   **(b)** 10   **(c)** 2

6. If the same 7473 of Problem 5 is used in a divide-by-ten ring counter, what will be the upper frequency limit of this counter?

7. Using the 7473, design a divide-by-ten ring counter. Show all waveforms.

8. Decode the decimal numbers 0–9 from the counter of Problem 7.

9. Determine the maximum counting modulus for ten flip flops if they are connected as a
   (a) Ripple counter   **(b)** Ring counter   **(c)** Johnson counter

10. Design a Johnson shift counter that will divide by 7. Show all waveforms.

11. The waveforms in figure 5–30 apply for a serial input four-bit latch. Fill in the Q_A, Q_B, Q_C, and Q_D outputs, and show the number stored at each clock pulse.

12. Draw a timing diagram, showing clock pulses, J_D and K_D programs, and all Q outputs for an eight-bit serial latch that is to store the binary number 10011010.

13. Repeat Problem 12 for a parallel latch to store the same binary number.

14. Given a clock frequency of 1 MHz, what is the minimum time required to store a 16-bit word in
    (a) A serial-loaded register
    (b) A parallel-loaded register

15. Using the 74193 counter, show all connections necessary to use the device as a count-up, divide-by-sixteen counter. Is this a ripple counter?

16. Using the 74193 counter, show all connections necessary when this chip is to

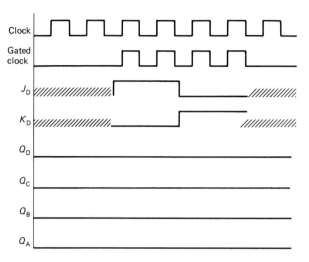

FIGURE 5–30 Waveforms for Problem 11

be used as a count-up, divide-by-ten counter. No decoding of the output is required. Indicate your output connection.

17. Using the 74193 counter, show all connections necessary for this chip to be used as a count-down, divide-by-twelve counter. Decoding of the output is desired.

18. Given the 74193 circuit shown in figure 5–31, determine the following parameters:
    (a) Output frequency
    (b) Count sequence
    (c) Output pulse width

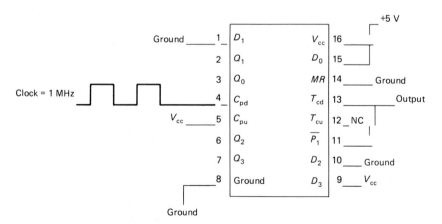

FIGURE 5–31 Circuit for Problems 18 and 19

Counters and Latches

19. For the circuit in Problem 18, what happens if C_{pd} is connected to ground?
20. Using two 74193 chips, design a divide-by-sixty counter, then use the output of the divide-by-sixty as the clock for a divide-by-ten ring counter. Show how you would display the output.

chapter 6

ECL and CMOS

6–1 INTRODUCTION

This chapter will introduce the two remaining major families of the three families of digital logic (TTL, ECL, and CMOS). The circuitry that comprises a standard gate for each of the families will be analyzed, and some interfacing circuits will be discussed. The advantages of each family will be discussed along with the costs of the advantages.

6–2 EMITTER-COUPLED LOGIC

The main advantage of the emitter-coupled logic (ECL) family is speed of response. This family is the fastest in operation. There are two series within this family. The first series out is called the 10k series; it generally has a typical propagation delay of 2–2.5 nsec per logic gate. The later series is the 100k series, which has a typical propagation delay of 1 nsec per logic gate. In addition to the increased speed of response, there is a change in the packaging of the two series. The 10k series comes in the 16-pin DIP package, whereas the 100k series generally comes in the 24-pin DIP package.

Whereas the TTL logic family requires $+5$ V for V_{cc}, with return being ground, the ECL family utilizes a negative voltage on the V_{ee}, and V_{cc} is returned to ground. The 10k series requires -5.2 V for V_{ee}, while the 100k series recommends that -4.5 V be used for V_{ee}. The logic levels are not the same as TTL logic levels. A logic 1 is nominally -0.7 V, and a logic 0 is nominally -1.6 V. This remains true to the definition for positive logic, which requires that the more positive of the two logic levels be considered the 1 and the less positive the 0.

A very convenient feature of ECL circuits has both the function and its complement as outputs from each of the logic gates. The inputs to the ECL gates have

ECL and CMOS

pull-down resistors to the V_{ee} and so do not require the designer to install an external pull-down resistor. If the input is left open, it will act as a low input. If you wish an unused input to be held in a 1 state, then a pull-up resistor to ground must be installed. Figure 6–1 is a schematic of a typical ECL gate.

Note that each gate has two output functions. Each function is the complement of the other. If the output function is a NOR gate ($y = \overline{A + B}$), then this function and its complement ($\overline{y} = A + B$) are available. The two inputs to this circuit are the standard logic levels that are required for the particular ECL family that is used. A high or 1 would require a -0.7 V level, while a low or 0 would require the level to be at -1.6 V.

The fast switching speeds of these devices are achieved by keeping the transistors out of saturation and maintaining the circuits near turn-on even when they are in the OFF state and by switching the current from one side of the device to the other and in effect holding the total current constant.

To understand the operation of the circuit, there are a couple of items to be reviewed before analyzing circuit operation. First, for a transistor to be in the ON or conducting state, a 0.7 V drop from base to emitter is required. This is a nominal voltage and may vary in actual practice. Second, the base voltage of Q_3 is held by an internal reference source to -1.2 V. If Q_3 is to be conducting, the emitter voltage must be at a nominal -1.9 V. The two inputs to our logic circuit are to

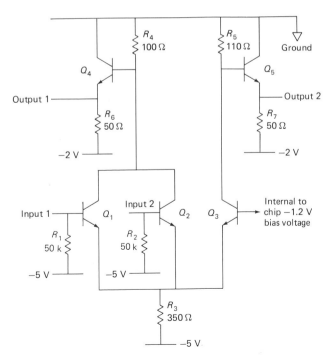

FIGURE 6–1 Basic ECL OR/NOR gate

the bases of Q_1 and Q_2. These two input transistors and the reference transistor Q_3 have all their emitters tied together, hence the name *emitter-coupled logic*.

For the circuit to function we assume that the inputs will be held at one of the two ECL logic levels: a high or $1 = -0.7$ V, and a low or $0 = -1.6$ V. If all the inputs are held in a low (-1.6 V) state, they will have an insufficient base to emitter voltage drop to cause Q_1 and Q_2 to conduct. Since Q_3 is holding the common emitters at -1.9 V, the base-to-emitter voltages of Q_1 and Q_2 will be limited to the difference of -1.9 and -1.6, or 0.3 V.

As was stated earlier, a transistor requires a 0.7 V base–emitter drop before the transistor will conduct; it follows that both Q_1 and Q_2 will be in the OFF state and so will not be conducting current through either transistor, and no current will flow through the R_4 resistor. This condition will cause the base of Q_4 to go to ground or 0 voltage. The emitter of Q_4 is returned to -2 V through R_6 (50 Ω). This condition will cause the base–emitter voltage of Q_4 to be sufficient to cause Q_4 to conduct.

For very rough current approximations, assume that the base of Q_4 is at ground and that there is a 0.7 V base–emitter voltage drop across Q_4. This will leave a voltage of 1.3 V across the 50 Ω resistor and will result in a maximum emitter current of $1.3/50 = 20$ mA. The ratio of emitter current to base current is defined as the β of the transistor and is roughly equal to 100. It follows that the actual base current will be approximately 0.2 mA. This 0.2 mA must flow through the R_4 100 Ω resistor and will cause a 0.02 V voltage drop across R_4. This verifies that the original assumption that the base of Q_4 will be at or near ground or 0 voltage was a good one.

The emitter output voltage of Q_4 will be at a -0.7 V level when transistors Q_1 and Q_2 are both OFF. The output will be a 1 (-0.7 V) for this condition. For the same input condition it follows that Q_1 and Q_2 are both OFF and that Q_3 is ON and conducting current. R_3 determines the amount of current that is being conducted through Q_3 and will control the current to a value of $(5 - 1.9)/350 = 8.9$ mA. This current will flow through the R_5 resistor, causing the base of Q_5 to be at a voltage of -0.97 V. Q_5 requires a 0.7 base–emitter voltage drop, resulting in an output 2 voltage of -1.7 V. For the stated 0/0 inputs, the outputs will be as follows: Output 1 = a high or 1 state, and output 2 = a low or 0 state. A distinct advantage of the ECL circuits comes from the fact that most outputs have both the function and the NOT of the function available.

By reviewing the transformations that can occur with the application of De Morgan's Theorems, we see that the NAND and NOR circuits can be caused to perform all of the logic functions. The above analysis was performed for the input conditions being 0/0. The truth table entry would be as shown in figure 6–2.

The second input condition to be examined is the 0/1 input. Take input 1 to the high state and hold input 2 low. By raising either of the two inputs to -0.7 V, the common emitter will be raised to -1.4 V. This will be sufficient to cause Q_3 to turn OFF, and in the case of input 1 going high, Q_1 will conduct. Current will flow through Q_1, and Q_3 will be cut off. Following the current paths through R_4 and R_5 will demonstrate that output 1 has now gone to a low state and output

ECL and CMOS

Inputs		Outputs	
2	1	1	2
0	0	1	0

FIGURE 6-2 Incomplete ECL gate truth table

2 switches to a high state. The truth table entry for this condition is line 2 of figure 6-3.

For the input condition 1/0 the current path simply switches from Q_1 to Q_2, and Q_3 remains in the OFF state. Outputs 1 and 2 will be unchanged. This will result in the output conditions for line 3 of the truth table of figure 6-4 being the same as line 2. If both inputs are taken high, the current through R_3 and R_4 will remain the same as it was when just one of the inputs was taken high. The only difference will be that Q_1 and Q_2 will share this current equally. Q_3 will remain in the OFF state. Line 4 of the truth table of figure 6-4 will thus indicate no change in the output states when going between these two input conditions. The complete truth table will be as shown in figure 6-4.

Output 2 is a standard OR gate, while output 1 is a NOR gate. The logic symbol for this circuit, shown in figure 6-5, must reflect the fact that both functions are available from this gate.

6-3 DESIGN CONSIDERATIONS FOR ECL

A look at the circuitry within the ECL chip indicates that the static input resistance of the logic elements is 50 kΩ. This input resistance is located on the chip and is returned to the -5.2 V V_{ee}. Using basic transistor approximations that suggest a minimum β of 100 for the input transistor, we can calculate the dynamic input resistance to these devices. The input voltage swing will be the standard ECL logic level swing of 0.9 V. This input voltage swing will cause the input transistor to

Inputs		Outputs	
2	1	1	2
0	0	1	0
0	1	0	1

FIGURE 6-3 Incomplete ECL truth table

Inputs		Outputs	
2	1	1	2
0	0	1	0
0	1	0	1
1	0	0	1
1	1	0	1

FIGURE 6-4 Completed ECL truth table

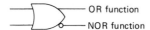

FIGURE 6–5 Symbol for ECL OR/NOR gate

conduct a current of 10 mA. A transistor with a β of 100 will require a base current of 10 mA/100 = 0.1 mA. The dynamic resistance will be on the order of the changing voltage divided by the changing current. For this case the value will be 0.9 V/0.1 mA = 9 kΩ.

The output resistance of the circuit will be determined by the emitter output resistance of the output transistors. This resistance will be low, much less than the 50 Ω lines it is specified as being capable of driving. A reason for the 50 Ω drive capability is to permit driving the readily available 50 Ω transmission lines and cables. A 50 Ω termination is to be provided externally by the designer. This terminating resistor must be returned to a negative voltage. It is suggested that the terminating resistor be returned to a -2 V reference supply. These circuits will be used on printed circuit boards, and it is possible to cause the traces on the board to have the desired characteristic 50 Ω impedance by proper selection of the trace width, height, and spacing to ground. This subject is covered in both the Fairchild and Motorola ECL product manuals. Formulas and charts are given to assist you in choosing the correct parameters for the circuit.

To reduce the power dissipation, the 50 Ω can be increased to 100 Ω. This change will require that the designer use the equivalent 100 Ω transmission lines on and off the board. The outputs of any two gates can be hard wired together and will result in an OR circuit. The emitter outputs are OFF when the output is a low. If either of the two hard-wired outputs is taken high, the other output can be OFF and suffer no effects when the other emitter is taken high. The equivalent output circuit is a common OR, which has the output high if either or both inputs are low. The hardwired output could also be considered a bubbled input NAND gate. (See figure 6–6.) The truth table would be the same for both circuits.

When the logic level shifts within the gate structure, there is a shift in the current path from one side of the gate to the other. It is important to note that there is no large change in current demand from the power supply. This lack of increased

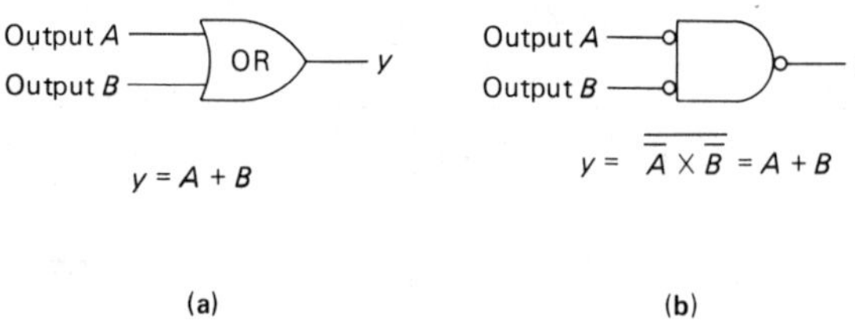

FIGURE 6–6 Equivalent hard-wired ECL output

ECL and CMOS

or decreased current demand will result in a system that has less noise on the common power supply. As a consequence, one should expect a more reliable system.

To extend this effect one step forward, even if only one of the two outputs is used, both outputs should be terminated in the same manner. If one is terminating with a 50 Ω resistor to −2 V, then both outputs should be terminated in this manner.

6–4 TTL TO ECL LOGIC LEVEL CONVERSION

Quite often a system is designed that uses more than one of the standard logic families. When this occurs, the logic levels of one family must be converted to the logic levels of the other. Should the two families be TTL and ECL, the following procedures can be followed. Reviewing the TTL logic levels, we note that an output high will be +2.4 V or greater (but less than 5 V). A TTL low will be less than 0.4 V (but not less than 0 V). An ECL high will be −0.7 V, and a low will be −1.6 V. The passive resistor combination shown in figure 6–7 is available in a 14-pin DIP package with six conversion circuits per package.

Using the resistor divider technique for determining the voltage output gives the results shown in table 6–1. These levels are not exactly those specified for the ECL logic family, but the circuit does work as a TTL-to-ECL converter. One must be careful when using the device. Do not try to drive more than one of these passive loads from a standard TTL output circuit. These passive converters are available with six converters per package. The package is a 14-pin DIP manufactured by Bourn and given the number 4114R-064. This conversion is good only when converting from TTL to ECL; the reverse conversion will not work with this chip.

A more conventional method for converting between the two logic families is to use the 10124 and 10125 chips that are listed in the ECL data catalogs. These chips were designed to convert one logic level to another. The 10124 chip converts a TTL level to an ECL level. This chip can convert four TTL levels to the equivalent ECL level. The 10125 can convert four ECL logic signals to the equivalent TTL logic level. These two chips require +5 V, −5.2 V, and common return connections.

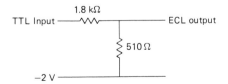

FIGURE 6–7 Passive TTL-to-ECL logic level converter

TABLE 6-1 TTL/ECL level comparisons

TTL Input Range	ECL Output
(Low) 0.0 V	−1.55 V
0.4 V	−1.47 V
(High) 2.4 V	−1.02 V
4.0 V	−0.30 V

6-5 THE TWO FAMILIES OF ECL

Within the ECL family there are two subfamilies. The earlier family is the 10k series; the faster 100k series came later. The logic levels of these two families are compatible. The 100k series has a propagation delay of 0.7 nsec when transitioning from a low logic level to a high level. The propagation delay for a high-to-low transition is specified as a maximum of 1.5 nsec. The 10k series is specified as a straight 2 nsec propagation delay for both transitions. The price that you pay for the increased speed of response is increased power dissipation. The 10k series dissipates approximately 35 mW per gate for the basic NOR circuit, whereas the 100k dissipates 50 mW for the same NOR function. It is recommended that the −5.2 V V_{ee} that is required for the 10k series be reduced to −4.5 V for the 100k series. This reduces the power dissipation. The −4.5 V is easily derived from the −5.2 V supply by simply inserting a silicon diode in series with the −5.2 V supply; the resultant 0.7 V drop across the silicon diode will produce the −4.5 V supply.

The 10k series comes in both a flatpak and a standard 16-pin DIP package. Ground is generally on pins 1 and 16. The −5.2 V is generally required on pin 8. The 100k series is packaged in both the flatpak and a 24-pin DIP. The V_{cc} (ground) pins are 6 and 7; the −5.2 V is required on pin 18.

Unused inputs to both the 10k and 100k series can be left unconnected if one wishes them to be in a low state. The 50k input resistors are returned to the −V_{ee} supply, and no pull-down resistors are required. Should you wish the unused inputs to be in the high state, they must be connected to an ECL high-level source.

6-6 CMOS

The CMOS logic family is used when low power dissipation is of primary interest. This family is the closest to being the "perfect" logic family. CMOS circuitry dissipates power only during the switching cycle. The CMOS family consists of two insulated gate field effect transistors connected in tandem. To understand the operation of a CMOS gate, it is necessary to understand the operation of the component FETs.

The FET in figure 6-8 is fabricated from a bar of P type silicon. This type of silicon has a deficiency of electrons, which is called an excess of holes. Implanted on opposite ends of this P type bar are two areas of N type silicon. N type silicon

ECL and CMOS

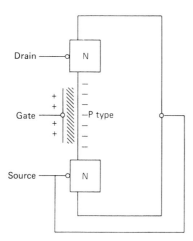

FIGURE 6–8 Enhancement-type MOSFET with N channel-insulated gate

has an excess of electrons available for conduction of electric current. Across the junction between the implanted regions and the P type bar is the same barrier that exists for a PN junction diode, and the response for this junction is similar to that of the diode. It will conduct if the P section is made more positive by 0.7 V than the N material and will block current if the N type is held or given a higher voltage than the P type material. With no voltage connected to the upper implanted region (called the *drain*) and the lower implanted region (called the *source*) tied electrically to the P type bar, there is the wrong combination of voltages to cause the conduction of current from the source to the drain. If a positive voltage is now connected to the drain, there remains the wrong combination of voltages on the device to allow it to conduct, so there is no current path from drain to source or from source to drain. If the gate is now forced positive, a path of available electrons will be created between the source and the drain, and a conduction path will now exist. If we locate the gate midway between the source and the drain, the voltage at which the gate goes more positive than the bar will be approximately half the voltage applied to the source. If we use the drain as an output and the gate as the input, the output will have a low-resistance conduction path through the induced N channel that exists within the P type bar. In our CMOS family this circuit will be used to cause the output to be held or clamped to ground when the output is to be in the low state.

A second complementary circuit using an N type bar with P type implantations is used to clamp the output to the V_{dd} supply voltage when the output is to be held in the high state. This circuit is shown in figure 6–9. In this circuit the gate must be taken to a voltage that is lower than the bar voltage. When this gate voltage is high, there is no induced P channel conduction path between the source and the drain. When the gate voltage is lowered, there is a path of current conducting holes between the source and the drain, and the device now has a low-resistance conducting path from the source to the drain. If the positive supply voltage V_{dd} is connected to the source and the output is taken from the drain, the output will be

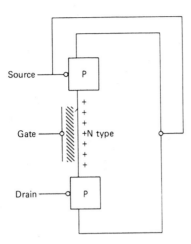

FIGURE 6-9 Enhancement-type MOSFET with P channel-insulated gate

tied to V_{dd} when the gate voltage is low via the low-resistance current hole-conducting path through the N type bar material. The output will now be connected to the V_{dd} supply and will be in the high state.

By the simple procedure of stacking these two devices, connecting the drains together and using that connection as your output, and tying the gates together and using that connection as the input, the circuit becomes an inverter. With the input low the output is connected to the V_{dd} power supply. If the gate input is high, the output is connected to ground. (See figure 6–10.)

6-7 A CMOS NAND GATE

For ease of representation, let us use a box with source, gate, and drain inputs to replace the circuitry shown in figures 6–8 and 6–9. To build a two-input NAND gate with this CMOS circuitry, the connections shown in figure 6–11 would be required. The truth table for this device is as shown in figure 6–12.

Treat the boxes as switches. When the P channel has a low logic level on the gate, it will be conducting; when it has a high logic level at the gate input, it will be open. The converse can be said for the N channel devices. They will be shorted switches when a high logic level is on the gate input and open when a low logic level is applied to the input. In the first case, both inputs are low, so both of the P channel devices will be conducting, and the output is tied to the V_{dd} supply voltage. For the second line, one of the parallel P channel devices will be turned OFF, and one of the lower N channel devices will come ON. Since the lower ones are in series, both must turn on to cause the output to be connected to ground. The output will thus remain connected to the V_{dd} supply. For the third line of the truth table, the only difference will be that the devices that were ON will now go OFF and the devices that were OFF go ON. The output will remain in the high state. For

ECL and CMOS

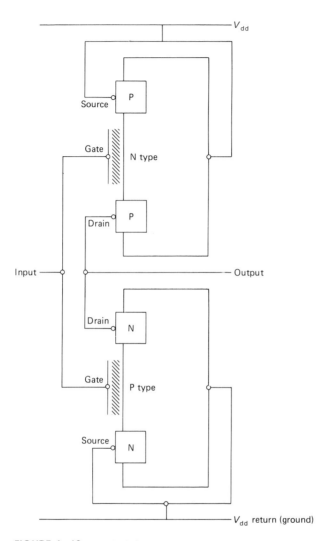

FIGURE 6-10 Typical CMOS inverter

the last line of the truth table the upper parallel P type devices will all be OFF, and the lower series N channel devices will all be in the ON state. The y output will consequently be connected to ground and thus be in the low state.

6-8 CMOS DESIGN CONSIDERATIONS

The input circuitry to the CMOS device is in reality a capacitor. As such, the input resistance is very high—on the order of 10^{12} Ω, the capacitance value being approximately 5 pF. When used in a circuit, with the logic levels changing, this input

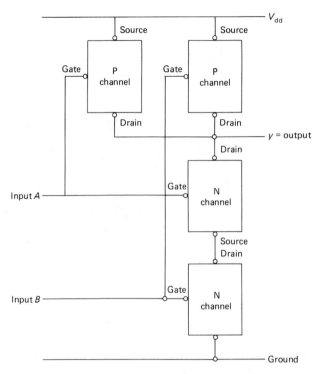

FIGURE 6–11 Typical CMOS NAND gate

capacitor must be charged and discharged during each of the logic level changes. This is when the circuit dissipates power. In any capacitor charge or discharge there is a power dissipation of $\frac{1}{2}CV^2$. It follows that if one is charging and discharging the capacitor at a very low repetition rate, there will be minimum power dissipation. As the repetition rate is increased, the power dissipation will of course increase.

With CMOS, no voltage value is specified for the V_{cc}; instead a maximum value is specified. Some of the specifications label this voltage V_{dd}, for drain voltage; and the return is called V_{ss}, for source voltage. Since no particular power supply voltage is specified, there can be no standard values for logic lows and logic highs. An investigation into the circuitry discloses that the logic gates will toggle at a

Inputs		Output
B	A	y
0	0	1
0	1	1
1	0	1
1	1	0

FIGURE 6–12 Truth table for NAND gate of figure 6–11

ECL and CMOS

voltage level that is the power supply voltage divided by 2. Should you choose a V_{dd} of +15 V, then you can expect your logic levels to be as follows:

0 or low < 7.5 V

1 or high > 7.5 V

One must be careful when working with CMOS devices; the chips must be stored with a conducting path between the inputs. Remember that if no external conduction path is maintained for the CMOS devices, the inputs are capacitive and so can accumulate a charge and develop sufficient voltage to cause them to self-destruct.

The propagation delays experienced in using CMOS gates should be on the order of 25–50 nsec. Rise and fall times tend to be 20–40% longer than the propagation delays. Since the system power dissipation is virtually zero when the devices are in either the ON or the OFF state, the power supplies for powering the CMOS circuitry will not require the current capacity that an equivalent TTL or ECL circuit would require. Since the inputs to the CMOS circuit are essentially capacitive, charging current will be required during level transitions.

6–9 CMOS TO TTL INTERFACING

A major problem that arises in attempting to drive TTL circuitry from CMOS outputs is the problem of current drive. Recalling the specs on the TTL inputs, we note that a standard TTL input gate would require that approximately 1.6 mA of current be sinked to ground through the output of the drive circuit. The spec for current sinking when the output is in the low state (0.4 V) allows 360 μA over a 50°C temperature range. If the temperature is held to 25°C, the current can reach 420 μA. If the voltage is allowed to climb to 0.8 V, then the current sinking capacity will quadruple to ~1.6 mA. This value is the standard sinking current requirement for a 1 U.L. TTL low input. This is marginally acceptable.

For the best design procedure, an interface chip designed specifically for this function is preferred. The National Semiconductor 74C909 and the DS1630 buffer can be used to buffer the CMOS family driving the TTL family.

6–10 TTL TO CMOS INTERFACING

The logic levels for TTL inputs are high > 2.0. The CMOS logic level change occurs at $V_{cc}/2$, or when CMOS is used with TTL, the V_{cc} would be 5 V, and the CMOS logic level change would occur at an input voltage of 2.5 V. Strictly speaking, the output from a TTL circuit is guaranteed to >2.4 V when in the high state. This leaves a region of 0.1 V where the interface circuitry could be confused. Realistically, the output from the TTL circuit would in almost all cases be greater than 2.5 V, and there would be no problem with allowing the TTL circuit to drive the CMOS family. What can cause the output voltages from the TTL circuits to drop in the case of a high output and to rise in the case of a low output is the requirement of large output currents. In the case of the TTL driving the CMOS, practically no current is required, so one can logically expect the level changes to be compatible. A problem might arise if one attempts to drive CMOS circuits in parallel with TTL circuits. This situation could cause the output of the TTL circuit to drop to the trouble region: $2.4 < V_{out} < 2.5$ V. If care is taken to limit the output current from the TTL circuit, minimum difficulty should be experienced with driving CMOS directly from TTL. The CMOS circuitry is not locked into using the 5 V for V_{cc}, and when V_{cc} is other than 5 V, a chip similar to the Motorola MC14504 hex level shifter would be required.

6–11 ECL TO CMOS INTERFACING

It is unlikely that you will experience these two families being used together in the same circuitry. The advantages of their being opposites of one another dictate that they not be used together. The very fast, high-power-dissipating ECL circuitry would not work well with the slower, power-saving circuitry of the CMOS family. If for some reason they are used together, one can always go to an intermediate TTL level and use the standard interfacing chips that have been previously discussed. The sequence could be from CMOS to TTL to ECL or the reverse.

PROBLEMS

1. Using an ECL 100114 differential line receiver, show all connections necessary to properly terminate a 50 Ω differential line.
2. What is the input resistance to a 10k series ECL device if no external components are added at the input of the device?
3. Using the resistor divider technique for changing TTL levels to ECL levels, calculate the current required from the TTL output for both the high and low output levels.
4. Using the 10125 level translator, show all chip connections to translate from an ECL logic level to a TTL level.
5. Make a list of time delays that you would expect to experience when using each of the following gates. (*Note:* Consult a manufacturer's databook.)
   (a) TTL   (b) LSTTL   (c) STTL   (d) FTTL   (e) CMOS
   (f) 10k ECL   (g) 100k ECL
6. If the outputs of two ECL circuits are hard wired together, draw the equivalent logic function that will result.
7. How many CMOS 3630 hex buffer chips could a standard 7400 NAND gate drive if we could accept the marginal operation of the device?

ECL and CMOS

8. What is the power dissipation of a circuit that utilizes 100 CMOS chips that have the same power dissipation as the 3630 device? What is the power dissipation of a circuit that utilizes 100 of the 100114 ECL chips?
9. Determine the highest-frequency signal that you would apply to the 3630 CMOS buffer circuit.
10. Design an interface that goes from ECL to CMOS logic levels.

chapter 7

chip survey and applications

7-1 INTRODUCTION

We have come to a point at which we understand the basic components of digital logic. These consist primarily of gates and flip flops. We have also learned Boolean algebra, the tool whereby we can design and analyze logic functions. Logic minimization techniques, either playing with algebraic factoring or mapping, help us to optimize a given task.

As a matter of fact, we could design complete digital systems with nothing else available but the dual-input NAND gate. As you have seen, the two-input NAND gate becomes

- An inverter if both inputs are tied together
- An AND gate if followed by such an inverter
- An OR gate by ANDing two inverted functions
- A latch by cross-feeding the outputs of two NAND gates
- A clocked flip flop by gating a clock with the inputs

If a two-input NAND gate fills all required functions, why complicate life by developing and marketing a myriad of chips, every one of which could be put together with 7400 chips? The question is rhetorical. The integrated circuit game started with four NAND gates in one chip, six inverters in one chip, and two flip flops in one chip. This in itself enormously reduced the chip count of a given design implementation. The number of functions on one chip is determined not so much by the real estate of transistors and resistors, but more by the required input/output pins. It soon became apparent that basic functions that are more complex than the simple NAND and the flip flop gate would further enhance design and packaging miniaturization.

Chip Survey and Applications

Entire families of TTL chips became available. In addition to the small-scale integrated circuit (SSI), the medium-scale integrated circuit (MSI) appeared. Combination AND/OR functions, multiplexors, eight-bit registers, arithmetic functions, buffers, tristate drivers, receivers, and others were also developed. As reliability and manufacturing yields improved further, many large-scale integrated circuit (LSI and VLSI) chips, the microprocessor, memories, and interface logic were developed.

- MSI: a chip that contains 12 or more gates as a single microcircuit.
- LSI: a chip that contains more than 100 gates as a single microcircuit.
- VLSI (very large scale integration): IC technology that is much more complex than LSI and involves a much higher equivalent gate count, though the minimum gate count has not yet been standardized.

No matter how large the variety of available functions, they cannot satisfy every need on the market. They are, however, very convenient building blocks for larger systems. Many designers find it necessary to squeeze an entire system into one chip. Another development fills that need—the *gate array*. As many as 200,000 gates are placed unconnected on one chip. The user specifies the desired interconnect pattern, which becomes the final step in chip manufacture.

In this chapter we will discuss, in survey fashion, commercially available chips, some typical applications, how to read the data sheets, and some pertinent characteristics.

7-2 DATA SHEETS AND SPECIFICATIONS

A *data sheet* of a particular chip usually contains a brief description of chip operation together with pertinent equations and/or a functional description in tabular form. It of course shows the package type and a description of the pin assignment.

In many cases this is all we look at. We know what the chip is expected to do, and we know how to wire it into our system. This presupposes familiarity with the particular chip family characteristics. A particular logic family shares many of the same operational parameters.

A list of parameters and definitions pertaining to digital logic follows. The symbols and terms used are in accordance with those currently agreed upon by the JEDEC council of Electronic Industries Association (AIE) for use in the United States and by the International Electrotechnical Commission (IEC) for international use.

- V_{IH}: High-level input voltage. An input voltage within the more positive of the two ranges of values used to represent the binary variables. *Note:* The input voltage is above the maximum switching level that guarantees proper operation by being recognized as a 1.

- V_{IL}: Low-level input voltage. An input voltage level within the less positive of the two ranges of values used to represent the binary variable. *Note*: The input voltage is below the minimum switching level that guarantees proper operation by being recognized as a 0.
- V_{OH}: High-level output voltage. The voltage at an output terminal with input conditions applied that, according to product specification, will establish a high level at the output. *Note:* The high-level output voltage must be higher than the maximum switching level.
- V_{OL}: Low-level output voltage. The voltage at an output terminal with input conditions applied that, according to product specification, will establish a low level at the output. *Note:* The low-level output voltage must be lower than the minimum switching level.
- V_{T+}: Positive-going threshold level. The voltage level at a transition-operated input that causes operation of the logic element according to specification as the input voltage rises from a level below the negative-going threshold voltage V_{t-}.
- V_{T-}: Negative-going threshold level. The voltage level at a transition-operated input that causes operation of the logic element according to specification as the input voltage falls from a level above the positive-going threshold voltage V_{t+}.
- I_{IH}: High-level input current. The current into an input when a high-level voltage is applied to that input. *Note:* Current into a terminal is considered positive.
- I_{IL}: Low-level input current. The current into an input when a low-level voltage is applied to that input.
- I_{OH}: High-level output current. The current into an output with input conditions applied that will establish a high level at the output. *Note:* This may be a negative current if current actually flows out.
- I_{OL}: Low-level output current. The current into an output with input conditions applied that will establish a low level at the output.
- I_{OZ}: Off-state output current (of a three-state output). The current flowing into an output having three-state capability with input conditions that will establish the high-impedance state at the output.
- I_{CC}: The current into the V_{cc} supply terminal of an integrated circuit.

The preceding terms define the voltage and current conditions. There are also the timing conditions:

- t_r: Rise time. The time interval between two reference points (typically 10–90%) on a waveform that is changing from the defined low level to the defined high level. (See figure 7–1.)
- t_f: Fall time. The time interval between two reference points (typically 10–90%) on a waveform that is changing from the defined high level to the defined low level. (See figure 7–1.)

Chip Survey and Applications

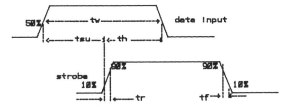

FIGURE 7-1 Timing waveform

- t_h: Hold time. The time interval during which a signal is retained at a specified input terminal after an active transition occurs at another specified input terminal. (See figure 7-1.)

- t_{su}: Setup time. The time interval between the application of a signal at a specified input terminal and a subsequent active transition at another specified input terminal. (See figure 7-1.) *Note:* The data input to a flip flop, for example, must be valid for at least the required setup time before a clock strobe is applied.

- t_{pd}: Propagation delay time. The time between the specified reference points on the input and output voltage waveforms with the output changing from one defined level to the other defined level. *Note:* This is defined under the two conditions that follow.

- t_{pHL}: Propagation delay time, high- to low-level output.

- t_{pLH}: Propagation delay time, low- to high-level output. (See figure 7-2.)

- t_w: Pulse duration. The time interval between two reference points on the leading and trailing edges of the pulse waveform. (See figure 7-1.)

Pertaining to tristate circuits the following definitions apply:

- t_{pHZ}: Disable time of a tristate output from high level. The time interval between the specified reference points on the input and output voltage waveforms with the three-state output changing from the defined high level to a high-impedance state.

- t_{pLZ}: Disable time of a tristate output from low level. The time interval between the specified reference points on the input and output voltage waveforms with the three-state output changing from the defined low level to a high-impedance state.

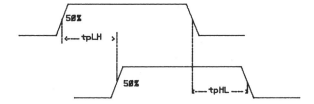

FIGURE 7-2 Timing waveform propagation delay

- t_{pZH}: Enable time of a tristate output to a high level. The time interval between the specified reference points on the input and output voltage waveforms with the three-state output changing from a high-impedance state to the defined high level.
- t_{pZL}: Enable time of a tristate output to a low level. The time interval between the specified reference points on the input and output voltage waveforms with the three-state output changing from a high-impedance state to the defined low level.

There is also a list of absolute maximum ratings. These usually pertain to

- The maximum allowable power supply voltage, V_{cc}
- The maximum allowable input voltage
- Operating free air temperature range
- Storage temperature range

The first of the four listed parameters, V_{cc}, is the one to watch. The input voltage is rarely exceeded, since the input comes from the output of another chip and is expected to be within specification. The temperature in our daily environment is well within the specified limits. These parameters become important when systems are designed for unusual environments such as outer space. V_{cc}, however, is under the engineer's control. Particularly when adjustable power supplies are used, it is very easy to exceed the maximum allowable voltage.

All operating characteristics are important. Depending on the application, some are more critical than others.

The high–low voltage input and outputs, the first through sixth items in the list, come into play when loading effects are considered. Either the chip is overloaded by too much fanout or there is a wiring error or bus contention. An oscilloscope picture easily indicates a suspiciously low high level or a high low level.

The high–low current input and output specifications, the seventh through twelfth items in the list, are useful to determine fanout and drive capability. The specified ranges are needed to calculate worst case conditions. For the design of a single system it usually suffices to use the typical values. If large numbers of the circuit are to be mass produced, then worst case conditions must be taken into account.

The timing specifications become important in working with high clock frequencies and/or short-duration pulses. Consider the case in which two waveforms are to be ORed and the transitions are assumed to occur at clock time as in figure 7–3. The expected output is at clock cycle 4, but because of waveform time delay there is also an unexpected voltage spike known as a *glitch* at clock cycle 5. That glitch might cause system problems. When two variables are expected to change state at the same time, but do not, there is a race condition.

Another case can be illustrated with a synchronous counter. Suppose we have a flip flop that is guaranteed to operate at 10 MHz. We build a counter as in figure

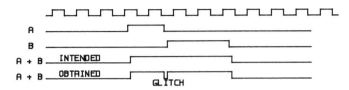

FIGURE 7-3 Glitch due to delay

10-10(a) (page 331) extended to ten stages. (We can look at the counter without having studied Chapter 10.) Each stage requires that the control input be available one setup time before the clock strobe arrives. There will be one flip flop propagation delay and eight gate propagation delays, or a total of nine delays before the input is available to the last flip flop stage. If the propagation delay is 10 nsec, the input to the last flip flop stage arrives 90 nsec after the last clock strobe. The cycle duration for 10 MHz is 100 nsec. This leaves 10 nsec for the setup time t_{su}. If the specified setup time is 30 nsec in the worst case, the last flip flop might not respond.

The circuit will work at 8.3 MHz but not at 10 MHz even if the flip flop specification is good to that frequency. The point is that we must look at the chip specification to ascertain proper operation. Moreover, marginal operation is worse than failure to operate at all.

The chips discussed in the remainder of this section are available in several TTL and MOS families. TTL is generally faster than MOS, but the latter uses less power. Functionally, TTL and MOS are the same in most cases. The choice of TTL or MOS depends on the application.

The main criteria in which the families differ are speed and power consumption. Table 7-1 gives a rough comparison of the various families.

It is difficult to talk about power consumption in CMOS logic. Standby power is negligible. Power is consumed when the circuit switches. How often a circuit switches depends on the particular operation of the system. It is clear, though, that

TABLE 7-1 Speed and power for TTL and CMOS families

Family	Propagation Delay (nsec)	Clock Frequency (MHz)	Power Dissipation per Gate (mW)
TTL (no subscript)	10	35	10
TTL H	6	50	22
TTL S	3	125	19
TTL AS	2.5	175	9
TTL L	33	3	1
TTL LS	10	45	2
TTL ALS	7	50	1 (at 1 MHz)
CMOS	60	10	1 (at 1 MHz)
HCMOS	8	20	1 (at 1 MHz)

in most cases a gate is not called upon to switch at the system clock frequency. Thus if the average switching rate is low, much power can be saved.

The TTL and HCMOS families are produced in the 74 and 54 series. The 74 series is for commercial use, and the 54 series for military application. The main difference is temperature range. Whereas the commercial series is specified between $-40°$ and $+85°C$, the military version is specified between $-55°$ and $+125°C$.

7-3 GATES

AND, NAND, OR, and NOR gates and the inverter are available in several TTL families, CMOS, and HCMOS. The part numbers and types are as follows.

- 7404: Hex inverter*
- 7400: Quad 2-input NAND gate*
- 7408: Quad 2-input AND gate*
- 7410: Triple 3-input NAND gate*
- 7411: Triple 3-input AND gate
- 7420: Dual 4-input NAND gate*
- 7421: Dual 4-input AND gate
- 7430: 8-input NAND gate*
- 74133: 13-input NAND gate
- 7402: Quad 2-input NOR gate*
- 7432: Quad 2-input OR gate
- 7427: Triple 3-input NOR gate
- 744075: Triple 3-input OR gate
- 744002: Dual 4-input NOR gate
- 744078: Eight-input OR/NOR gate

These are the workhorse gates of logic. Asterisks indicate the most frequently used chips. Designers favor NAND over NOR logic. The family used depends on the rest of the system. Mixing families is possible but not advisable. Attention must be given to voltage level, current drive, and speed incompatibility.

7-4 BUFFERS, DRIVERS, AND TRANSCEIVERS

In addition to the NAND, NOR, and AND/OR chips, which are used for standard logic configurations, there is a need for special-purpose logic chips to interface with other devices and systems. These are buffers, drivers, and transceivers, which are used for

Chip Survey and Applications

- Higher current capability
- Higher voltage capability
- Wired OR and wired AND connections
- Driving signal over longer distances
- Bus interfacing

Open Collector Buffers

Here the load resistance is not part of the chip proper but part of an external circuit. Current requirements for devices other than standard logic chips often require larger currents and are driven from ground to V_{cc} rather than to TTL level. Such is the case when driving LEDs or relays. A typical device for that purpose is an inverting 7406 and 7416 or a noninverting 7407 and 7417 open collector driver. These have a 30 mA drive capability. Moreover, the collector of the 7406 and 7407 can be operated at 30 V. Figures 7–4 and 7–5 are typical applications.

N inverting open collector circuits can be used for a wired NOR function and N noninverting open collector circuits can be used as a wired AND function as shown in figures 7–6 and 7–7.

Open collector dual input NANDs allow input control, such as the 7438 with 48 mA or the 7439 with 80 mA current capability. These are widely used as bus drivers or flat cable drivers. The characteristic impedance of cables is on the order of 180 Ω terminated with a 330–390 Ω combination. Such terminations are also found on computer bus lines. Bus driving will be discussed in Chapter 12. Examples are shown in figures 7–8 and 7–9.

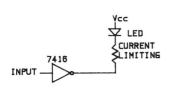

FIGURE 7–4 Open collector LED driver

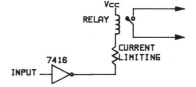

FIGURE 7–5 Open collector relay driver

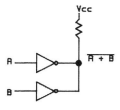

FIGURE 7–6 Wired NOR

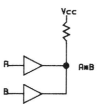

FIGURE 7–7 Wired AND

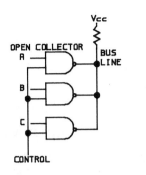 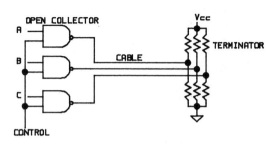

FIGURE 7-8 Open collector bus driver

FIGURE 7-9 Open collector cable driver

Line Drivers and Receivers

To transmit clean waveforms over distances longer than 30 ft, it is often necessary to use 50 or 75 Ω transmission cables. The TI 74128 is a NOR line driver for that purpose.

The National Semiconductor DS26LS31 is a differential line driver for twisted pair or parallel wire transmission lines. The signal is then received with a differential line receiver such as a DS78LS120 as in figure 7-10.

Other driver and receiver types are available to match requirements for bus-oriented transmission systems, various bus standards, noise and crosstalk problems, data terminal equipment, and so on. Important characteristics to look for are

- Current capability
- Maximum output voltage
- Open collector or tristate output
- Switching characteristics
- Differential or single ended
- Technology: CMOS or bipolar, standard, Schottky, etc.

A few examples are given in table 7-2. Some additional examples appear in Chapter 13.

Transceivers

Bus-oriented systems transmit signals to and receive signals from the bus. On data and address buses an entire word is transferred with only two control bits, one to determine the direction of signal flow and the other to enable the chip. These are 20-pin chips available in slim dual in-line, open collector, or tristate. Typical part numbers are 74245 for noninverting, 74620 with inverting tristate outputs, and 74621 and 74622 with open collector outputs. The 74646 and 74647 include input

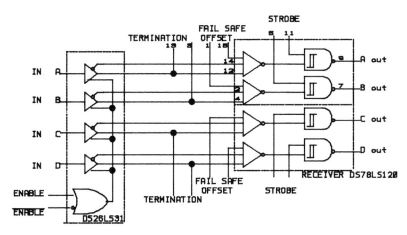

FIGURE 7-10 Differential line driver DS26LS31

as well as output registers with multiplex control to transfer signals directly or via the registers. Figure 7-11 shows the logic for one bit.

There are six control signals:

- $\overline{G}$: Chip enable
- DIR: Low = B data to A bus, high = A data to B bus
- CAB: Store A data
- CBA: Store B data

TABLE 7-2 Examples of buffers, drivers, and receivers

Part No.	Manufacturer	Type	Output	Technology	Current (mA)	Voltage	Speed (nsec)
7416	TI*	Hex Inv	OC	Std	30	15 V	10
7417	TI	Hex NonInv	OC	Std	40	15 V	6
7438	National*	Quad NAND 2-Input	OC	LS	48	15 V	22
74241	TI	Octal Dr	Tr St	HC–TTL	7.8	V_{cc}	15
74245	Motorola	Octal Transc	Tr St	HC–TTL	24	V_{cc}	24
DS1686	National	Relay Dr	OC	CMOS	300	65 V	1
DS26LS31	National	Quad Diff Dr	Tr St	LS	30	V_{cc}	15

Abbreviations: Inv: inverter; Dr: driver; Transc: transceiver; Diff: differential; OC: open collector; Tr St: tristate; HC: high-speed CMOS; Std: standard; LS: low-power Schottky.

* Most devices are available from several other manufacturers.

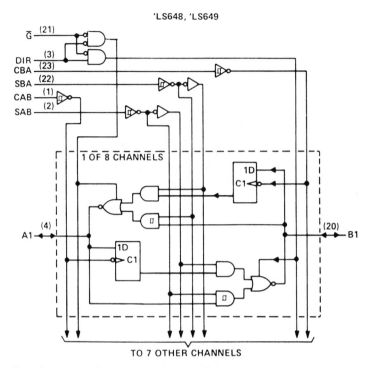

FIGURE 7-11 Octal bus transceiver with storage (Courtesy of Texas Instruments Incorporated)

- SAB: Low = real time A data to B bus, high = stored A data to B bus
- SBA: Low = real time B data to A bus, high = stored B data to A bus

It is not practical to suggest which particular chip to use for a given purpose, since requirements and needs vary widely. Each of the available chips has somewhat different characteristics, and the user must judge which one most closely meets the need. Chip properties also vary with the logic family. Some part numbers are produced in standard TTL; others are produced in the LS, L, S, or HC family. Most are available from Texas Instruments, National Semiconductor, Motorola, RCA, American Micro Devices, and other manufacturers.

7-5 DECODERS, ENCODERS, SELECTORS, AND MULTIPLEXORS

It is often necessary to select one of a number of data sources. The data source can be one data bit—a control line, for example—or an entire word. Tristate gates or registers are used to gather data onto a common output bus, and a selector enables one set of data at a time.

Chip Survey and Applications

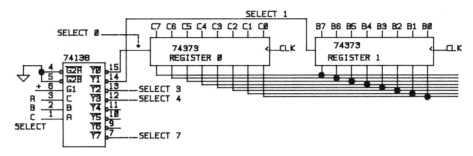

FIGURE 7–12 Address select logic

A 74138 decodes three lines to eight lines. One of eight output lines becomes true as a function of the three-bit input. The 74139 is a dual two-line to four-line decoder, and the 74154 is a four-line to sixteen-line decoder with tristate outputs. These find application in memory address decoders and selectors. Figure 7–12 shows a typical application of a data selector.

Another way to select data is to use a two-, four-, eight-, or sixteen-line to one-line decoder. The 74157 is a quad two to one decoder. The 74153 is a four to one, the 74152 is an eight to one, and the 74150 is a sixteen to one decoder. A QUAD 2-line to 1-line data selector.

An example of an encoder is a priority encoder. The binary output of the priority encoder reflects the number of true inputs starting with the least significant

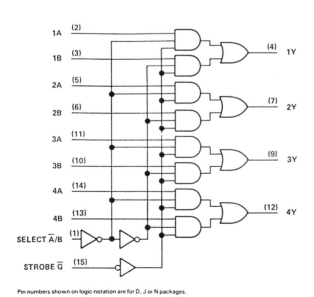

FIGURE 7–13 Data selector (Courtesy of Texas Instruments Incorporated)

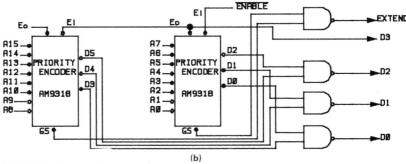

FIGURE 7-14 16-bit input priority encoder

bit. This is best illustrated in terms of a truth table for the 74148, as shown in figure 7-14(a). Notice that the output binary number is equal to the number of 1's at the input. The input listed in the truth table is the only permitted input. Any other input pattern is an error indicated by a 11 status output. An enable input E_i, enable output E_o, and G_s overflow allow external expansion without the need for external circuitry. An eight-bit priority encoder expanded to sixteen bits is shown in figure 7-14(b). The 74147 and 74148 have BCD and octal outputs, respectively. An application of a priority encoder to a flash converter is discussed in the chapter on *D/A* and *A/D* conversion (Chapter 12).

In some instances designers with special requirements must design their own logic. Not every conceivable combination can be made available. However, it is helpful to find the most common types. A few examples are given in table 7-3.

7-6 ARITHMETIC CIRCUITS AND PROCESSOR ELEMENTS

The most often used arithmetic chip is the 74283 four-bit binary full adder. This chip will be discussed in Chapter 8. A more complete unit is the 74181, a four-bit parallel binary accumulator that includes subtraction, shifting, complementing comparison, and logic functions.

Chip Survey and Applications

TABLE 7-3 Examples of encoders and decoders

Part No.	Manufacturer	Type	Technology	No. of Pins
74152	TI	8 to 1 decoder	LS	24
74138	Motorola	1 of 8 decoder	HC	16
74139	TI	Dual 2 to 4	ALS	16
74C48	National	BCD to 7-segment LED driver	CMOS	16
CD22105	RCA	Four-digit LCD decoder-driver	CMOS	40
AM9318	AMD	8-input priority encoder	Std	16

To speed up arithmetic operations, there are carry look-ahead generators such as the AM9342 by AMD. This device accepts four pairs of carry-generating signals and provides anticipated active carries over four groups of binary adders.

Hardware multiplication, using binary full adders, will be discussed later. High-speed multipliers, however, accomplish multiplication in one clock cycle. The 74284 will give a 4×4 product in 40 nsec, as is shown in figure 7-15.

Ultra-high-speed multipliers with ECL logic can multiply 32×32 bits in 10 nsec. Such devices find application in sophisticated array processors and systems with advanced mathematical operations.

Comparators can also be considered arithmetic elements. The 7485 four-bit comparator (also discussed in Chapter 8) is the most often used. It performs comparisons of two four-bit nibbles, and three outputs tell whether nibble $A > B$, $A = B$, or $A < B$. The chips can be cascaded or interconnected as a tree for higher speed.

Eight-bit comparators in the ALS family, 74518 to 74522, are 20-pin chips with either open collector or totem pole output. There is, however, only one identity output, and these comparators are therefore not as general purpose as the 7485. The 74526 is a sixteen-bit and the 74528 a twelve-bit identity comparator.

Successive approximation registers perform a very useful mathematical algo-

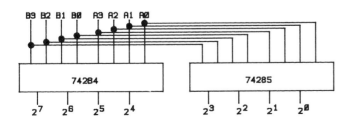

FIGURE 7-15 4×4 bit binary multiplier

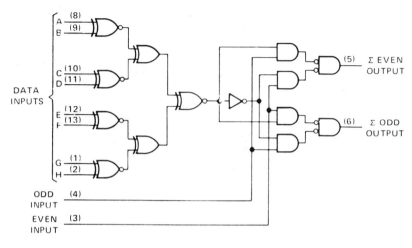

Pin numbers shown on logic notation are for J or N packages.

FIGURE 7–16 Parity generator (Courtesy of Texas Instruments Incorporated)

rithm. They are used in *A/D* converters, data compressors, and other fast-converging processes. These will be described in Chapter 10. National Semiconductor Corporation makes the eight-bit DM2502 and the twelve-bit 54C905.

Error detection circuits are used with data transmission. They fall into two classes, odd/even parity generators and error detection and correction circuits. Discussion on that subject can be found in the chapter on peripherals (Chapter 14). The 74180 and 74280 are four-bit parity generators with expansion capability. The two outputs indicate whether the sum of all 1's is even or odd. The circuit is shown in figure 7–16.

Error detection and particularly error correction are complex networks. The

TABLE 7–4 Examples of arithmetic chips

Part No.	Manufacturer	Type	Technology	No. of Pins
74283	TI	4-bit binary full adder	LS	16
AM9341	AMD	4-bit arithmetic logic unit	LS	24
54C905	National	Successive approx. register	C	24
74180	TI	9-bit odd/even parity generator	Std	14
74284	TI	4 × 4 multiplier	Std	16
67558	Monolithic M	8 × 8 multiplier	S	40
74C85	National	4-bit comparator	CMOS	16
74682	TI	8-bit comparator	LS	20

74632 is a 52-pin chip that detects and corrects single-bit errors with a modified Hamming Code.

A few examples of adders, parity generators, and comparators are given in table 7–4.

7–7 FLIP FLOPS, REGISTERS, AND LATCHES

Two types of flip flops, the *J/K* and the *D*, are available. The *T* flip flop is a *J/K* flip flop with *J* and *K* tied together. Synchronous *R/S* flip flops are not needed, since the *J/K* flip flop fills that function. If nonsynchronous Set and Reset are needed, the *D* as well as the *J/K* include these as independent inputs.

One of the first *J/K* flip flops to appear on the market was the 7476. It has nonstandard V_{cc} and ground connections at pins 5 and 13, respectively. It is still available, but the preferred version now is the 74112 with standard pin connections. The various versions differ from each other mainly in the number of flip flops per chip. Other differences are the availability of Preset and Clear (this is really the same as Set and Reset) and whether or not the flip flops share a common clock or can be independently strobed. Flip flops are available in all families and most manufacturers. (See table 7–5.)

Quad and octal flip flops are useful as byte registers. Often the edge-triggered flip flop is not needed, and latches will do. They are used as temporary storage elements in I/O application or indicator units. The *D* input is transferred to the *Q* output when the Enable input is true. The output will follow the input as long as the Enable input is true. When the Enable goes false, the data is retained. The 74259 is an eight-bit addressable latch that allows input access to a selected location while leaving the other seven data bits undisturbed.

An application of the octal *D* flip flop with tristate outputs is shown in figure 7–17. Since the outputs can be selectively enabled, it serves as a data selector.

TABLE 7–5 Examples of flip flops and registers

Part No.	Type	No. of Pins
7476	Dual *J/K* flip flop with Preset and Clear	16
74112	Dual *J/K* flip flop with Preset and Clear	16
7473	Dual *J/K* flip flop with Clear	14
74113	Dual *J/K* flip flop with Preset	14
7474	Dual *D* flip flop with Preset and Clear	14
74175	Quad *D* flip flops with common clock and Clear	16
74174	Hex *D* flip flop with common clock and Clear	16
74273	Octal *D* flip flop with common clock and Clear	20
74374	Same as 74273 but tristate outputs, no Clear, but an Enable input	20
74375	Quad *D* latch	16
74573	Octal *D* latch	20
74259	8-bit addressable latch	16

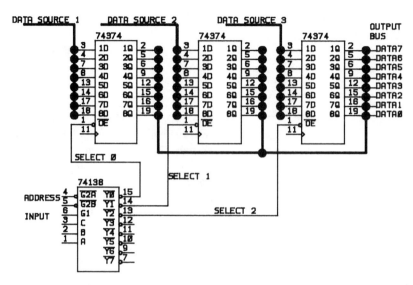

FIGURE 7–17 Selector

7–8 COUNTERS

Counters are needed in many digital systems. The basic function of a counter is simple, yet available chips differ from each other in several respects:

1. Number of bits—most counters are four bits long, some are eight bits, and some are longer.
2. Synchronous or ripple.
3. Modulus: binary count to 2^n, or BCD count from 0 to 9, or special counters that allow choice of moduli 2, 5, 6, 8, 10, 12, 16, or N-position Johnson counters with one active output position/cycle, or programmable divide-by-N counters.
4. Synchronous or nonsynchronous Load and Clear.
5. Synchronous cascade capability with count enable input.
6. Counter with input register or tristate output register.

Table 7–6 lists several popular counters.

The 74193 and 74163 are both synchronous counters, but the 74193 does not have a count enable input. Therefore cascaded stages are ripple connected; that is, the second four-bit counter must wait for the carry of the first to be true before receiving a signal at the clock input. The 74163 is synchronous for all bits. This is shown in figures 7–18(a) and 7–18(b).

Down counters are useful for generating a strobe after a given number of clock cycles. The Preset inputs are loaded with a desired binary value and the counter counts down to zero. The borrow output indicating an all zero becomes the strobe.

Chip Survey and Applications

TABLE 7-6 Examples of counters

Part No.	Description	Features	No. of Pins
74193	4-bit binary up/down	Asynchronous Clear and Load; no count enable	16
74393	Dual 4-bit binary up counter	Asynchronous Clear; no Preset inputs	14
74163	4-bit binary up counter	Synchronous Clear and Load; count enable input	16
74590	8-bit binary up counter	Synchronous Clear with tristate output registers	16
74192	Decade up/down	Asynchronous Clear; no count enable	16
7490	Decade or biquinary	Asynchronous Clear; no Preset inputs	14
CD4059	CMOS divide-by-N	Programmable	24
CD4020	CMOS 14-stage up counter	Ripple counter	16

Note: A counter may be synchronous count, but asynchronous Clear and Load.

The same strobe can also serve to load the Preset value into the counter for a repeat as in figure 7–19.

7-9 SHIFT REGISTERS

All a shift register does is move a given bit pattern one position when strobed. The shift register is characterized by a number of features:

- Length
- Shift in one direction only or left as well as right
- The input may be serial only or serial and parallel

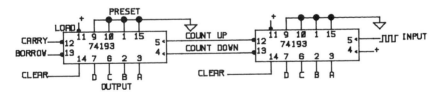

FIGURE 7-18(a)   Ripple carry counter

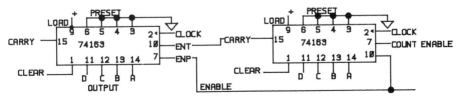

FIGURE 7-18(b)   Synchronous counter

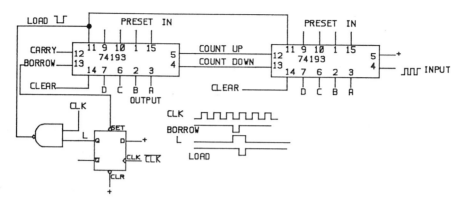

FIGURE 7-19 Down counter with repeat Preset

- The output may be serial only or parallel
- The output may be tristate

Several types are listed in table 7-7.

Shift registers have many applications. A few of these are time delay circuits, CRT refresh memory, Johnson counters, multipliers, logic functions, parallel-to-series conversion, and series-to-parallel conversion. The last two applications are shown in figures 7-20(a) and 7-20(b), respectively.

TABLE 7-7 Examples of shift registers

Part No.	Description	Features	No. of Pins
74195	4-bit shift register	Parallel input and output; shift from A to D only	16
74194	4-bit shift register, bidirectional	Serial/parallel input, parallel output	16
74165	8-bit shift register	Parallel input, serial output; shift from Y_0 to $32Y_7$ only	16
74589	8-bit shift register with input register	Serial parallel input, 3-state parallel output	16
74673	16-bit shift register	Serial input, 3-state parallel output; shift from Y_0 to Y_{15} only	24
CD40100	CMOS 32-stage shift register, bidirectional	Serial input, serial output	14
CD4062	CMOS 200-stage shift register	Serial input, serial output; shift from Y_0 to Y_{199} only	14

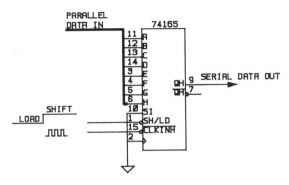

FIGURE 7–20(a) Parallel-to-series conversion

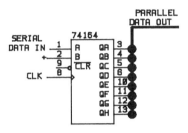

FIGURE 7–20(b) Series-to-parallel conversion

7–10 TIMING CHIPS

Most systems require a clock generator or oscillator. If frequency stability is not an issue, then any R/C controlled oscillator such as a 555 timer will do. Otherwise, a crystal oscillator is needed. A wide variety of oscillators is available from Motorola, Sprague, Nortec, Mitsubishi, Intersil, RCA, National Semiconductor, Q-tech, and other manufacturers. You can get them, for example, at a fixed single frequency in three-pin cans from 1 Hz to 80 MHz from Q-tech or use an RCA 21-stage counter CD4045 with crystal input or a Motorola MC14450 16-stage divider buffer for low-frequency and low-power applications.

One shots are also used as timers as well as single-pulse generators. Table 7–8 shows a few available one shot oscillators and timers. Examples of the 555 timer and two single shots as an oscillator are given in figures 7–21 and 7–22, respectively.

The one shot is often used as a switch debouncer. Depression of a switch triggers a retriggerable one shot. As long as the switch bounces, the output waveform stays true. As soon as the switch contact settles, the one shot will time out and reset. When the switch is released, the contact will bounce again. A second one shot will lock out the first and time out when the contact potential has stabilized. However, a switch bounce eliminator chip such as the Motorola MC14490 uses an

TABLE 7–8 Examples of timing chips and one shots

Part No.	Manufacturer	Description	Technology	No. of Pins
555	Signetics	Timer	Linear Integr.	8
QT1	Q-tech	Oscillator	Crystal	3
74S124	TI	VCO	S	16
MM53107	National	Oscillator divider	MOS	8
560	Signetics	Phase locked loop	Linear Integr.	16
MC14490	Motorola	Hex contact bounce eliminator	CMOS	16
74123	TI	Dual monostable	Dual input, retriggerable	16
74121	TI	Dual monostable	Dual input, nonretriggerable	16

internal oscillator to generate a clean pulse that is four clock cycles delayed. It features six contact bounce eliminators in one chip.

Another type of oscillator is the voltage-controlled oscillator (VCO). The output frequency is a function of a dc control input. Without any dc input the oscillator operates at some predesigned center frequency. The input is used to shift the frequency to a desired operating point. The TI 74124, 325, and 327 are examples.

One well-known application of the VCO is the phase-locked loop used in frequency multiplication, voltage-to-frequency conversion, data synchronization, and frequency discriminators. The 741297 is TTL, and the RCA 4046 is a CMOS low-power phase-locked loop that includes a VCO, a phase comparator, and buffers. A frequency multiplier application is shown in figure 7–23.

In this example the VCO operates near 512 kHz. The output is divided by 32 to 16 kHz and compared to a 16 kHz incoming frequency. The two frequencies, the incoming frequency and the counted down frequency from the VCO, are compared in the phase comparator. A dc error signal develops at the output of the phase comparator if the two frequencies are not equal and out of synchronization. The dc voltage output from the phase comparator becomes the input to the VCO and regulates the VCO frequency until the error signal is minimum. The VCO output at 512 kHz is thus locked to the incoming 16 kHz, hence the name *phase-locked loop*.

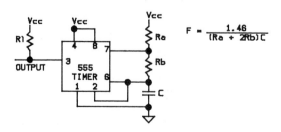

FIGURE 7–21 Astable operation with 555 timer

Chip Survey and Applications

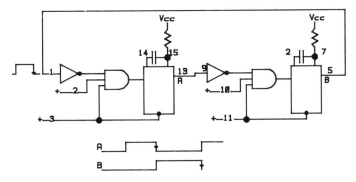

FIGURE 7-22 Astable operation with two one shots

7-11 MEMORIES

Large memories are part of general-purpose or special-purpose computing systems, and separate manuals describe their characteristics. Details will be discussed in Chapter 9.

Static memories are available from TI, Harris, Motorola, MMI, RCA, Raytheon, Intel, Signetics, National, Fairchild, IDT, MOSTEK, AMI, AMD, and other manufacturers. They typically range from 1K to 64K capacity in various architectures. Most of the same manufacturers also produce dynamic RAMs, ROMs, PROMs, and EPROMs.

Small memories such as 74870, which is a dual 16 × 4 register file, often form an integral part of a logic design. The RCA 40208 is a 4 × 4 expandable, addressable array used as a multiport register that is used as a scratch pad memory or data storage. Many other miniarrangements fit the designer's needs. The 74170 is shown in block diagram form in figure 7-24.

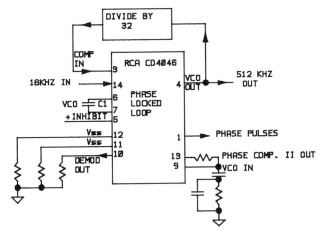

FIGURE 7-23 Frequency multiplier with a phase-locked loop (Courtesy of GE Solid State)

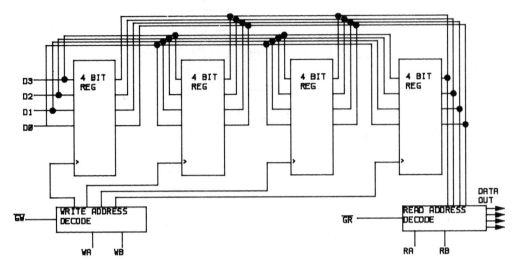

FIGURE 7–24 4 × 4 register file

Another useful register-type memory is the first-in–first-out (FIFO) memory. It is used as a buffer memory to synchronize data output between two subsystems operating at different speeds or even with different clocks. Data output from System 1 is stored in the FIFO and falls down to the last unoccupied stage. System 2 will remove the data as soon as it is ready to do so. All data stored in previous stages falls to the bottom, ready to be removed. This type of memory will be discussed in Chapter 9. The TI 74225 is a five-bit-wide, sixteen-bit-deep register. The 74232 and 74233 are four bits wide and 64 bits deep. The Monolithic Memories 57402 is five bits wide and 62 bits deep. A few examples are given in table 7–9.

TABLE 7–9 Examples of memory chips

Part No.	Manufacturer	Type	Technology	No. of Pins
SN74S201	TI	256-bit RAM	S	16
AM2102	AMD	1K RAM	NMOS	16
MWS5114	RCA	4K RAM	CMOS	18
IDT6116	IDT	16K RAM	CMOS	24
HM76641	Harris	64K PROM	Fusible link	24
M2716	Intel	16K EPROM	Ultraviolet erasable	24
57401	MMI	64 × 4 FIFO	S	16

Chip Survey and Applications

7–12 HYBRID DEVICES

Other devices that are not strictly digital compose the entire class of hybrid chips. These include *D/A* and *A/D* converters, sample and hold circuits, analog comparators, analog switches, and multiplexers. This course does not deal with hybrid chips, yet digital circuits must often use these devices when dealing with real systems.

Manuals to consult are from Analog Devices, Datel-Intersil, PMI, Siliconix, RCA, National, Motorola, and Micro Power Systems.

D/A and *A/D* Converters

D/A and *A/D* converters will be discussed in Chapter 12. A large variety of *D/A* and *A/D* converters are listed in the manuals ranging from six-bit to sixteen-bit conversion. Conversion times range from 10 µs to 60 nsec. A few are listed below. A garden-variety inexpensive *D/A* converter is the DAC08 from PMI or the DAC0830 from National with a settling time of 1 µs and 20 mW power dissipation when operated from a +5 V single power supply. A good *A/D* converter to use is the ADC0800 from National with 35 µs conversion time if speed is not essential. Faster successive approximation *A/D* converters are also available. The Analog Devices AD5240 converts 12 bits in 5 µs.

For higher speeds, one must turn to the flash converter. Micro-Power Systems makes the MP7683 eight-bit converter with a 200 nsec conversion speed. RCA's 41051 has a 67 nsec conversion speed and includes autobalancing between cycles.

Important characteristics to look for in *D/A* converters are

- Number of input bits
- Settling time
- Percent nonlinearity
- Power dissipation

Important characteristics to look for in *A/D* converters are

- Number of output bits
- Conversion speed
- Integral linearity
- Differential nonlinearity
- Power dissipation

Figure 7–25 shows an application of a *D/A* converter in a ramp generator. The input to the *D/A* is a binary count generated with a binary counter. The *D/A* output must be a stepladder following the counter. The linearity of the output will be as

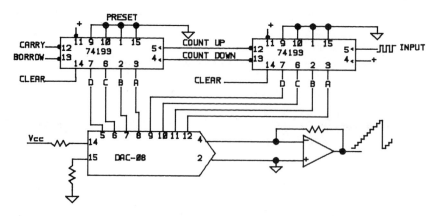

FIGURE 7–25 Ramp generator

good as the D/A converter resolution. When the count returns to zero, the ramp return is limited only by the op-amp slew rate. The ramp frequency is as good as the stability of the clock driving the counter.

Sample and Hold Amplifiers

As the name implies, these devices "sample" an analog signal and then "hold" the peak of the sample. They find application where time-varying inputs cannot be tolerated as in fast successive approximation A/D converters, data acquisition systems, analog delay lines, or wherever the input must be frozen for measurement. Some manufacturers are PMI, Analog Devices, Burr Brown, National, and Datel-Intersil. PMI's SMP10, for example, has a 50 nsec aperture time, a 3.5 μs acquisition time, and a droop rate of 120 μV/ms. This means that the sample is 50 nsec wide, the output reaches the sample value in 3.5 μs, and the signal decays at a 120 μV/ms rate. An application is shown in figure 7–26.

Important characteristics to look for are

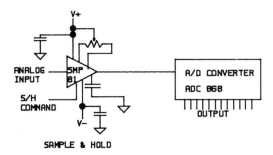

FIGURE 7–26 A/D converter with sample and hold

Chip Survey and Applications

- Acquisition time
- Aperture time
- Droop rate
- Slew rate
- Power bandwidth
- Signal transfer nonlinearity

A few examples are given in table 7–10.

Analog Switches

An analog switch acts as an open or closed contact. It is functionally similar to a relay. In the ON state a FET conducts current equally well in either direction and will transfer an analog signal (or for that matter a digital signal). In the OFF state the FET switch is open and blocks signal transfer. Unlike bipolar devices, a FET does not offset the transferred signal and is in that sense equivalent to a mechanical switch. Whereas a mechanical switch has essentially zero resistance when closed, the FET switch does have an ON resistance r_{ds} (on) of 5–500 Ω, depending on the device chosen. In the OFF state, though, the resistance is on the order of megaohms.

Digital switches require only that the output be high or low as a response to an input control. Analog switches require that the signal shape be transferred with fidelity.

Analog switches are used for multiplexing, commutating, sample and hold, A/D and D/A conversion, chopping, and other applications in which low resistance switches without offset are needed. A typical application might be the monitoring of temperature, pressure, supply voltage, and current in a satellite system. The

TABLE 7–10 Examples of D/A, A/D, and sample and hold chips

Part No.	Manufacturer	Type	No. of Pins	Speed	
AD7523	Intersil	8-bit D/A	16	100	nsec settling time
MP7523	Micro-Power Systems	8-bit D/A	16	150	nsec settling time
DAC0800	National	8-bit D/A	16	100	nsec settling time
AD7520	National	12-bit D/A	16	600	nsec settling time
AD571	Analog Devices	10-bit A/D	18	25	μs conversion time
ADC868	Datel	12-bit A/D	Hybrid package	500	nsec conversion time
41051	RCA	Flash 8-bit A/D	24	66	nsec conversion time
5021	OEI	Sample and hold	Hybrid package	100	MHz bandwidth
SMP10	PMI	Sample and hold	14	3.5	μs acquisition time
LF198	National	Sample and hold	8	10	μs acquisition time

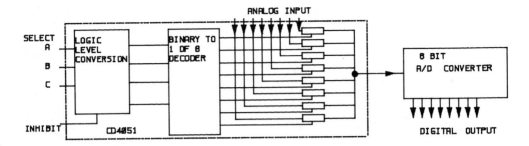

FIGURE 7-27 Analog multplexor and A/D

sampled values are multiplexed into an A/D converter, then sent to the telemetry system, and transmitted to ground together with other scientific data. Figure 7-27 shows the monitor subsystem.

Important characteristics to look for are

- ON resistance: $R_{ds}(on)$
- Drain source cutoff current: $I_d(off)$
- Gate source breakdown voltage: $B_V(GSS)$
- Gate source cutoff voltage: $V_{GS}(off)$
- Speed parameters—for example, $t_d(on)$, turn-on delay time

A few examples are given in table 7-11.

7-13 COMPARATORS

Analog comparators are high-gain amplifiers that provide a logic output indicating the amplitude relationship between two analog signal inputs. They are designed to swing from a low to a high (or reverse) output level—ground to V_{cc}, for

TABLE 7-11 Examples of analog switches

Part No.	Type	Manufacturer	r_{ds} (on)	I_d (off)	t_d (on)
U291	N-Channel JFET	Siliconix	3 Ω	1 nA	15 nsec
P1087	P-Channel JFET	Siliconix	75 Ω	2 nA	15 nsec
CD4051	8-switch MUX	RCA	470 Ω	0.01 nA	120 nsec
CD4529	8-switch MUX	National	165 Ω	0.001 nA	200 nsec
AD7510	Quad switch	Analog Devices	75 Ω	0.5 nA	1 μs
SW-06	Quad switch	PMI	80 Ω	30 nA	340 nsec

example—with minimum rise time. Linear response is not the aim, but speed of response is. In that sense, comparators differ from operational amplifiers. Moreover, a comparator accepts a much larger differential input voltage than does an operational amplifier.

Several parameters of importance are

- *Input slew rate:* The rate of change of the output waveform as a response to large input voltage step.
- *Input to output, high propagation delay:* The time between a small differential input polarity change and the 50% low-to-high transition point of the output.
- *Input to output, low propagation delay:* The time between a small differential input polarity change and the 50% high-to-low transition point of the output.
- *Sensitivity:* The smallest input change that causes the output to switch.
- *Input offset voltage:* The DC level around which a minimum +/− input change will cause output switching.

Four applications are shown in figures 7–28(a), 7–28(b), 7–28(c), and

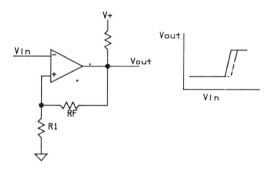

FIGURE 7–28(a) Hysteresis circuit

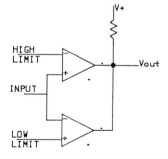

FIGURE 7–28(b) Limit detector

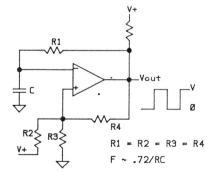

FIGURE 7–28(c) Square wave oscillator

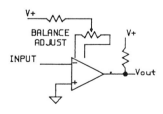

FIGURE 7–28(d) Zero crossing detector

TABLE 7–12 Examples of comparators

Part No.	Type	Manufacturer	Response Time (μs)	Slew Rate (V/μs)	Sensitivity (V/mV)	Offset (mV)
LM339	Quad comparator	Motorola	1.3	17	200	1
CMP01	Single comparator	PMI	0.18	92	500	0.4
CMP05	Single comparator	PMI	0.06	200	16	0.4
529	Single comparator	Signetics	0.01	300	5	6
AM1500	Single comparator	AMD	0.15	30	200	0.7

7–28(d). The use of comparators in *A/D* conversion will be discussed in Chapter 12. A few examples of comparators are given in table 7–12.

7–14 OPTOELECTRONIC COMPONENTS

Light-sensitive or light-emitting devices are found in most electronic systems. Displays for calculators and computers, appliances, optical couplers for isolation, signal transmission, light-sensitive memories, CCDs for imaging, and light-sensitive diodes for camera light meters are some examples.

Displays

A very familiar optoelectronic component is the display. The simplest visible indication of an electronic signal level is the light-emitting diode (LED). Four LEDs at the output of a BCD counter, for example, (with proper drive) display the output value in BCD code. For laboratory breadboard indicators this is often a quick and easy way to inspect the operation of a circuit. Data General NOVA and Eclipse minicomputers display the accumulator contents on the front panel in octal form with a string of LEDs. A more convenient way to display decimal or hexadecimal values is done with an LED display chip. The diodes are arranged in seven lines as needed to show the number 8. Activating selected LEDs within the chip displays a choice of 10 decimal digits or 16 hexadecimal digits.

Of course, the input for the proper LED combination must be decoded. A BCD or hex value consists of four bits. The LED chip requires seven inputs. The decode chips are four-line to seven-segment decoders. A variety of display chips are available. They come as single-digit or multiple-digit displays, with or without decoding included. (See Chapter 13 for illustrations.)

Light-emitting diodes consume considerable current, typically 5–20 mA per section. Displaying the number 8 would take 140 mA. Liquid crystal displays (LCDs) consume as little as 5 μA per section and are recommended for battery-powered systems.

Seven segments are not sufficient to display letters. Dual alphanumeric chips have 13 light-emitting lines, as shown in figure 7–29, which include two additional

Chip Survey and Applications

FIGURE 7–29 Alphanumeric displays

vertical lines in the middle of the chip as are needed for the letter T and four diagonal lines as are needed for K, M, and other letters.

Dot matrix displays are more commonly used for alphanumeric displays. Instead of 13 diodes arranged in dashed lines, the diodes are used as dots in matrix fashion, either 7×9 or 5×7. An example of an alphanumeric display driven from a character generator is given in Chapter 14.

Couplers

Other components are infrared emitters, optical signal detectors and amplifiers, optically coupled isolators, and fibers.

Signals can be transmitted in the visible frequency range, either modulated or in the digital ON–OFF fashion. One example of the latter is the case in which two systems are electrically isolated from each other yet must communicate. This can be done with pulse transformers. The logic signal can also be changed to an ON–OFF light beam with an infrared transmitter. A light fiber guides the light beam to a receiver. The signal is then amplified, and the electrical logic signal is reproduced.

Optical signal transmitters and receivers are designed for frequencies at the near-infrared wavelength. Basically, they have the same construction as any junction photodiode but are optimized for low noise and high speed. An optical isolator is shown in figure 7–30. A few component examples are given in table 7–13.

7–15 LOGIC ARRAYS

The complexity of digital systems has grown enormously over the last few years. As the availability of complex chips has increased, so has the variety and complexity of the users' applications. It is difficult for the semiconductor industry to supply a large-scale integrated (LSI) chip for every conceivable application. The cost of

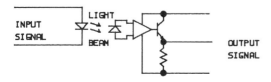

FIGURE 7–30 Optically coupled ac linear coupler

TABLE 7-13 Examples of optoelectronic devices

Part No.	Type	Manufacturer	Package	Comment
TIL23	P/N GA light source	TI	0.06 × 0.122"	1 mW output
MFOE100	Infrared emitter	Motorola	To18 can	0.55 mW output
TIXL55	Detector and amplifier	TI	0.122 × 0.143"	1 GHz gain bandwidth
MRD500	Photo diode	Motorola	5 mm × 5 mm	1 nsec response
MOC5003	Optical isolator	Motorola	6-pin dual in-line	10^{11} Ω isolation
TIL301	7-segment display	TI	14-pin dual in-line	Common cathode
TIL302	7-segment display	TI	14-pin dual in-line	Common anode
5082-7731	7-segment display	HP	14-pin dual in-line	Common anode
5082-7440	9-character, 7-segment display	HP	17-pin 0.72" × 2.1" PC board	Common anode
5082-7010	5 × 7 LED matrix BCD display	HP	8 pin 0.58" × 1.05"	Package with built-in decoder
LR1704	5 × 7 alphanumeric	IEE	14-pin dual in-line	
LR3784	Dual alphanumeric	IEE	14-pin dual in-line	
3920-365	6 character LCD	HAMLIN	PC board	For edge connector

producing an LSI chip is still considerable unless nationwide usage is foreseen. The need therefore exists for custom integrated circuits.

Such custom integrated systems can be built with the usual monolithic technique or interconnected from a set of prefabricated gates on a wafer level.

The manufacturer fabricates a wafer that is about the size of a fifty-cent piece with as many as 200,000 NAND gates. These gates are left unconnected. The user, with a particular application in mind, need only specify the interconnections. The manufacturer then completes the design with the user's specifications. Present technology allows the interconnection of as many as 30,000 gates. The entire wafer will yield a number of copies. The typical custom chip will consist of 500–5000 gates.

Such wafers can be built with bipolar as well as CMOS technology. CMOS is attractive because of the low power consumption. Moreover, the frequency limitations of slow CMOS were recently overcome. High-speed CMOS is as fast as LS–TTL. The latest wafers use a 2 micron process (this refers to the speed-critical well sizes), and a 1 micron process will soon be available.

Chip Survey and Applications

The finished product of such a set of custom interconnected gates is known as a *logic array*.

The manufacturer is not familiar with the functional specifications and performance requirements of the user. In some cases the user supplies the manufacturer with all the details, then sits back and waits for the finished product to be delivered. Usually, though, close cooperation between the two parties is needed. An often preferred method is for the manufacturer to supply the user, who is more familiar with the specific circuit, with a computer terminal at the manufacturer's plant. LSI Corporation of Milpitas, California, for example, provides the following procedure for the user.

Logic Array Design Procedure

1. Enter the network into the computer—that is, specify all connections. Depending on the available computer system, this might take the form of entering all connections on the keyboard. An example of an instruction for the circuit in figure 7–31 would be as follows. Chip U_1 is a two-input NAND gate by the name ND2; chip U_2 is an inverter by the name IVA. The instruction

$$U_1 = IVA(U_1(Z))$$

means that the input of U_2, which is an inverter, is equal to, that is connected with, the output Z of U_1. In similar fashion the entire network is described or *net-listed*. In some graphic systems the network is schematically entered at the terminal. The computer creates its own wire list from the schematic. The computer will store and display the net-list.

2. Call a verify program. This checks that no basic design rules were violated, such as outputs connected to outputs, fanout exceeding the drive capability of a given output, or inputs left unconnected. The computer will respond with a list of errors to be corrected.

3. Simulate system performance. A set of variable input bits are specified for a given number of clock cycles, and the corresponding outputs are computed. Logic is simulated in blocks of time, using an input test pattern provided by the designer. These are iteratively simulated until it becomes clear that the circuit is functioning properly. The states of all input and output signals and selected internal nodes are saved in a file. A display program lists the results. The designer checks the performance on a clock cycle by clock cycle basis.

4. Run a fault simulation program. To ensure that the logic network is adequately tested, this program is run after simulation is completed. This resimulates

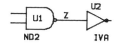

FIGURE 7–31

the same input patterns as above and logs any nodes in the network that were not toggled during the previous simulation. Additional tests might be needed.

5. Specify the package. When performance is deemed satisfactory, the user chooses an appropriate package with an adequate number of input, output, and power pins. The I/O connections are specified, and the computer generates the final wire list.

6. Generate a test. Finally, a test program must be generated to allow the computer to test the final product.

Logic Array Components

Logic arrays are also called *gate arrays*. The term gate array is somewhat misleading, since the base array is not an array of gates but an array of components, N and P transistors, which may be interconnected to form inverters, NAND gates, NOR gates, flip flops, and so on. The more complex combinations of gates are called *macrocells*. The user need only call the desired macrocell, a two-input or three-input NAND gate, for example, and the computer provides the standard interconnections. In other words, the internal connections of the gate within the macrocell are transparent to the designer.

A gate consists of two N and two P transistors and a macrocell of one or more gates. The complexity of a final system is measured by the number of gates used. Each macrocell may consist of one or more equivalent gates. An inverter is one gate, a two-input NAND is also one gate, a flip flop is five gates, and a 16-bit fast binary adder is 248 gates.

The designer can estimate the total number of gates required for the system and choose the appropriate type and size of the masterslice.

Figures 7–32, 7–33, and 7–34 show a few examples of LSI macrocells.

Suppose we call for a two-input NAND gate as in figure 7–32(a). What we get is the circuit in figure 7–32(b). When inputs A and B are both high, transistors T_1 and T_2 are OFF, and T_3 and T_4 are ON. The output is at ground level. If A and B are both low, T_1 and T_2 are ON, and T_3 and T_4 are OFF. The output is at V_{dd} level.

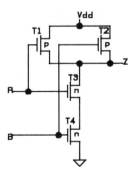

FIGURE 7–32(a) Two-input NAND (Courtesy of LSI Logic Company)

FIGURE 7–32(b) Equivalent transistor structure (Courtesy of LSI Logic Company)

Chip Survey and Applications

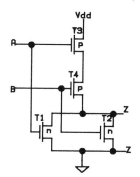

FIGURE 7–33(a) NOR gate (Courtesy of LSI Logic Company)

FIGURE 7–33(b) Equivalent transistor structure (Courtesy of LSI Logic Company)

A NOR gate is shown in figures 7–33(a) and 7–33(b).

A more complex arrangement of an AND OR invert function is shown in figures 7–34(a) and 7–34(b).

The switching speed of the gates is simply the charging and discharging of load capacitance through the ON resistance of the transistors. Therefore the propagation delay will increase linearly with load capacitance (proportional to fanout) and the number of series transistors in the gate. NOR gates are slower than NAND gates. The ON resistance of P transistors is about twice the ON resistance of N transistors. Notice the NAND gate has two N transistors in series for the output = 0 state and two P transistors in parallel for the output = 1 state. The charging and discharging resistances are about equal. This is not true for the NOR circuit, in which two P transistors in series to V_{dd} have four times the resistance of one N transistor to ground. The NAND version is therefore the preferred component. A typical speed versus fanout plot is shown in figure 7–35.

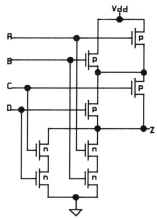

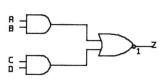

FIGURE 7–34(a) AND OR inverter (Courtesy of LSI Logic Company)

FIGURE 7–34(b) Equivalent transistor structure (Courtesy of LSI Logic Company)

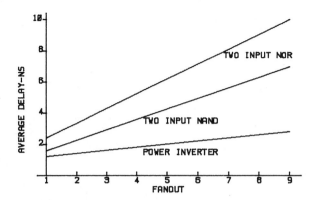

FIGURE 7–35 Average gate delay versus fanout (Courtesy of LSI Logic Company)

Frequent use is also made of a component that has not been discussed previously, the transmission gate. It is a simple CMOS switch. When it is in the ON state, signals flow in both directions; when it is turned OFF, the switch is an open circuit.

The basic flip flop macrocell is the D version. These are designed with transmission gates and inverters as in figure 7–36.

The concept is simple. Assume that gate 4 is ON and gate 3 is OFF. The logic level at D appears to the input of G_3 after some time delay. Next G_4 is turned OFF, and G_3 is turned ON. The level of D is maintained at the input of G_5 even though G_4 is an open circuit. The assumption, of course, is that the switchover between G_4 and G_3 occurred in a make-before-break fashion.

Two identical circuits are used to make a master/slave D flip flop. The clock input to the slave is inverted.

Some timing constraints must be observed: One-input D must be stable while G_4 is ON. The signal from D must propagate through G_4, G_5, and G_6 before reaching G_3. The setup time is the propagation delay through these three gates. However, the clock signal must go through G_1 and G_2 before enabling G_3. Therefore the setup time for the D flip flop is equal to the sum of G_4, G_5, and G_6 propagation delays minus the sum of the G_1 and G_2 propagation delays. The equation is

$$T_{\text{setup}} = (T_{pd}G_4 + T_{pd}G_5 + T_{pd}G_6) - (T_{pd}G_1 + T_{pd}G_2)$$

To meet the hold time requirement, the data pin D must not change state before G_4 is disabled. Since the clock signal must propagate through G_1 and G_2 before disabling G_4, the hold time is equal to the sum of the G_1 and G_2 propagation delays. Thus

$$T_{\text{hold}} = (T_{pd}G_1 + T_{pd}G_2)$$

The typical setup time for the 2 micron technology is 4 nsec, and the typical hold time is 2 nsec.

Chip Survey and Applications

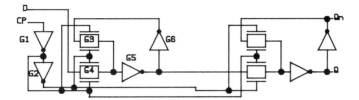

FIGURE 7–36 Master/slave D flip flop (Courtesy of LSI Logic Company)

REVIEW QUESTIONS

1. Define MSI, LSI, and VLSI.
2. What is a switching level of a gate? How do high- and low-level voltage input and output specifications relate to the switching level?
3. How do current input/output specifications relate to fanout?
4. What is a glitch? What is a race condition?
5. Is system speed related to propagation delay?
6. Is mixing of chip families advisable? Is it possible?
7. How do buffers differ from ordinary gates?
8. Where are transceivers used?
9. Where would you use a 7447 four-line to seven-segment decoder?
10. What chip would you use for binary addition? For comparison?
11. What is the difference between a latch and a flip flop?
12. Explain the difference between a ripple counter and a synchronous counter.
13. List a few shift register applications.
14. When would you use a 555 timer? When would you use a crystal oscillator?
15. What are a RAM, a ROM, and an EPROM?
16. What is the purpose of a sample and hold circuit? List applications.
17. Where would you use analog switches?
18. Explain the difference between an analog and a digital comparator.
19. How are characters formed in a dot matrix display?
20. What is a macrocell? What role do logic arrays play in logic systems?

chapter 8

arithmetic

8-1 INTRODUCTION

We are familiar with base 10 arithmetic simply because we have ten fingers. Computers, in that sense, have only two fingers and operate in base 2. We must therefore adjust our thinking processes to base 2 to understand and design digital arithmetic with logic gates. Boolean algebra is the tool that we use. There is no fundamental expression in Boolean algebra that deals with addition. Rather, the fundamental concepts are the AND, OR, and NOT functions. The AND function implies simultaneous events, and the OR function implies the occurrence of at least one event. At first glance there does not seem to be any connection between event occurrence and ordinary arithmetic addition; yet with the aid of truth tables the logic for addition can be defined. All other arithmetic operations, such as subtraction, multiplication, and division are mere extensions of the ability to add.

8-2 ADDITION

Let us start by adding two binary numbers A and B, each of which could be either 0 or 1. There are four possibilities, which are best listed in a truth table, as shown in table 8-1. Note that the left side of the truth table lists the possible combinations of B and A, while the right side gives the desired responses. The Boolean functions for the carry and sum are read from the table. Let the carry be called C_o for "output carry" and use the Greek letter Σ for sum.

The half adder logic circuit in figure 8-1 comes from table 8-1. The circuit is called a *half adder* because the addition of B and A alone is not sufficient to carry out a practical addition of two numbers longer than one bit. As soon as you move from the least significant column to the next higher column, you must consider the possibility of a carry from the previous column.

Arithmetic

TABLE 8–1 Half adder truth table

B	A	Carry (C_o)	Sum Σ
0	0	0	0
0	1	0	1
1	0	0	1
1	1	1	0

EXAMPLE 8–1

Add 5 + 7. In binary:

```
   (1) (1) (1)    carries
    0   1   0   1
  + 0   1   1   1
  ─────────────
    1   1   0   0
```

■

If we developed a truth table for a three-bit addition using C_i, B, and A, where C_i is the carry from the previous column, then we could add two binary numbers. What if we wanted to add more than two binary numbers of n bits? This is an easy operation in decimal. Computers are designed to add two numbers at a time, store the subtotal, then add the next number to the subtotal, and so on until the addition is complete. The circuitry becomes unnecessarily complex if addition is done otherwise. Thus it suffices to develop an adequate method to add two numbers with carry. Again we must consider all possible combinations of the three variables C_i, B, and A. Table 8–2 lists the conditions, the needed equations are directly read from the table, and the logic implementation is shown in figure 8–2.

$$C_o = \overline{C_i}BA + C_i\overline{B}A + C_iB\overline{A} + C_iBA$$

$$\Sigma = \overline{C_i}\overline{B}A + \overline{C_i}B\overline{A} + C_i\overline{B}\overline{A} + C_iBA$$

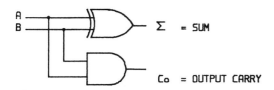

FIGURE 8–1 Half adder logic: C_o and $\Sigma = BA = \overline{B}A + B\overline{A} = B \oplus A$

TABLE 8–2 Full adder truth table

C_i	B	A	C_o	Σ
0	0	0	0	0
0	0	1	0	1
0	1	0	0	1
0	1	1	1	0
1	0	0	0	1
1	0	1	1	0
1	1	0	1	0
1	1	1	1	1

$$C_o = BA + (B \oplus A)C_i$$
$$\Sigma = \overline{C_i}(B \oplus A) + C_i(\overline{B \oplus A})^* = C_i \oplus (B \oplus A)^\dagger$$

Note that the full adder is implemented with two half adders and a NAND gate. The entire full adder circuit can now be represented by one functional block as in figure 8–3.

An entire string of bits representing a binary number is usually written with the same letter but increasing subscripts. The least significant bit has a subscript 0, meaning that the bit is multiplied by $2^0 = 1$ or is of weight 1. The fourth bit, for example, has a subscript 3 meaning that the bit is of weight $2^3 = 8$. Thus a four-bit binary number (A) is written as A3 A2 A1 A0.‡ To add two four-bit numbers (B) and (A) electronically, we just connect four binary full adders as shown in figure 8–4. Figure 8–4 uses $B = 9$ and $A = 10$ to illustrate the function of a four-bit adder. Figure 8–5 combines the four binary full adders into one chip such as the 74283.

This architecture of a binary full adder for four bits is operational but too slow. Notice that the carry of each bit location must ripple through the entire system until it reaches the output. That limits the speed of the device, particularly if long words

---

*It is easy to show that the complement of the Exclusive OR function is

$$\overline{BA + \overline{B}\overline{A}} = B\overline{A} + \overline{B}A$$

$_A$ B	0	1
0	0	1
1	1	0

If the 1 entries in the Exclusive OR map represent the Exclusive OR function, then the 0's represent the complement function, which is the Inclusive OR function, also known as the Exclusive NOR function.

†If $B + A$ is replaced by M, then the function can be rewritten as $C_i \overline{M} + \overline{C_i} M$, which is an Exclusive OR function. Substituting $B + A$ for M, we get an Exclusive OR of an Exclusive OR.

‡A four-bit number is called a *nibble*; an eight-bit number is called a *byte*.

Arithmetic

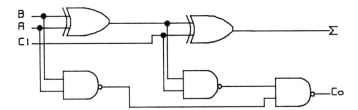

FIGURE 8–2 Full adder logic

are to be added. (A group of bits is called a *word*.) There is a method whereby this delay can be reduced, called *carry look-ahead*. A logic network looks at all inputs and predicts the carry. The conditions for the prediction are as follows:

- A position within the adder will generate a carry at that position if the input = 1 1.
- A position within the adder will propagate a previous carry if one of the inputs at that position is a 1.
- A position within the adder will absorb a carry if both inputs at that position are 0.

Thus a *carry* will exist at the output of a four-bit adder if

- the input to the most significant position is a 1 1 (generates its own carry), or
- the second lesser significant position generates a carry and the most significant position propagates it, or
- the third lesser significant position generates a carry and the two more significant positions propagate it, or
- the least significant position generates a carry and the three more significant positions propagate it, or
- there is an input carry from a lesser significant adder and all four positions propagate it.

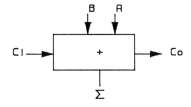

FIGURE 8–3 Binary full adder functional block

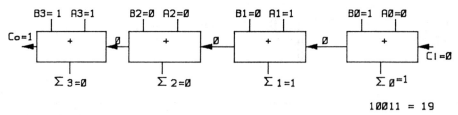

FIGURE 8–4 Four-bit binary adder

This can be expressed in Boolean terms as follows:

$$
\begin{aligned}
C_o = \ & B3 * A3 \\
+ \ & B2 * A2(B3 \oplus A3) \\
+ \ & B1 * A1(B3 \oplus A3)(B2 \oplus A2) \\
+ \ & B0 * A0(B3 \oplus A3)(B2 \oplus A2)(B1 \oplus A1) \\
+ \ & C_i(B3 \oplus A3)(B2 \oplus A2)(B1 \oplus A1)(B0 \oplus A0)
\end{aligned}
$$

This is the carry look-ahead equation.

The Exclusive OR function and the AND function are available within the adder structure (see figure 8–6). Logic solutions are usually not unique. The 74283 also includes carry look-ahead to the sum outputs. This concept can be extended to larger groups and to groups of groups.

The last term of the carry look-ahead equation,

$$C_i(B3 \oplus A3)(B2 \oplus A2)(B1 \oplus A1)(B0 \oplus A0)$$

has an interesting property of its own. It says that if one of the bits in each input pair is a 1 and the input carry $C_i = 1$, then there will be an output carry. In other words, the last term of the carry look-ahead equation can be called a *carry bypass enable function* (Fcbe). The input carry C_i can bypass the group B3 A3 → B0 A0 when tested with Fcbe.

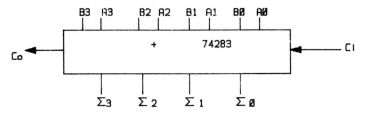

FIGURE 8–5 Four-bit binary full adder

Arithmetic

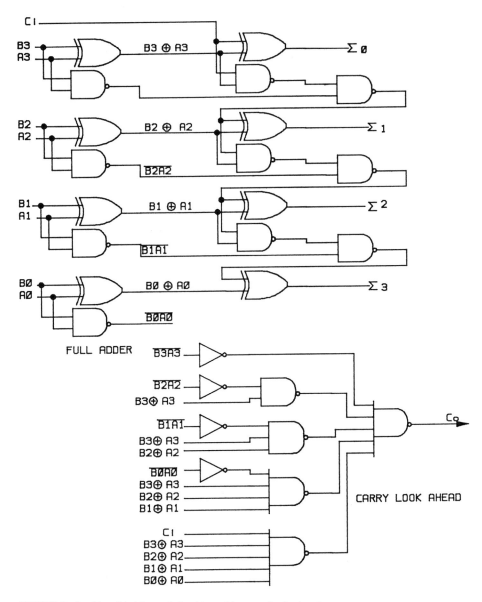

FIGURE 8-6 Four-bit binary full adder with carry look-ahead

The Exclusive OR functions for the *BA* pairs are usually not available as output pins and must be externally implemented. Figure 8-7 shows the logic for the carry bypass enable function Fcbe and its application for a string of binary full adders. The bypassed carry must be ORed with any carry produced by the bypassed group.

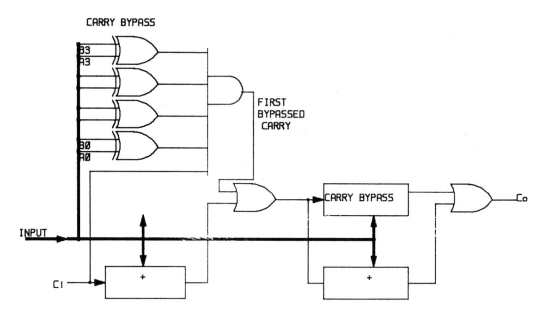

FIGURE 8-7 Adder with carry bypass

8-3 THE ACCUMULATING ADDER

As was mentioned earlier, if more than two words are to be added (we might as well get used to the term "word" rather than "number"), the subtotal or interim result is stored and then added to the next word.

Instead of using the adder as shown in figure 8-4, in which two independent words are added, one can add only one word to an initially cleared accumulating system. The output of the adder is entered into a set of *D* flip flops. The output of each flip flop is fed back to one of the input sets of the adder. The other set of adder inputs is reserved for words to be entered. Figure 8-8 illustrates the arrangement.

The accumulator output is now called the *B* word and becomes the second input to the adder. At first the accumulator is cleared. To begin with, when word *A* is applied to the adder, that word + 0 (since the accumulator is empty) appears at the output of the four-bit adder. A clock pulse to the accumulator enters word *A*. Now word *A* exists at the *B* input to the adder. Another word *A* can now be entered, and word *A* + word *B* appear at the adder output. A second clock pulse enters the new sum into the accumulator. A third word *A* can now be applied to the *A* inputs. The process is limited only by an accumulator overflow.

Note that the accumulator is not a special device. It is just a holding register from one add operation to the next. The name comes only from the particular application.

Arithmetic

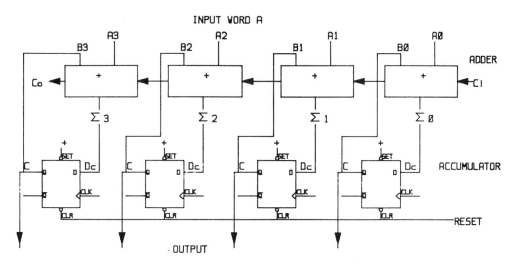

FIGURE 8-8 Accumulating adder

8-4 SUBTRACTION BY COMPLEMENT ADDITION

One could, of course, generate truth tables for subtract functions that are similar to the add truth table in table 8–1 and design a binary full subtractor. It is most inconvenient to do add and subtract operations with separate hardware. It is possible to do addition as well as subtraction with the same binary adder, using complement addition techniques.

For the moment, let us revert to decimal subtraction. For example, subtract 3 from 9. We know that the answer is 6. Instead of subtracting 3, we could add the complement of 3 with respect to 10, which is 7.

―――――――――――――――――――――――――――――― EXAMPLE 8-2

Subtract 3 from 9. Instead of $9 - 3$, we have $9 + 7 = 16$. The 1 is ignored, and the 6 is our answer. ∎

―――――――――――――――――――――――――――――― EXAMPLE 8-3

Subtract two decimal digits: $78 - 26 = 52$. The complement of 26 with respect to 100 is 74. We get $78 + 74 = 152$. If the 1 is ignored, the answer is 52. ∎

These two simple examples illustrate that the process works. In reality it is somewhat more complicated. When subtracting two numbers the answer may be positive or negative, there may or may not be an end carry, and the numbers to be subtracted may themselves be positive or negative numbers. We will have to construct a table listing all possible combinations, investigate the results, and see whether or not some corrective action has to be taken to get the right answer. The process also works when numbers are expressed in nine's complement form, except that the rules whereby one obtains the right answer differ somewhat. We will shortly develop such tables for binary addition and subtraction. (The decimal version is of no further interest.)

Complement addition avoids subtraction. One does either addition of two positive values, addition of two negative values, or addition of a positive and a negative value.

A sign convention must be adopted. The computer does not understand a plus or a minus sign. The convention is that a signed number is recognized as positive if the most significant bit is a 0 and as negative if the most significant bit is a 1. There is only one way to represent a positive number: a 0 followed by the binary number (the magnitude). There are three ways to represent a negative number:

1 followed by the magnitude

1 followed by the one's complement

1 followed by the two's complement

EXAMPLE 8-4

Suppose we have the number 3, represented with four bits where the most significant bit is the sign bit:

0 011	+3	
1 011	-3	sign and magnitude
1 100	-3	sign and one's complement
1 101	-3	sign and two's complement

■

Notice the two's complement is the one's complement + 1. There are advantages to each of the complement representations. The one's complement is simply the magnitude where each bit is inverted. This is easy to implement. An Exclusive OR gate will conditionally invert a number as shown in figure 8–9.

If the sign bit is a 0, the other bits pass through the Exclusive OR gate unchanged. If the sign bit is a 1, all other bits are reversed.

The two's complement is more difficult to obtain because a 1 must be added to the result. This implies either an additional adder (or an add operation with an existing adder) or some logic that encodes the number incremented by 1. The two's complement is the more common application because it is a faster way to carry out

FIGURE 8–9 The Exclusive OR as a conditional inverter

Arithmetic

additions, as will be shown later. Such a two's complementer is easily encoded from table 8–3. Let D_i, B_i, C_i, and A_i be the input bits and D_o, C_o, B_o, and A_o be the two's complement output bits (see table 8–3). Reading the Boolean output functions from the four maps in figure 8–10 yields

$$A_o = A_i \quad \text{(This is obvious.)}$$

$$B_o = B_i \overline{A_i} + \overline{B_i} A_i = B_i \oplus A_i \quad \text{(This is also obvious.)}$$

$$C_o = \overline{C_i} B_i + \overline{C_i} A_i + C_i \overline{B_i} \overline{A_i}$$

$$= (B_i + A_i)\overline{C_i} + \overline{(B_i + A_i)} C_i$$

$$= (B_i \oplus A_i) + C_i \quad \text{(Less obvious but still easy to see.)}$$

$$D_o = \overline{D_i} A_i + \overline{D_i} B_i + \overline{D_i} C_i + D_i \overline{C_i} \overline{B_i} \overline{A_i}$$

$$= (A_i + B_i + C_i)\overline{D_i} + \overline{(A_i + B_i + C_i)} D_i$$

$$= (A_i + B_i + C_i) \oplus D_i \quad \text{(Even less obvious.)}$$

Notice the repetitive pattern one obtains. One can thus expand the pattern without further derivation to more bits. The implemented logic version is shown in figure 8–11. (The variables D_i, C_i, B_i, and A_i and D_o, C_o, B_o, and A_o should have been called A_{3in}, A_{2in}, A_{1in}, A_{0in} and A_{3out}, A_{2out}, A_{1out}, A_{0out}. It would be in keeping with the convention of Word A in and Word A out. However, reading the maps and equations would have been confusing.)

TABLE 8–3 Two's complement truth table

D_i	C_i	B_i	A_i	D_o	C_o	B_o	A_o
0	0	0	0	0	0	0	0
0	0	0	1	1	1	1	1
0	0	1	0	1	1	1	0
0	0	1	1	1	1	0	1
0	1	0	0	1	1	0	0
0	1	0	1	1	0	1	1
0	1	1	0	1	0	1	0
0	1	1	1	1	0	0	1
1	0	0	0	1	0	0	0
1	0	0	1	0	1	1	1
1	0	1	0	0	1	1	0
1	0	1	1	0	1	0	1
1	1	0	0	0	1	0	0
1	1	0	1	0	0	1	1
1	1	1	0	0	0	1	0
1	1	1	1	0	0	0	1

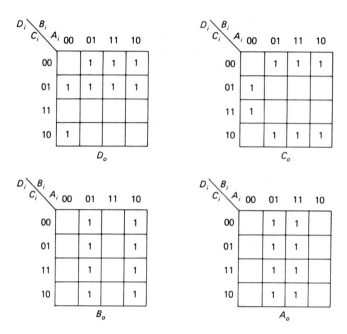

(the '2' complement of 0000 = 10000. the overflow bit '1' is ignored)

FIGURE 8–10 Two's complement map

The circuit shown in figure 8–11 reverts to the ordinary convention of labeling the input and output Word A_{in} and Word A_{out}, respectively. Furthermore, some provision must be made to prevent the two's complement output if the sign = 0. Notice the sign bit controlling the output with the AND gates to inhibit inversion if $S = 0$. The repetitive pattern also makes it obvious how to provide two extra terminals for expansion to lower- and higher-order bits.

As was stated previously, complement addition requires appropriate corrections to get the right answers. These corrections are well defined when all the cases for

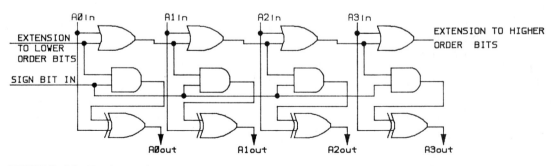

FIGURE 8–11 Two's complementer

Arithmetic

adding positive and negative numbers are investigated. Let us first look at a one's complement adder. We are talking about numbers without decimal points or, more appropriately, binary points or exponential multipliers. Such numbers are called *fixed point numbers*. Numbers with fractional parts and/or exponential multipliers are called *floating point numbers*.

Fixed Point One's Complement Adder

There are six cases:

- Case 1: Add two positive numbers, and *no* overflow carry occurs into the sign position.
- Case 2: Add two positive numbers, and *an* overflow carry occurs into the sign position.
- Case 3: Add a positive and a negative number, and *no* end carry occurs past the sign position.
- Case 4: Add a positive and a negative number, and *an* end carry occurs past the sign position.
- Case 5: Add two negative numbers, and *no* overflow carry occurs into the sign position, but there is an end carry.
- Case 6: Add two negative numbers, and *an* overflow carry occurs into the sign position as well as *an* end carry past the sign position.

This is best illustrated with actual numbers as shown in table 8–4. (This illustration is limited to sign and three bits.) Note that the sign is treated like all other bits, that is, it is entered into the adder as if it were part of the number.

In analyzing the corrections, *two* different carries must be considered: a carry past the sign position and a carry from the number into the sign position.

The conclusions are as follows:

1. An overflow alarm must be activated when both of the signs are positive (sign = 0) but the result is negative or when both signs are negative and the result is positive.

Let the sign of word $A = S_a$ and the sign of word $B = S_b$ and let the output sign of the adder $= S_o$. We get

$$F_{\text{overflow}} = S_a S_b \overline{S_o} + \overline{S_a}\, \overline{S_b} S_o$$

2. Any time there is an end carry it must be added to the result.

3. The result is complemented for readout when the signs of words A and B are different and there is no end carry or when both signs are negative and there is an end carry. We get

$$F_{\text{complement}} = (S_a \overline{S_b} + \overline{S_a} S_b)\overline{EC} + S_a S_b EC$$

TABLE 8-4 One's complement addition

Case 1	(+3) + (+2)		
	0011		No end carry (call the end carry *EC*)
	0010		No carry into sign position
	0101		The answer is +5 as expected and requires no correction.
Case 2	(+5) + (+4)		
	0101		No end carry but an overflow carry into the sign position
	0100		
	1001		The sign = 1 indicating a negative result. This cannot be. One cannot add two positive numbers and obtain a negative result. Obviously, there is an overflow, since the sum cannot be represented with three bits. No correction is possible. Activate an overflow alarm.
Case 3	(−6) + (+4)		
	1110	This becomes 1001	No end carry past the sign position
	0100	This remains 0100	
		1101	Complement the result.
		1010	= −2 as expected
			Correction: Complement the result but only for final readout, since the complement form is appropriate for continued operations.
Case 4	(+6) + (−4)		
	0110	This remains 0110	There is an end carry past the sign position.
	1100	This becomes 1011	
		10001	Add the end carry.
		──1	
		0010	= +2 as expected
			Correction: Add the end carry.
Case 5	(−2) + (−3)		
	1010	This becomes 1101	There is an end carry past the sign position.
	1011	This becomes 1100	
		11001	Add the end carry.
		──1	
		1010	Complement the result.
		1101	= −5 as expected
			Correction: Add the end carry and complement the result (but only for readout).
Case 6	(−5) + (−4)		
	1101	This becomes 1010	There is an end carry past the sign position and an overflow carry into the sign position.
	1100	This becomes 1011	
		10101	
		──1	
		0110	We added two negative numbers and got a positive result. Correction: Activate an overflow alarm.

Arithmetic

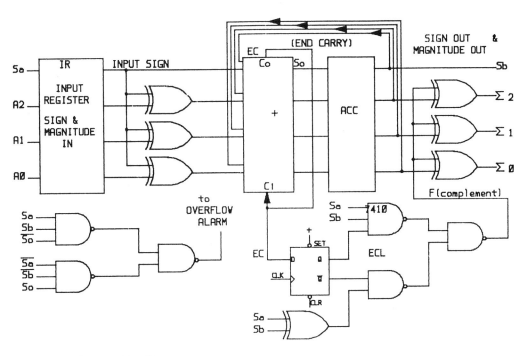

FIGURE 8-12 Four-bit adder

The end carry, *EC*, must be valid after the accumulator was clocked. It is therefore necessary to hold the *EC* in a flip flop (carry flag).

We are now in a position to implement an adder-subtractor unit in figure 8-12 by modifying figure 8-8. It is also appropriate to use an input register (IR) with the adder unit.

Fixed Point Two's Complement Adder

A similar derivation for the six cases yields table 8-5.

The conditions for overflow and complementation are the same as those in the one's complement method; end carries, however, need not be added to the result. This is a sigificant difference. The time for the end carry to ripple through the adder may be as long as the initial addition, thus doubling the add time. The two's complement avoids the end around carry, thus speeding up the add operation. Figure 8-13 is a repeat of figure 8-12 for the two's complement adder.

When two maximum numbers (all 1 pattern) are added, the accumulator will overflow by one bit. Repeated additions cause further overflows to more than one bit. To avoid that difficulty, the entire complement adder can be lengthened by n bits. It is not sufficient just to lengthen the accumulator, since the subtotal is fed back to the adder.

Thus the adder and accumulator blocks of figure 8-13 can be expanded to accommodate an output word of any length. Figure 8-14 shows an eight-bit adder

TABLE 8–5 Two's Complement Addition

Case 1	Same as one's complement	No correction.
Case 2	Same as one's complement	Activate an overflow alarm.
Case 3	$(-6) + (+4)$ 1110 This becomes 1010 0100 This remains 0100 ──── 1110 Two's complement 1010 $= -2$ as expected	Correction: Two's complement the result but only for readout.
Case 4	$(+6) + (-4)$ 0110 This remains 0110 1100 This remains 1100 ──── 10010	Ignore the end carry; the answer = $+2$ as expected. Correction: None.
Case 5	$(-2) + (-3)$ 1010 This becomes 1110 1011 This becomes 1101 ──── 11011	Ignore the end carry and two's complement the result; 1101 = -5 as expected. Correction: Two's complement the result.
Case 6	$(-5) + (-4)$ 1101 This becomes 1011 1100 This remains 1100 ──── 10111	The result is positive. Correction: Activate an overflow alarm.

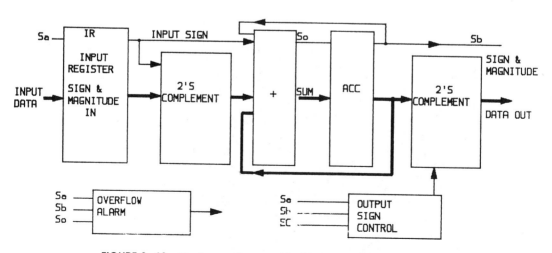

FIGURE 8–13 Two's complement adder block diagram

Arithmetic

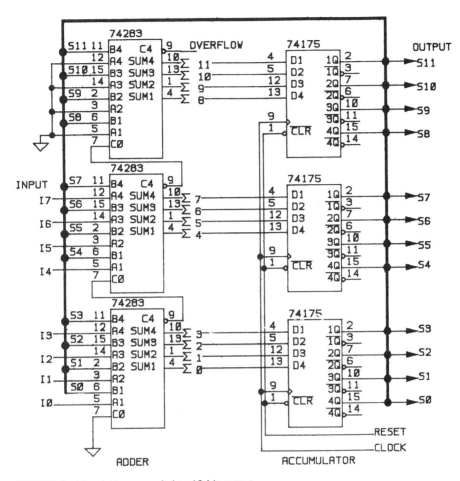

FIGURE 8–14 Adder expanded to 12-bit output

with a 12-bit accumulator. The expansion is obviously at the more significant end of the word. The upper four bits have the accumulator feeding back in the conventional fashion, but there are no corresponding input bits. Commercially available four-bit binary full adders such as the 74283 provide for four sets of input pairs. The accumulator feedback occupies one input of each pair; the other input is simply grounded.

The upper four bits really do not require a binary full adder, since one of the inputs does not exist. There are therefore a few wasted gates in the chip. Ordinarily, the convenience of the chip availability does not warrant the design of a special logic section for the upper four bits. In many instances, though, such an adder may be part of a logic gate array. In that case a number of unused gates can be eliminated and the carry look-ahead structure modified to suit the system requirements.

Suppose the number of bits per word is m and the adder is extended by n bits. The maximum value the adder can hold without overflow is $2^{m+n} - 1$. The maximum value of the input word is $2^m - 1$. The maximum number of additions (S) that the adder can handle without overflow is

$$S = \frac{2^{m+n} - 1}{2^m - 1}$$

An eight-bit adder, when extended to 16-bit, for example, can handle $(2^{16} - 1)/(2^8 - 1) = 257$ additions of an all 1 pattern before an overflow occurs.

Sums that are larger than the size of the accumulator can also be handled in a different manner. An eight-bit microprocessor with an eight-bit accumulator simply does not have the convenient accumulator length to accommodate overflows. In this case the microprocessor sets an overflow carry bit in a flag register. The addition of two words can yield only a single carry bit. One of the bits in the flag register is assigned for the carry bit. This information is available to the programmer, who can use it to increment some other register with a subsequent program step. More details on the subject of flag register use are left to courses on microprocessor applications.

Many fixed point arithmetic units consist of a complement adder only. If the accumulator has the additional ability to shift the data left and right, multiplication and division can be done in software. It is a slow process by implication, but it is adequate for many purposes. If a fast hardware version of a multiply-divide unit is needed, it can be done by modifying the circuit of figure 8–12 with an additional register and a few control gates.

8–5 MULTIPLICATION

First let us define an algorithm and then see how to implement it. We take our multiplication skills for granted, but let us review some of the fundamental concepts. We learned to multiply in decimal mode by memorizing the small multiplication table from 0 to 100 and then proceeded by multiplying one decimal digit of the multiplier times the multiplicand. The resulting subtotals are added in shifted fashion, since each successive subtotal carries a weight differing by 10 from the previous one. The shifted subtotal is shifted left if one starts the multiplication with the least significant digit or is shifted right if the multiplication starts with the most significant digit.

EXAMPLE 8–5

Multiplicand times multiplier:

27 × 83	Start with LSB	27 × 83	Start with MSB
81		216	
216	Left shifted	81	Right shifted
2241		2241	

∎

Arithmetic

The same concept is extended to binary multiplication. In a way it is much simpler, since the multiplication table to be memorized goes only from 0 to 1 (not from 0 to 100). A number in binary mode is either multiplied by 1—in other words, it remains as is—or multiplied by 0, resulting in 0. Each shifted subtotal differs by a weight of 2 from the previous one.

EXAMPLE 8-6

Take the same values as in Example 8-5:

```
     11011 × 1010011      Start with LSB
     ─────────────────
          11011
         11011
        00000
       00000               Partial products shifted left
      11011
     00000
    11011
    ─────────────
    100011000001  = 2241


     11011 × 1010011      Start with MSB
     ─────────────────
    11011
     00000
      11011
       00000              Partial products shifted right
        00000
         11011
          11011
    ─────────────
    100011000001  = 2241
```

The computer can follow the same algorithm except for the fact that an adder can add only one subtotal at a time. It becomes obvious that a convenient way to use the adder to do multiplication is as follows:

1. Starting from the MSB (or LSB), multiply the multiplicand by the MSB (or LSB) of the multiplier (either 1 or 0) and store the partial product, that is, add it to the cleared accumulator.
2. Shift the partial product as well as the multiplier one position.
3. Multiply the multiplicand by the next most (or least) significant multiplier bit and add it to the shifted partial product.
4. Shift the new partial product as well as the multiplier.

5. Again multiply the multiplicand by the next multiplier bit and add it to the shifted partial product. Note that Steps 4 and 5 are the same as Steps 2 and 3.
6. Repeat Steps 4 and 5 until the last bit of the multiplier has been used.
7. The resulting product is in the accumulator.

It takes twice as many clock cycles to do the multiplication as there are bits in the multiplier. The accumulator must be long enough to accommodate the result, which is the sum of the bits in the multiplicand and multiplier. Since computers work with fixed word length, such that multiplicand and multiplier are of the same length, the accumulator must be of double length.

If the accumulator is not of double length, as is the case in many microprocessors, the addition techniques mentioned before are applicable for the partial sums.

In hardware terminology the multiplication procedure is translated into the following steps:

1. Load a word, the multiplicand, into the input register (this is equivalent to a 0-shift step) and load a word, the multiplier, into a special register called the M-Q register (this the second word with a 0-shift). M stands for multiplier and Q for quotient. At this point we are only interested in multiplication.

2. Start with the MSB (in this example). If it is a 1, add the contents of the input register to the accumulator. If it is a 0, add all 0's to the accumulator. (One must go through the dummy operation of adding 0's to obtain the following shift and to keep order in terms of constant timing.)

3. Shift the contents of the accumulator as well as the M-Q register one position to the left.

4. Again, if the MSB = 1, add the contents of the input register to the accumulator by adding the bits as they appear properly lined up after shifting. If the MSB = 0, perform the dummy operation of adding all 0's; that is, the contents of the accumulator remain unchanged.

5. Repeat the shift add routine until all bits are exhausted.

In Step 4 the rule was to shift left, yet in the second part of Example 8–6 we shifted right when starting with the MSB. In the pencil and paper operation we shifted each new partial product to the right with respect to the previous partial product. In the computer operation it is more convenient to shift the previous partial product to the left with respect to the next addition. The result is the same. The shift is only relative as long as it is consistent.

EXAMPLE 8–7

A four-bit simple example (see figure 8–15) of 5×9 will illustrate the procedure, and an eight clock cycle time exposure illustrates the data flow in the arithmetic unit.

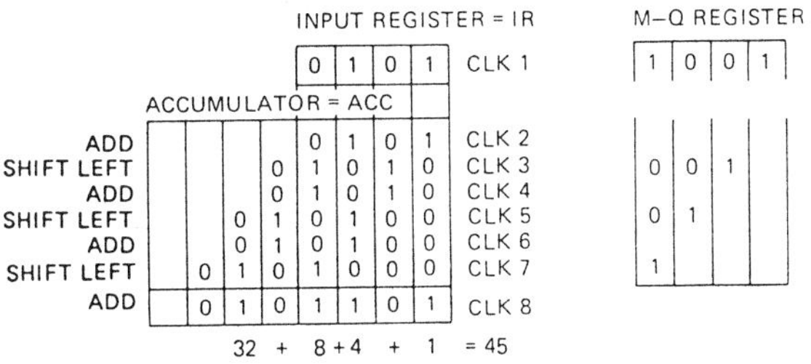

Arithmetic

```
                          INPUT REGISTER = IR    M-Q REGISTER
                          | 0 | 1 | 0 | 1 | CLK 1   | 1 | 0 | 0 | 1 |
            ACCUMULATOR = ACC
ADD                 | 0 | 1 | 0 | 1 | CLK 2
SHIFT LEFT      | 0 | 1 | 0 | 1 | 0 | CLK 3       | 0 | 0 | 1 |
ADD             | 0 | 1 | 0 | 1 | 0 | CLK 4
SHIFT LEFT  | 0 | 1 | 0 | 1 | 0 | 0 | CLK 5       | 0 | 1 |
ADD         | 0 | 1 | 0 | 1 | 0 | 0 | CLK 6
SHIFT LEFT  0 | 1 | 0 | 1 | 0 | 0 | 0 | CLK 7     | 1 |
ADD         0 | 1 | 0 | 1 | 1 | 0 | 1 | CLK 8
            32 +    8 + 4 +     1 = 45
```

FIGURE 8–15 Multiplication data flow

The accumulator left shift is more practical than a right shift, as would be necessary if one started the operation with the LSB, because right shifts would imply that the word initially entered into the accumulator would have to be left justified (that is, the entire word would be moved one word to the left before starting). Shift left is easily done in a microprocessor where some registers have add but no shift capability. A left shift doubles the magnitude. An add to itself accomplishes a left shift, since a word added to itself also doubles the magnitude.

The underlined ACC row implies that the content of the input register is added column by column to the shifted bits above the line because a 1 appeared in the most significant position of the M-Q register. For clock cycle 1 (CLK 1 in Example 8–7) the 1 in the MSB position in the M-Q register caused a simple transfer of the bits in the input register to the accumulator because the accumulator was previously cleared (that is, an add to all 0's).

EXAMPLE 8–8

See figure 8–16.

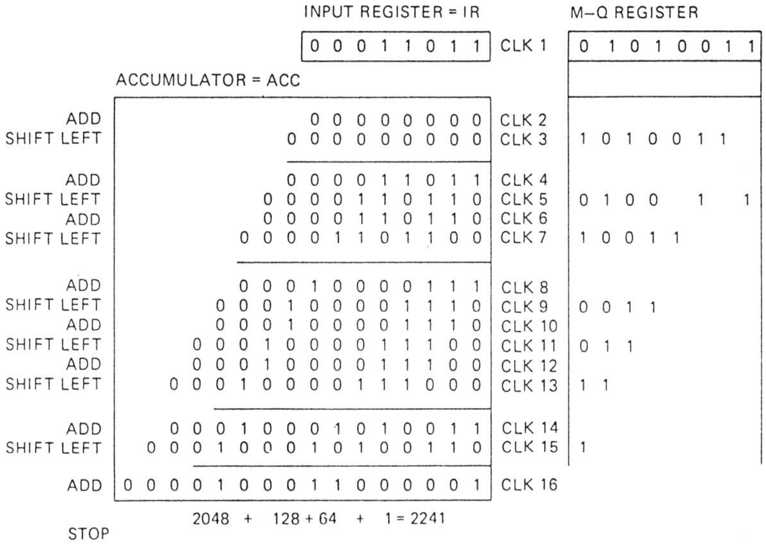

```
                               INPUT REGISTER = IR              M-Q REGISTER
                               0 0 0 1 1 0 1 1   CLK 1          0 1 0 1 0 0 1 1
                ACCUMULATOR = ACC
ADD                            0 0 0 0 0 0 0 0   CLK 2
SHIFT LEFT                    0 0 0 0 0 0 0 0    CLK 3          1 0 1 0 0 1 1

ADD                            0 0 0 1 1 0 1 1   CLK 4
SHIFT LEFT                    0 0 0 1 1 0 1 1 0  CLK 5          0 1 0 0   1   1
ADD                           0 0 0 1 1 0 1 1 0  CLK 6
SHIFT LEFT                  0 0 0 1 1 0 1 1 0 0  CLK 7          1 0 0 1 1

ADD                          0 0 1 0 0 0 0 1 1 1 CLK 8
SHIFT LEFT                 0 0 1 0 0 0 1 1 1 0   CLK 9          0 0 1 1
ADD                        0 0 1 0 0 0 1 1 1 0   CLK 10
SHIFT LEFT               0 0 1 0 0 0 1 1 1 0 0   CLK 11         0 1 1
ADD                      0 0 1 0 0 0 1 1 1 0 0   CLK 12
SHIFT LEFT             0 0 1 0 0 0 1 1 1 0 0 0   CLK 13         1 1

ADD                      0 0 0 1 0 0 0 1 0 1 0 0 1 1 CLK 14
SHIFT LEFT             0 0 0 1 0 0 0 1 0 1 0 0 1 1 0 CLK 15     1

ADD            0 0 0 0 1 0 0 0 1 1 0 0 0 0 0 1   CLK 16
STOP            2048  +   128 + 64    +    1  =  2241
```

FIGURE 8–16 Multiplication data flow

In Example 8–7, eight clock cycles were required, two for each bit. Another example should make the reader feel comfortable with this algorithm. Let us choose the values of Example 8–8: $27 \times 83 = 2241$.

The sign of a number has no bearing on the result. Positive as well as negative numbers are treated as products of magnitudes. If a number is signed, the sign is stripped by masking the sign bit with a 0. The sign of the product is reconstructed by an Exclusive OR function of the two input signs. If both signs are equal, the sign of the product is positive = 0. If the two input signs differ, the output sign is negative = 1.

The complement adder of figure 8–13 can now be modified to figure 8–17 to include multiplication capability. Modification and additions consist of

- An M-Q register to store and shift the multiplier
- A set of control gates ahead of the binary full adder to selectively allow addition of the input register contents to the accumulator if the control bit of the multiplier is 1 or mask the input register content if the control bit of the multiplier is 0 (to perform a dummy all 0 add) (where the control bit of the multiplier is either the most or least significant bit in the M-Q register after the last shift, depending on which multiplying algorithm was chosen)
- An Exclusive OR gate for output sign determination
- Some mode control logic
- Changes in the accumulator to be double length and capable of shifting left

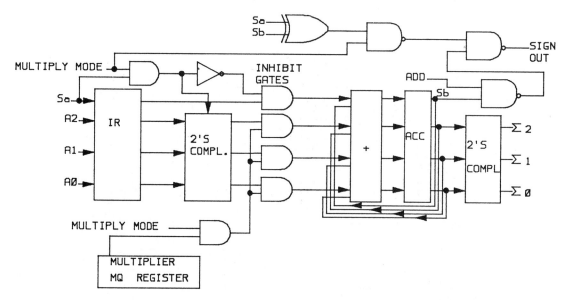

FIGURE 8–17 Complement adder and multiplier

Incorporating these requirements into figure 8–13 results in the logic network of figure 8–17.

The accumulator can handle only one multiplication. The next multiplication causes an overflow. Every multiplication requires a system expansion to n more bits, where n is the number of bits in the multiplier. This is not practical or even necessary. Repeated hardware multiplication can be handled with floating point multiplication. (The system of figure 8–17 is more tolerant of repeated additions.)

The above discussion covers one reasonable method of multiplication. Many variations are possible by using the right shift method or combinations of both in which the accumulator is shifted left but the M-Q register is shifted right. Another method uses table look-up techniques. In principal, at least, one could store all the possible results of all multiplications in a memory. All eight-bit by eight-bit multiplication, for example, has as many as 65,536 sixteen-bit products. If the eight-bit multiplicand as well as the eight-bit multiplier are applied as address to such a memory, the product could be found at that address. The availability of small, inexpensive memory chips makes such a solution attractive in many cases. Of course, when larger sets of products are needed, it is indeed preferable to compute the product. An interesting combination of table look-up and computation is also possible. Assume that we have the numbers $27 \times 83 = 2241$. A straightforward table look-up for two-digit multiplication (in decimal) would require a table of 10,000 entries. The number 27 can be split into the sum of two numbers $2 \times 10 + 7$ and 83 into $8 \times 10 + 3$. The multiplication now looks like

$$(A + B)(C + D)$$
$$= AC + AD + BC + BD$$
$$= (2 \times 8 \times 100) + (2 \times 3 \times 10) + (7 \times 8 \times 10) + (7 \times 3)$$
$$= (16 \times 100) + (6 \times 10) + (56 \times 10) + 21$$

where $A = 2 \times 10$, $B = 7$, $C = 8 \times 10$, and $D = 3$.

If we ignore the factors of 100 and 10, the partial products are 16, 6, 56, and 21. These fit nicely into a table of only 100 entries (not 10,000). The only operation needed after the table look-up of the partial products is to shift and add. The advantage of such a scheme is speed. A table look-up is faster than a multiply. Take an eight-bit binary word multiplied by another eight-bit binary word. Let us split each word into two nibbles; we have nibble A, nibble B, nibble C, and nibble D. It now requires four partial product look-ups = four clock cycles. Since the algorithm is fixed, the four partial products can be directly loaded into the proper shifted position through an appropriate gating network, saving shift clock cycles. This leaves four additions. The entire multiply operation thus needed 8 rather than 16 clock cycles. The time saving is greater for longer words.

Another speed-up technique is to eliminate dummy 0 adds. Techniques can be developed to sense that condition and essentially shift across a 0 or a string of 0's, eliminating the dummy add. This saves clock cycles but causes multiply time variations depending on the 1, 0 pattern of the multiplier.

8-6 DIVISION

The last of the four arithmetic operations to be discussed is division. By mere intuition we could guess that if multiplication is done by shifting and adding, then division is done by shifting and subtracting. This is indeed true. However, there is a distinction to be made between proper and improper fractions. In a proper fraction the numerator is smaller than the denominator. Let us consider that case first. The numerator is called the *dividend*, the denominator is called the *divisor*, and the result is the quotient. Again let us take two decimal numbers to illustrate the procedure.

EXAMPLE 8-9

$$\frac{52}{73} = 0. + \frac{52}{73}$$

The first question we ask is: How many times does 73 fit into 52? The answer is 0 times. Thus we write down a 0 followed by a decimal point. Next we shift the 52 one position to the left and obtain 520. Again

$$\frac{520}{73} = 7 + \frac{9}{73}$$

The remainder is 9 (really 9/730). We shift the 9 one position to the left and obtain 90. Again

$$\frac{90}{73} = 1 + \frac{17}{73}$$

The remainder is 17 (really 17/7300). We shift the 17 one position to the left and obtain 170. Again

$$\frac{170}{73} = 2 + \frac{24}{73}$$

The remainder is 24 (really 24/73,000). One must decide how far to go. The division can go on forever. In this example the answer is 0.712 (the three leading digits of the three above divisions). ∎

Arithmetic

When we asked the question "How many times did 73 fit into 520?", what we really asked was "How many times can we subtract 73 from 520 before one more subtraction yields a negative answer?" The same questions can be asked in the binary case. For the decimal example it was a number between 0 and 9 + a remainder. In the binary case it is 0 or 1 + a remainder. If the answer is once, a 1 is entered in that quotient position, and the difference of the shifted dividend and the divisor is used to continue the solution. If the answer is not at all, it means that one subtraction gave a negative answer, in which case 0 is entered in that quotient position, and the unchanged shifted dividend is used to continue the solution. In terms of a step-by-step procedure:

1. Enter the dividend into the accumulator (ACC).
2. Enter the divisor into the input register (IR).
3. Assume that the quotient register was previously cleared. Also assume that we have a proper fraction, which implies a binary point in the least significant position of the quotient register.
4. Shift the accumulator and M-Q register left. Temporarily enter an unknown X into the shifted in position of the M-Q register. (The next subtraction will determine whether it should be a 1 or 0.)
5. Subtract the divisor from the shifted dividend. If the difference is positive, replace X with 1 and the difference of the dividend and divisor into the accumulator. If the difference is negative, replace X with 0 and leave the accumulator value unchanged.
6. Repeat Step 5 for as many binary positions as desired.

───────────────────────────────── EXAMPLE 8–10

See figure 8–18.

$$\frac{7}{11} = \frac{0111}{1011} = 0.63636$$

The quotient is

$$\frac{1}{2} + \frac{1}{8} + \frac{1}{128} = 0.6328125$$

When compared to 0.6363636, the relative accuracy is

$$\frac{(0.6363636 - 0.6328125)}{0.6363636 \times 100\%} = 0.355\%$$

FIGURE 8–18 Division data flow

		IR							M–Q REG								
		1	0	1	1	11	divisor										
		ACC															
0			0	1	1	7	dividend									.X	X = as yet unknown
1	SH L	0	1	1	0	14									.1		X = 1 subtr. was +
2	SUBTR			1	1	14	− 11 = 3							.1	X		X = as yet unknown
3	SH L		1	1	0	6								.1	0		X = 0 subtr. was −
4	SUBTR		1	1	0	6							.1	0	X		
5	SH L	1	1	0	0	12							.1	0	1		X = 1 subtr. was +
6	SUBTR				1	12	− 11 = 1					.1	0	1	X		
7	SH L			1	0	2						.1	0	1	0		X = 0 subtr. was −
8	SUBTR			1	0	2					.1	0	1	0	X		
9	SH L		1	0	0	4					.1	0	1	0	0		X = 0 subtr. was −
10	SUBTR		1	0	0	4				.1	0	1	0	0	X		
11	SH L	1	0	0	0	8				.1	0	1	0	0	0		X = 0 subtr. was −
12	SUBTR	1	0	0	0	8			.1	0	1	0	0	0	X		
13	SH L	1	0	0	0	16			.1	0	1	0	0	0	1		X = 1 subtr. was +
14	SUBTR		1	0	1	16	− 11 = 5		.1	0	1	0	0	0	1	X	
15	SH L		1	0	1	10			.1	0	1	0	0	0	1	0	X = 0 subtr. was −
16	SUBTR		1	0	1	10											

When the fraction is larger than 1, the process must be preceded by an initializing operation. The dividend is to be shifted right until it is smaller than the divisor. In Step 0 of Example 8–10 it was assumed that the fraction was proper, and a binary point was placed (mentally) in the least significant location of the M-Q register. If this cannot be assumed, then perform as many shift right operations of the dividend, ahead of Step 0, as are necessary to obtain the first negative answer, that is, the dividend is now smaller than the divisor. For every shift right operation until the negative difference is obtained, increment a counter by 1. That number is remembered until the proper division is finished. At the end of the divide operation the quotient is shifted left by that same number (across the binary point). Or else the fraction is left as is, and the stored number in the counter becomes the exponent of 2, a multiplier of the quotient. The choice is one of representation. The first gave the result in integer and fraction, the second in scientific notation. Let's try it.

EXAMPLE 8–11

See figure 8–19. $\quad \dfrac{11}{3} = \dfrac{1011}{0011} = 3.666$

IR (Divisor): 0 0 1 1 → 3

COUNTER:
0	0	0	0
0	0	0	1
0	0	1	0

ACC (Dividend)

0	SUBTR	1 0 1 1	11	Difference is + increment the counter by 1.
1	SH R	1 0 1|1		
2	SUBTR	1 0 1|1	5	Difference is +; increment the counter by 1.
3	SH R	1 0|11		
4	SUBTR	1 0|11	2	Difference is −; proceed with division.
				Note that the shifted bits are not discarded.

FIGURE 8–19 Division data flow

Arithmetic

Now start shifting left (see figure 8–20).

```
                                    M–Q REG
                                                    0 .
 5  SH L    1 0 1  1                              0 . X
 6  SUBTR   0 1 0  1       2                      0 . 1     Difference was +
 7  SH L    1 0 1          5                     0 . ! 1    X !
 8  SUBTR     1 0       5 – 3 = 2                 0 . 1 1   Difference was +
 9  SHL     1 0 0          4                    0 . 1 1 X
10  SUBTR       1       4 – 3 = 1               0 . 1 1 1   Difference was +
11  SHL       1 0          2                   0 . 1 1 1 X
12  SUBTR     1 0          2                   0 . 1 1 1 0  Difference was –
13  SHL     1 0 0          4                 0 . 1 1 1 0 X
14  SUBTR       1       4 – 3 = 1             0 . 1 1 1 0 1 Difference was +
15  SHL       1 0          2                 0 . 1 1 1 0 1 X
16  SUBTR     1 0          2                 0 . 1 1 1 0 1 0 Difference was –
17  SHL     1 0 0          4              0 . 1 1 1 0 1 0 X
18  SUBTR       1       4 – 3 = 1         0 . 1 1 1 0 1 0 1 Difference was +
```

FIGURE 8–20 Division data flow

Now we need to remember the two shifts stored in the counter. Shift left twice:

$$| 1 1 . 1 0 1 0 1 \, X \, X |$$

The answer is

$$3 + \frac{1}{2} + \frac{1}{8} + \frac{1}{32} = 3.62597656$$

The answer would approach 3.666 if the process were carried out to several more places. ∎

The arithmetic unit of figure 8–17 can now be extended to give a complete fixed point system as in figure 8–21.

The only components added to figure 8–17 are

1. A flip flop to sense the binary full adder output sign and temporarily store it just prior to the subtract cycle [This presents some difficulty. The sign must be known when the subtract cycle is executed. If the sign is 1, then subtraction of the divisor must be suppressed by shutting off the inhibit gates. If the sign is 0, then subtraction of the divisor proceeds normally. But how can we know the sign unless the subtraction has been performed? After all, we cannot perform the subtraction and simultaneously suppress it. There are two ways to handle that. The first is to add the divisor back into the accumulator if the sign is 1. We speak of a restoring cycle. The other is to get the sign directly from the binary full adder or comparator output (after the settling time is over) and store it in a flip flop. The accumulator has not been disturbed at this time. Then proceed with the subtract cycle controlled by the known sign. If the carry look-ahead is fast, one can determine and store the

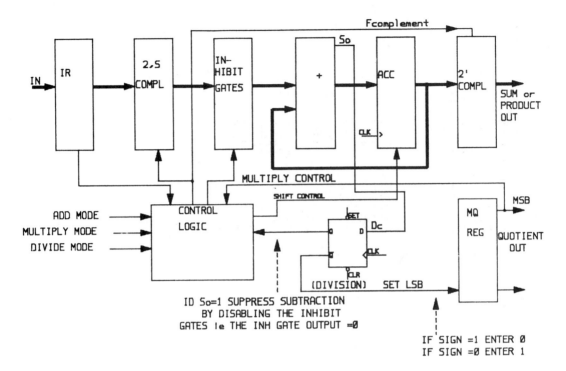

FIGURE 8-21 Fixed point add, multiply, and divide unit

sign half a cycle after the shift, shut the inhibit gates, and obtain the accumulator contents + 0 at the adder output during the next half-cycle. If the clock rate is too high and a fast compare is not possible, a restore cycle must be inserted.]

2. An input to the LSB position of the M-Q register from stored sign (inverted from the Q output) to be strobed in with the subtract clock

3. Some additional mode control

Complementary addition and multiplication were carried out in fixed point notation. Only integer numbers were considered. Division in fixed point notation will yield a 0 answer when the numerator is smaller than the denominator and the integer portion only when the opposite is true; 3/11 would be 0, and 11/3 would be 3. In that sense the system in figure 8-21 does a floating point operation for the division.

The question arises as to what can be done to operate in floating point for all four arithmetic modes.

8-7 FLOATING POINT OPERATIONS

Complement Addition

Numbers that have greater range than can be represented with available word length are split into a signed integer coefficient and a signed exponential portion. The two values are stored in separate registers, a coefficient register and an exponential register. When two words A and B are added, it must be made certain that the two exponents are equal. If they are, then the two integers can be added as shown before while the exponent remains unaltered. If the exponents are not equal, either one or both of the exponents must be changed and the coefficients shifted accordingly.

There are three possibilities:

1. Shift coefficient A to the left and decrement the exponent of A until the exponent of A equals the exponent of B. This might run into the difficulty of not reaching equality of exponents before the coefficient of A overflows its most significant position.

2. Shift coefficient B to the right and increment the exponent of B until the exponent of B equals the exponent of A. This has the shortcoming of losing the least significant bits of coefficient B and impairs computational accuracy.

3. A compromise would be to first shift one coefficient, the one with the larger exponent, to the left until the exponents match or a 1 is detected in the most significant position, and then shift the coefficient of B to the right as much as is necessary to line up the exponents. This method guarantees the least loss of accuracy.

Once the exponents are lined up, complement addition proceeds in the normal manner.

The left-justified coefficients may overflow by one bit after addition. The overflow bit is temporarily stored. The sum is then shifted one position to the right, and the exponent is incremented by one.

── EXAMPLE 8–12

$$\begin{array}{ll} 0011\ 1010 \times 2^4 & \text{in IR} \\ +\ 0100\ 0011 \times 2^6 & \text{in ACC} \end{array}$$

Shift the ACC one position to the left and the IR one position to the right:

$$\begin{array}{l} 0001\ 1101 \times 2^5 \\ +\ 1000\ 0110 \times 2^5 \\ \hline =\ 1010\ 0011 \times 2^5 \end{array}$$ ∎

── EXAMPLE 8–13

$$\begin{array}{ll} 0011\ 0110 \times 2^{-3} & \text{in IR} \\ +\ 0010\ 0001 \times 2^1 & \text{in ACC} \end{array}$$

Shift the ACC two positions to the left and the IR two positions to the right:

$$\begin{aligned}
&\ 0000\ 1101\ \times\ 2^{-1} \\
&+\ 1000\ 0100\ \times\ 2^{-1} \\
\hline
&=\ 1001\ 0001\ \times\ 2^{-1}
\end{aligned}$$

Notice that the least significant bit of the value in the IR was lost. ∎

EXAMPLE 8–14

$$\begin{aligned}
&\ 0011\ 1101\ \times\ 2^{4} \quad\text{in IR} \\
&+\ 1010\ 1101\ \times\ 2^{3}
\end{aligned}$$

Shift the IR one position to the left:

$$\begin{aligned}
&\ 0111\ 1010\ \times\ 2^{3} \\
&+\ 1010\ 1101\ \times\ 2^{3} \\
\hline
&=\ 10010\ 0111\ \times\ 2^{3}
\end{aligned}$$

Shift the result in the ACC one position to the right and increase the exponent by one:

$$=\ 1001\ 0011\ \times\ 2^{4}$$

Again the LSB was lost. A nine-bit result cannot be expressed with an eight-bit byte, but no more least significant bits were lost than was necessary. ∎

Multiplication and Division

No exponent lineup is required. The two coefficients are multiplied or divided in the normal manner, while the contents of the exponent registers are added or subtracted.

The operation can be carried out in two steps. First, the product or quotient is determined in a multiply or divide mode, then the exponents are processed by the same adder in a complement add mode. If higher speed is desirable, a separate adder for the exponent registers can carry out the add operation in parallel with the multiply or divide operation.

The choice of the logic for an arithmetic logic unit (ALU) is certainly not unique. The variety of ALUs is as bountiful as there are different computers, processors, and calculators on the market. The designer chooses in terms of complexity of requirements, desired speed, economy of gates, and, as is often the case, new ideas he or she wants to include.

As a summary of the entire discussion a version of an arithmetic unit is shown in figure 8–22.

Arithmetic

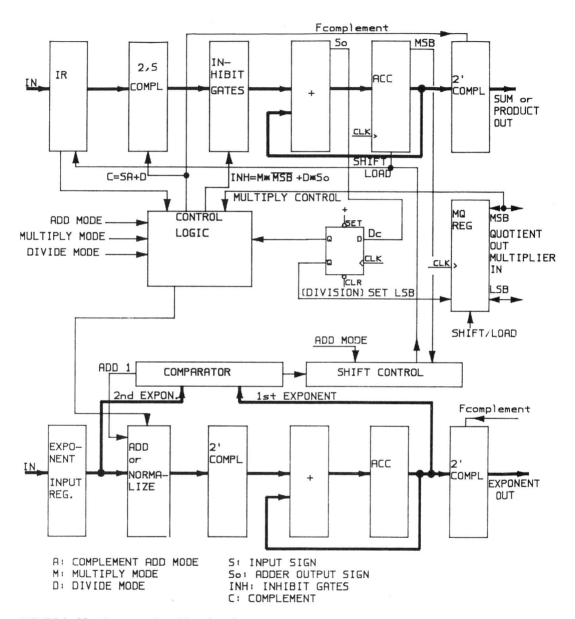

FIGURE 8–22 Floating point arithmetic unit

Logic Functions

Arithmetic logic units usually include the ability to perform the logic functions AND, OR, and Exclusive OR. These can be used in a variety of ways.

ANDing two words is a way of masking. If, for example, we want to strip a word of all its bits except the LSB, then word *A* is ANDed with all 0's except a 1 in the LSB position. The expected result is obtained:

    Word *A*   0111 0011
    AND   0000 0001
                    0000 0001

Now suppose we have two words that were previously stripped of all bits but one and we want to reassemble a new word:

    Word *A*   0000 0001
    OR Word *B* 0000 1000
                      0000 1001

EXAMPLE 8–15

Suppose we have an eight-bit byte that represents two BCD numbers, say, 57 = 0101 0111. For some subsequent operation the two BCD numbers must be separated. Here they are said to be in packed format, and for the operation we need them in unpacked format. We AND the byte in question with 0000 1111 to separate the lesser significant BCD digit; then we AND the byte with 1111 0000 to separate or unpack the more significant BCD digit

                0101 0111   0101 0111
    AND 1111 0000   AND 0000 1111
                0101 0000   0000 0111 = 7

Shifting right four positions, we get

    0000 0101 = 5   ■

ANDing a word with an all 1 pattern tests the word for an all 0, that is, the only way an all 0 will result is if the tested word itself is all 0. Such a result sets a zero flag in the flag register of the arithmetic logic unit, which can be used by the computer to make decisions. ORing a word with an all 0 pattern accomplishes the same result. Exclusive ORing a word with itself will set the word to 0, that is, clear the particular register. These are just a few simple examples, the power of which can be appreciated when they are applied to computer programming. The ability to shift and perform logic operations allows much freedom in bit manipulation and decision making.

An Example of a Commercially Available ALU

Arithmetic logic units (ALU) that are available in chip form typically do not include all the functions of the block diagram of figure 8–22. One example is the Advanced Micro Devices 2901 ALU, which is reproduced in figure 8–23. It is a four-bit adder

Arithmetic

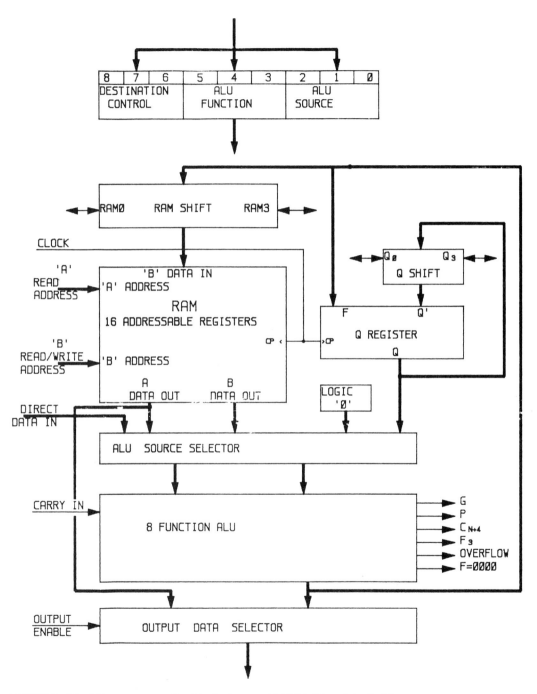

FIGURE 8–23 Microprocessor slice block diagram AM2901 (Copyright © Advanced Micro Devices, Inc., 1983. Reprinted with permission of copyright owner. All rights reserved.)

with a double-length accumulator. It has the capability of shifting the sum so that multiply-divide operations can be performed with external instructions. Such four-bit units are called *bit slice* architecture. As many chips as are needed can be put in parallel to form longer words. The chip also includes internal random access memory (RAM) for multiple storage of operands. In addition to the basic binary arithmetic functions it includes logic functions.

8–8 DECIMAL ARITHMETIC

Arithmetic with logic is most efficiently performed in binary form. Since humans are born with ten fingers, we are used to decimal arithmetic, or numbers to the base 10. It certainly is not convenient to read results in binary or enter data into a machine in binary. The human interface requires decimal-to-binary or binary-to-decimal conversion unless, of course, we could operate the logic in some decimal form to begin with.

Conversion is usually accomplished in software so that the binary nature of the computer becomes transparent to the user. Hardware conversion can be done with SN74184 or SN74185 converter chips and requires conversion trees. The conversion is relatively slow, as much as 320 nsec for a 16-bit conversion requiring 16 chips. In many instances in which speed is not too important, conversion can be accomplished with counters. Simply load the binary number into a binary counter and down-count. Simultaneously up-count a decimal (BCD) counter and stop both counters when the borrow of the binary counter indicates 0. The same two counters can be used for BCD-to-binary conversion by loading and down-counting the BCD counter while up-counting the binary counter.

To bypass this inconvenience, it is possible to design BCD arithmetic units. They are obviously less efficient, since the BCD code uses only 10 out of 16 possible states, but entry and exit are more direct. When doing additions, we depend on the generation of a carry for the next column. A four-bit binary number generates a carry when the number 1111 overflows. In BCD that value is never reached because any number above 1001 is illegal. The way to get around this problem is to shift the operating range 0000–1001 to 0011–1100. In other words, place the operating range in the middle of the 0000–1111 range by adding 0011 to every BCD number entered. This new number is now in *excess 3 code*.

If, for example, we add 5 + 5, we expect 0 and a carry of 1. As was mentioned before, a binary 1010 will not yield a carry, but if we add 5 + 3 to 5 + 3 in excess 3 code, we get 0000 and 1 for a carry. If there is no carry, the result will be in *excess 6 code;* therefore 0011 must be subtracted to put the result back into excess 3 code for further computation. If there is a carry, the result is back in BCD, and 0011 must be added to put it back in excess 3 code for further computation.

To repeat:

If a BCD excess 3 sum has a carry, add 3 to the sum.

If a BCD excess 3 sum has no carry, subtract 3 from the sum.

Arithmetic

── EXAMPLE 8–16

```
            BCD            Excess 3
            2 + 3 = 5      5 + 6 = 11    Subtract 3 = 8    No carry
                                                           11 is less than 16
                           0101
                           0110
                           ────
                           1011
 Subtract                  0011
                           ────
                           1000  = 8
```

Indeed 5 in excess 3 = 8

```
            9 + 7 = 16     12 + 10 = 22     There is a carry
                           1100             22 is larger than 16
                           1010
                           ──────
                         1 0110
                     Add   0011
                         ──────
                         1 1001 = 9 and a carry
```

Indeed 6 in excess 9 = 6   ■

Initial rules for BCD addition are:

1. Enter operands in excess 3 code.
2. Subtract 3 from sum if there is no carry.
3. Add 3 to sum if there is a carry.

Two binary full adders are needed to carry out the operation. The second adds or subtracts 3. Subtracting 3 is the same as adding the complement of 3 with respect to 15, which is 12. (The one's complement of 0011 = 1100.) We get the circuit of figure 8–24.

Now we want to find out how to handle words that are longer than 1 BCD digit and how to use the basic adder above to do complement addition.

Six cases can be illustrated, similar to the binary method, in which each BCD number is represented by four binary bits and the sign is 1 or 0 for negative and positive, respectively. This is shown in table 8–6.

With the rules established, a BCD adder can be drawn as in figure 8–25. The numbers entered into the adder of figure 8–25 must be in excess 3 as well as complement form. The BCD excess 3 adder must have the *end around carry* around the entire adder, while the excess 3 restore adder only has local end around carry for each bit. Let us try some examples.

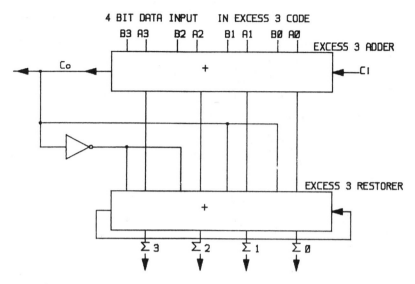

FIGURE 8–24 One BCD digit adder

TABLE 8–6 BCD addition

Case 1	+ 2 + 3	
	0 0010	
	0 0011	BCD
	0 0101	Excess 3
	0 0110	
	0 1011	The sign is 0; there is no overflow
	− 11	Subtract 3 (12 is the one's complement of 3)
	0 1000	
	1000	8 = 5 in excess 3
		Correction: For further computation—none.
		For readout: Subtract 3.
Case 2	+ 9 + 4	
	0 1001	
	0 0100	BCD
	0 1100	
	0 0111	Excess 3
	1 0011	The 1 in the sign bit signals an overflow. Two positive numbers cannot have a negative sum.

Arithmetic

TABLE 8–6 (Continued)

	$\underline{+11}$	Add 3
	0110	6 = 3 in excess 3
		Correction: Set overflow alarm.
Case 3	+ 3 − 8	
	0 0011	
	1 1000	BCD
	0 0110	
	1 1011	Excess 3
	0 0110	
	$\underline{1\ 0100}$	One's complement
	1 1010	There is no carry; subtract 3
	$\underline{-11}$	
	1 0111	−7 = complement of −5 in excess 3
	1 1000	Correction: None.
		For readout: Complement and subtract 3.
	1 0101	−5
Case 4	+ 9 − 4	
	0 1001	
	1 0100	BCD
	0 1100	
	1 0111	Excess 3
	0 1100	
	$\underline{1\ 1000}$	One's complement
	10 0100	First the end carry must be added
	$\underline{\quad\ \ 1}$	
	0101	
	$\underline{+11}$	There is a carry: add 3
	1000	8 = 5 in excess 3
		Correction: None.
		For readout: subtract 3.
Case 5	− 2 − 5	
	1 0010	
	1 0101	BCD
	1 0101	
	1 1000	Excess 3

continues

TABLE 8-6 (Continued)

	1 1010	
	1 0111	One's complement
	11 0001	The end carry must be added
	———1	
	1 0010	There is a carry: add 3
	+11	
	1 0101	= −7 in excess 3
		Correction: For further computation—None.
		For readout: complement and subtract 3.
Case 6	− 4 − 9	
	1 0100	
	1 1001	BCD
	1 0111	
	1 1100	Excess 3
	1 1000	
	1 0011	One's complement
	10 1011	
	———1	
	0 1100	Add 3
	11	
	0 1111	The 0 in the sign position signals an overflow
		Correction: Set overflow alarm.
		(The two's complement version does not work because the symmetry between 0 and 15 is shifted.)

EXAMPLE 8-17

$$91 - 37 = 54$$

Step 1	0 1001 0001	
	1 0011 0111	BCD
Step 2	0 1100 0100	
	1 0110 1010	Excess 3
Step 3	0 1100 0100	
	1 1001 0101	One's complement

Arithmetic

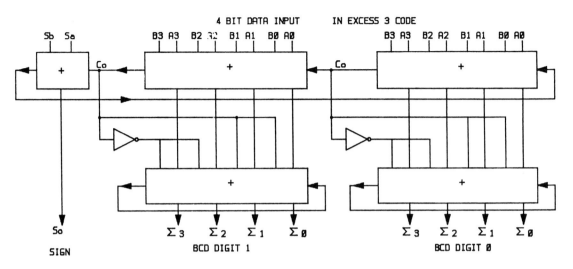

FIGURE 8-25 BCD adder

Step 4

```
 10 0101 1001
_____1          End carry must be added
  0 0101 1010
    +11  -11       Restore excess 3
  0 1000 0111      54 in excess 3
```

━━━ EXAMPLE 8-18

$37 - 91 = -54$

```
0 0011 0111
1 1001 0001        BCD
0 0110 1010
1 1100 0100        Excess 3
0 0110 1010
1 0011 1011        One's complement
1 1010 0101        Restore excess 3
   -11  +11
1 0111 1000        Complement -54 in excess 3
1 1000 0111
```

Another method for decimal addition that is used with microprocessors is to add two BCD words in the ordinary binary fashion and perform a *decimal adjust* on the obtained sum. Each BCD digit consists of a four-bit nibble.

- If the sum of two BCD nibbles does not exceed 1001 = 9, no correction is needed.
- If the sum of two BCD nibbles exceeds 9, a special nibble carry is set and 0110 = 6 is added to the nibble.
- The nibble carry is added to the next most significant nibble.

The following examples illustrate the method.

EXAMPLE 8-19

Let us add 35 + 43 = 78:

```
  0011 0101
+ 0100 0011
  ─────────
  0111 1000  = 78    No correction is needed
```

Neither nibble of the sum exceeded 9, and there are no nibble carrys. ∎

EXAMPLE 8-20

Add 87 + 16 = 103:

```
    1000 0111
+   0001 0110
    ─────────
    1001 1101    Lower nibble > 9, and nibble carry is set
         0110    Add 6
    ─────────
           1     Nibble carry
    1010 0011    Upper nibble > 9
    0110         Add 6 to upper nibble
  ───────────
  1 0000 0011  =  103
```
∎

Decimal Multiplication

Decimal multiplication is more complicated than binary multiplication. The simplest way to accomplish multiplication is to do repeated addition for each BCD digit. That is, if the most significant digit of the multiplier digit is 1, add eight times; otherwise, shift the multiplier left to the next least significant bit. If it is a 1, add

Arithmetic

four times; otherwise, shift to the next bit, and so on. The maximum number of additions for each digit is obviously nine. Additions must be carried out with a BCD adder. Shifting across zeros saves clock cycles. After each BCD digit multiplication, shift the accumulator four positions, which is equivalent to a multiplication of 10. Higher-speed methods exist, but the additional logic is warranted only if speed is critical.

EXAMPLE 8-21

Multiply 371 × 297 = 110,187. Note that 2 * means "two times" or "twice" (see figure 8-26).

	ACC						M-Q REG	
			IR					
			0011 0111 0001	371				
Initial state			0000 0000 0000			297	0010 1001 0111	
						ShL	010 1001 0111	
						ShL	10 1001 0111	
Add 2*			0111 0100 0010		742			
						ShL	0 1001 0111	
			0111 0100 0010		742			
Shl 4*		0111	0100 0010 0000		7420	ShL	1001 0111	
Add 8*	0001	0000	0011 1000 1000		10,388 =			
						371 × 8 + 7420	ShL	001 0111
						ShL	01 0111	
Add 1*						ShL	1 0111	
	0001	0000	0111 0101 1001		10,759 = 10,388 + 371			
Shl 4*	0001	0000	0111 0101 1001 0000		107,590	ShL	0111	
						ShL	111	
Add 4*	0001	0000	1001 0000 0111 0100		109,074 =			
						371 × 4 + 107,590	ShL	11
Add 2*	0001	0000	1001 1000 0001 0110		109,816 =			
						371 × 2 + 109,074	ShL	1
Add 1*	0001	0001	0000 0001 1000 0111		110,187 =			
						371 × 1 + 109,816			

FIGURE 8-26 Decimal multiplication

Decimal Division

Division follows the reverse order of multiplication: repeated subtraction while testing the sign bit and shifting four bits for each BCD digit. Similar to the binary case the fraction must be normalized by shifting the dividend to the right four positions while adding 1 to the exponent counter for every four shifts until the dividend is smaller than the divisor. The number entered into the M-Q register is equal to the number of subtractions before a minus sign is detected. A sign test with a digital comparator avoids the additional subtraction to find the minus sign, which would require an add of the divisor, that is, a restore cycle. This is shown in Example 8-22.

EXAMPLE 8-22

See figure 8-27.

$$\frac{193}{17} = 11.3529$$

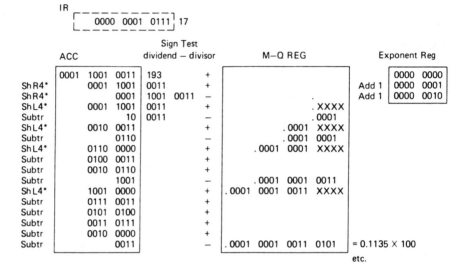

FIGURE 8-27 Decimal division

8-9 SERIAL ARITHMETIC

Addition

When speed is not important and circuitry is to be minimized, serial adders become attractive. The method is simple. Two numbers are stored in two separate shift registers for words A and B. The two least significant bits that are available at the serial output port are added in a binary full adder, while the carry for the next add is stored in a flip flop. Next, the shift registers are shifted to the right one position, while the last obtained sum is shifted into the most significant position of the shift register B. The process is repeated until all bits are exhausted. The sum of the two words is found in shift register B. Shift register A is empty and can be reloaded with another value for continued accumulation. See figure 8-28.

It is useful to follow the step-by-step process for the serial addition of two positive bytes.

Arithmetic

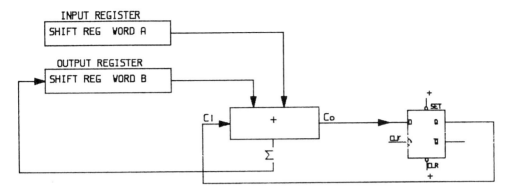

FIGURE 8-28 Serial adder

EXAMPLE 8-23

Assume that registers A and B are cleared. A byte = 00110101 is serially shifted into register A. Register B is clocked and shifted at the same time. Nothing but 0's are reentered into register B during the first eight shift cycles.

Another byte = 10011110 follows with another eight shift cycles. The contents of A are added to the contents of B, which is 0. Thus the contents of register A just transfer to register B.

By now 16 clock cycles have elapsed, and the addend and augend are in registers A and B, respectively. The next eight shift cycles will accomplish the desired addition. We expect

$$\begin{array}{r} 10011110 \quad (+158 \text{ decimal}) \\ + 00110101 \quad (+53 \text{ decimal}) \\ \hline 11010011 \end{array}$$

Register A	Register B	C_i	Σ Sum	C_o	Clock
10011110	00110101	0	1	0	Initial clock 1
01001111	10011010	0	1	0	
00100111	11001101	0	0	1	Clock 2
00010011	01100110	1	0	1	Clock 3
00001001	00110011	1	1	1	Clock 4
00000100	10011001	1	0	1	Clock 5
00000010	01001100	1	1	0	Clock 6
00000001	10100110	0	1	0	Clock 7
00000000	11010011 *Answer*	0	1	0	Clock 8

Notice that

- Register *A* simply shifts right.
- Register *B* shifts right, too, but the most significant bit is the previous sum from the adder.
- The input carry C_i is the adder output carry delayed by one cycle. ■

Complement Addition

Negative numbers are represented in two's complement form. The complementation for a number that is loaded into shift register *A* as sign and magnitude can be done simultaneously with the addition. Let us take a second look at a binary number and its two's complement:

$$\begin{array}{r} 1\ 10001100 \text{ Sign and magnitude} \rightarrow 1\ 01110011 \quad \text{One's complement} \\ +\qquad 1 \\ \hline 1\ 01110100 \quad \text{Two's complement} \end{array}$$

Compare the sign and magnitude and the two's complement and notice that several of the lesser significant bits appear to be unchanged. When a 1 is added to the one's complement, the 1 will ripple through the one's complement changing all 1's to 0's and then is stopped when the first 0 appears. That first 0 is changed to a 1, but no further changes occur for the higher-order bits. This implies that some of the lesser significant bits of the one's complement are changed back to what they were in sign and magnitude form. The result of all this is that a sign and magnitude number can be encoded into the two's complement by the following algorithm, starting with the LSB and working our way up to the MSB:

1. All trailing 0's are left unchanged.
2. The first 1 encountered is unchanged.
3. All bits thereafter are complemented up to and including the MSB of the number *except the sign bit*.

How can this algorithm be implemented? The problem is sequential in nature; therefore the methods to be discussed in Chapter 10 are applicable. The derivation will be given as an example in Chapter 10. For now it suffices to explain the logic in figure 8–29.

The sign of *A*, while it is part of the byte in register *A*, is also strobed into the *hold sign flip flop* and remains there for the duration of the addition.

If the sign of *A* is positive, the input *S* to the AND gate is 0, and therefore the AND gate output is 0. It follows that the output of the *complement control flip flop*

Arithmetic

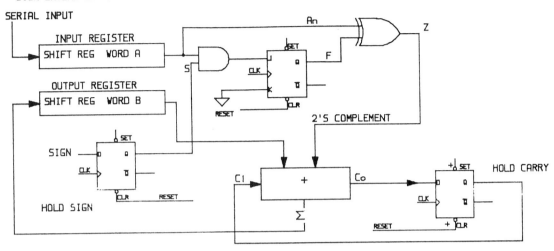

FIGURE 8–29 Serial complement adder

stays at 0. Therefore the output of the Exclusive OR gate Z is always equal to the input A_n. In this case the addition does not differ from Example 8–23.

If, however, the sign of A is negative, S becomes 1. As long as $A_n = 0$, the input to the complement control flip flop remains 0, and so does the output F. Thus $Z = A_n$ as long as 0's arrive from register A.

The first 1 to exit from register A changes the J input to the complement control flip flop to 1. The output F remains at 0 for one additional clock cycle. Thus the first $A_n = 1$ leaves $Z = A_n = 1$ as stated in the above algorithm for two's complementation.

With the next clock cycle, though, the situation changes. The complement control flip flop output F changes to a 1. All following bits from register A, up to the sign bit, are inverted with the Exclusive OR gate. F will return to 0 after clock cycle 7; this satisfies the requirement of *no* sign bit inversion.

Example 8–24 illustrates the cycle-by-cycle progress of a serial subtract (or complement add) operation.

-----EXAMPLE 8–24

Assume that the two bytes in registers A and B were previously entered:

01101110	Positive byte in register B	(+ 110 decimal)
10110100	Negative byte in register A	(− 52 decimal)
00111010	Answer	(+ 58 decimal)

Register A	Register B	S	J	K	F	Z	C_i	Sum	C_o	Clock
10110100	01101110	1	0	0	0	0	0	0	0	Initial clock 1
01011010	00110111	1	0	0	0	0	0	1	0	
00101101	10011011	1	1	0	0	1	0	0	1	Clock 2
00010110	01001101	1	0	0	1	1	1	1	1	Clock 3
00001011	10100110	1	1	0	1	0	1	1	0	Clock 4
00000101	11010011	1	1	0	1	0	0	1	0	Clock 5
00000010	11101001	1	0	1	1	1	0	0	1	Clock 6
00000001	01110100	1	1	X	0	1	1	0	1	Clock 7
00000000 ↗ 00111010 *Answer*		1	0	X	X	X	1	X	X	Clock 8

The most significant bit of the answer in Register B is the sign; the other seven bits are the sum in two's complement form. ∎

In Example 8–24 the answer was positive and the two's complement was the same as the magnitude. No further correction was needed for readout. If the answer had been negative, then it would have been necessary to obtain the two's complement again before sending the result to a display. The two's complement form, however, is the desirable form for further operations.

The serial adder can be extended to include multiplication and division in the same manner as in the parallel case.

8-10 THE DIGITAL COMPARATOR

An important element in digital arithmetic is the magnitude comparator. One way to achieve a comparison is to subtract two numbers. The resulting sign will, of course, indicate which of the two numbers is the larger value. It will also give the exact difference, which is not needed for a mere comparison; that is, the adder gives more information than required and suggests a waste of circuitry.

When two numbers are compared, you need only look at the MSBs of two words. Whichever has a 1 while the other has a 0 must be the larger value. There is no point in investigating the remaining bits. If, however, the two MSBs are equal, then you must look at the next lower significant bit. The same reasoning applies as for the MSB. Eventually, you might have to look at all bits. If they are all equal, then the two values are equal; if not, one of the bit comparisons will indicate which of the two dominates.

An Exclusive OR will tell whether two bits are equal or not. By itself it cannot indicate which is the 1 and which the 0. In combination with two NAND gates this can be determined as shown in figure 8–30.

It is convenient to compare four bits simultaneously. Four one-bit comparators are interconnected with successive inhibit gates in decreasing order, starting with the MSB. If the MSBs are unequal, all lesser significant outputs are held at 0, and

Arithmetic

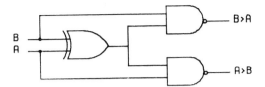

FIGURE 8-30 One-bit pair comparator

the comparator output takes the result from the MSB position. If the MSBs are equal, its output is 0, and the next lower position is enabled. This method continues down the line to the LSBs. Since all upper-position "equal" results have a 0 output and all lower positions past an unequal position are inhibited, there can only be one "true" output for the four positions, for either nibble A or nibble B. If none are true, the two nibbles are equal. One need only OR the four position outputs to find $A > B$, $A < B$, and $A = B$. Figure 8-31 shows the scheme.

We can examine the function of the comparator with an example.

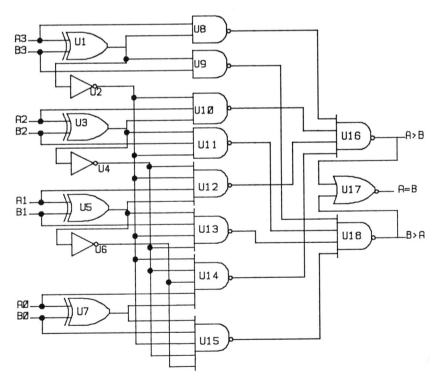

FIGURE 8-31 Four-bit comparator

EXAMPLE 8-25

Nibble $A = 0110$ and Nibble $B = 0101$. We expect $A > B$ output $= 1$.

Since $A_3 = B_3 = 0$, the output of the Exclusive OR gate $U_1 = 0$. The output of the inverter $U_2 = 1$, meaning that no decision can be made from the output of U_1. The outputs of U_8 and $U_9 = 1$.

Since $A_2 = B_2 = 1$, the output of the Exclusive OR gate $U_3 = 0$. The output of the inverter $U_4 = 1$, meaning that no decision can be made from the output of U_4. The outputs of U_{10} and $U_{11} = 1$.

Since $A_1 = 1$ and $B_1 = 0$, the output of the Exclusive OR gate $U_5 = 1$. The output of the inverter $U_6 = 0$, which inhibits any further investigation by forcing the outputs of U_{14} and $U_{15} = 1$.

Since $A = 1$, the output of $U_{12} = 0$. Since $B = 0$, the output of $U_{13} = 1$.

Notice that only one of the eight NAND gates U_8–U_{15} has a 0 output. This forces the output of U_{16} $A > B$ to 1. ∎

The SN7485 four-bit comparator also includes a fifth position comparison, which allows expansion. The same reasoning that applied to bit-by-bit comparison also applies to comparison of groups of four. Thus words of any length can be compared. One drawback is the trickle time if the word is too long. Another way to speed up the comparison at a slight cost in chip count is to build a comparison tree. (See figures 8-32 and 8-33.)

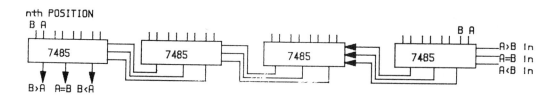

FIGURE 8-32 Serial comparator

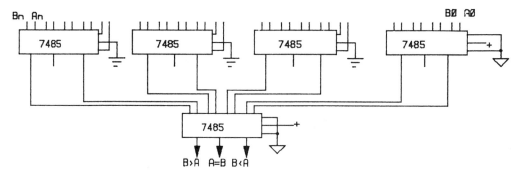

FIGURE 8-33 Parallel tree comparator

Arithmetic

REVIEW QUESTIONS

1. How does a full adder differ from a half adder?
2. What shall one do with the C_i input of the lowest-order adder? Discuss options.
3. What is an accumulator?
4. Why is it not necessary to build a special subtractor for an arithmetic unit?
5. Discuss one's complement addition.
6. What is meant by an end around carry?
7. What is the advantage of two's complement addition?
8. Discuss a two's complement encoder.
9. What is meant by fixed point arithmetic?
10. Review the required corrections for the six cases of one's and two's complement addition.
11. Why should the accumulator exceed the input word length and how does a microprocessor get around that problem with a one-byte accumulator?
12. Why must the binary full adder length equal the accumulator length (if a double-length accumulator is available)?
13. What is an M-Q register?
14. Why is a left shift multiplying scheme preferable to a right shift scheme?
15. Discuss the relationship between word length and clock cycles to carry out an addition. To carry out a multiplication.
16. Discuss the sign test method for the division process.
17. Discuss proper fraction and improper fraction division.
18. What is meant by a floating point number?
19. What additional register is needed in the arithmetic unit when doing floating point calculations?
20. Discuss normalization of exponents for the add process. For the multiply-divide process.
21. How is the sign of a number handled in addition?
22. How is the sign of a number handled in multiplication or division?
23. What is meant by bit slice architecture?
24. Discuss the BCD code.
25. Why is it necessary to encode BCD into an excess 3 code to do decimal addition?
26. The primary end around carry of a BCD adder is differently connected than the end around carry of the local correction adder. Discuss the reasons.
27. Why do decimal multiplication and division require bursts of four shifts?
28. What is the advantage of serial arithmetic?

29. Discuss the two's complement method for a serial adder.
30. Why is the two's complementer not needed for the output shift register in a serial adder?
31. Discuss digital comparison.
32. Why is a comparison tree faster than a comparison string?

PROBLEMS

1. Apply all four possible input combinations to the half adder of figure 8–1 and find the corresponding outputs.
2. Apply all eight possible input combinations to the full adder of figure 8–2 and find the corresponding outputs.
3. Apply two four-bit numbers to the input of figure 8–4 and enter all sum and carries at the appropriate nodes.
4. Apply three four-bit numbers to the accumulating adder of figure 8–8 and enter all sum and carries at the appropriate nodes for each clock cycle.
5. Redesign the binary full adder of figure 8–6 without carry look ahead. Compare the gate delay for the output carry for the two versions.
6. For convenience, copy the logic of a 74283 four-bit adder onto larger paper. Identify the full adder sections. Identify the carry look-ahead logic. Add 1110 + 1001. Follow the logic flow by identifying 1's and 0's of each gate output. Check the resulting sum and output carry.
7. Compare the delay of the carry bypass enable logic to a binary full adder without carry look-ahead.
8. Draw a block diagram for a 16-bit adder with carry bypass enable and count the gate delay for an input carry.
9. Design an eight-bit incrementer, that is, add 1 to an eight-bit number. Hint: write a truth table for four bits, find the logic equation, and then expand to eight bits.
10. Design a nine-bit adder with carry look-ahead in groups of three.
11. Represent the numbers 5, 17, 237, in sign and magnitude form, in one's complement form, and in two's complement form.
12. Apply a four-bit number to the circuit of figure 8–11. Let the sign = 1. Enter each gate response.
13. Add in binary using one's complement:
    (a) $+37 + 48$
    (b) $+53 - 27$
    (c) $+27 - 53$
    (d) $-19 - 85$
14. Repeat Problem 13 in two's complement.

Arithmetic

15. Multiply in binary:
    (a) 6 × 8. Use four bits to represent each value.
    (b) 9 × 12. Use four bits to represent each value.
    (c) 13 × 3. Use four bits to represent each value.
    (d) 37 × 61. Use eight bits to represent each value.
    (e) 241 × 129. Use eight bits to represent each value.
16. Repeat Problem 15 but follow the arithmetic unit flow format of Examples 8–7 and 8–8.
17. Divide:
    (a) 1/7. Follow the flow format of Example 8–10.
    (b) 3/5
    (c) 5/11
    (d) 13/15. Carry the division to six binary places.
18. Divide:
    (a) 21/8. Normalize and follow the flow format of Example 8–11.
    (b) 8/7
    (c) 123/14
    (d) 255/31
19. Carry out the following logic operations:
    (a)
    ```
             1111 0000
    AND      0000 1111

             1111 0000
    OR       0000 1111

             1111 0000
    EX-OR    0000 1111
    ```
    (b)
    ```
             1001 0110
    AND      0001 0000

             1001 0110
    OR       0111 1111

             1001 0110
    EX-OR    1111 0000
    ```
20. Use a logic function that would test the following words and set a flag:
    (a) Is the MSB = 1? in 1000 0011
    (b) Are the first and third bits = 1? in 0011 0001 (MSB is the first bit.)
    (c) Are the first or third bits = 1? in 0011 0001
    (d) Clear a register of its contents.
    (e) Strip the upper four bits of the word 11000111
    (f) Strip the lower four bits of the word 11000111

(g) AND a word with itself. What happens?
(h) OR a word with itself. What happens?
(i) EX-OR a word with itself. What happens?

21. Add in BCD and excess 3. Indicate corrective actions.
    (a) $+5 \ +7$
    (b) $+3 \ -2$
    (c) $+2 \ -3$
    (d) $-6 \ -1$
    (e) $-6 \ -5$

22. Add in BCD and excess 3. Indicate corrective actions.
    (a) $+91 \ -38$
    (b) $+37 \ +12$
    (c) $-89 \ -99$
    (d) $-38 \ -61$

23. Multiply in BCD and excess 3:
    (a) 123×85
    (b) 234×06
    (c) 83×77
    (d) 469×111

24. Divide in BCD and excess 3:
    (a) 4/17
    (b) 35/17
    (c) 147/27
    (d) 200/25

25. Add in serial:
    (a) $+9 + 5$. Construct your own flow format.
    (b) $-9 + 5$. Use figure 8–29 to determine the result after each clock cycle.

26. Apply two four-bit words to the comparator of figure 8–32. Enter all gate responses.

27. Apply two 16-bit words to the inputs of both comparators of figures 8–32 and 8–33. Enter all gate responses.

chapter 9

memory

9–1 INTRODUCTION

The memory is as essential to computer systems as the arithmetic logic unit. Any device or logic circuit that is capable of holding information can be looked at as a memory cell. The flip flops that we used in counter or control circuits are memory elements. In a larger sense we talk about memory when many bits or words are to be stored for any length of time, to be written into a chosen location and to be retrieved from that location when needed.

Two questions arise:

1. How does one find the information of interest within a large memory?
2. How soon must that information be available?

We talk about "addressing" and "access time." Addressing a memory will be discussed throughout this chapter. Access time is a function of the type of memory. Computers use three types of memories: tapes, disks, and RAMs (random access memory).

Tapes are the slowest, with several seconds access time, since data are serially stored along the tracks. To find a particular file, the tape might, in the worst case, have to be wound from the beginning to the end of the spool. Tapes are too slow to effectively interact with a computer. They are used for permanent, inexpensive mass storage.

Disks are of medium speed, with several milliseconds access time. Data are serially stored along circular tracks, but the read-write heads go directly to the track of interest. Disks are used to store and retrieve currently used data files, application programs, and system programs.

RAM is the fastest memory, with microseconds (and below) access time. The medium is either magnetic cores or solid-state devices. Any data location within the memory is directly and immediately addressable.

The subject of tapes and disks is treated in the chapter on peripherals (Chapter 14). In this chapter we will concentrate on RAMs.

RAM is the most expensive in terms of storage cost per bit and space. The size of RAM needed is determined by the quantity of interactive data and programs needed to carry out a task. All other information that is not immediately needed is stored on disk and tape. Until recently, core memories dominated the market for RAM applications, with access times of about one microsecond and capacity that typically did not exceed one megabyte. Bipolar and CMOS memories have surpassed core memories in performance and cost and will therefore be the primary target of this discussion.

9-2 MEMORY COMPONENTS

In bipolar technology the D type latch is used as the basic memory element. One bit of information is applied to the D control input and written into the latch with a Write Enable signal $\overline{W}$. In this case it is not the falling edge but the low level of the Write Enable signal that sets Q of the latch to the value of the input. The input data must therefore remain constant for the duration of the

Write Enable = True

Two properties of the memory element should be mentioned:

- When power is turned OFF, the content of the memory element (or cell) is lost. The memory is said to be *volatile*.
- When data from the cell is read, the data remains undisturbed. Memory data readout is said to be *nondestructive*. (See figure 9–1(a).)

The CMOS version is similar, but the circuit arrangement needs a few words of explanation.

When the cell is enabled from the write address decoder and the input data is high, T_2 is closed and T_4 is open. Thus the output is at 0 level. The feedback loop to T_1 and T_3 also opens T_1 and closes T_3.

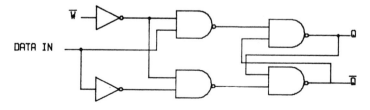

FIGURE 9–1(a) Bipolar latch with Write Enable

Memory

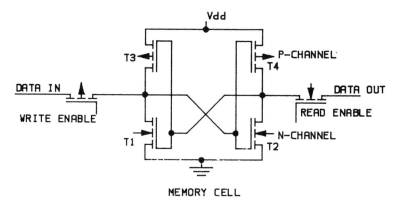

FIGURE 9-1(b) CMOS latch with Write Enable (Courtesy of GE Solid State)

When the input signal is removed, the input = 1 level is maintained by the feedback loop. (See figure 9-1(b).)

If the input is *low* while the cell is enabled, T_1 and T_4 are closed and T_2 and T_3 are open, holding the output high and the feedback loop low.

When the input signal is removed, the input = 0 level is maintained by the feedback loop. T_1 and T_3 as well as T_2 and T_4 are complementary symmetry inverters.

The output taken at the drain of T_2 is inverted with respect to the input data. A common tristate inverter for all memory cells feeding a common data bus line restores the polarity.

Magnetic core storage is based on the property of magnetic remanence. When current flows in a wire that passes through the core, it produces a magnetic flux in the core, say in the clockwise direction in figure 9-2, which remains even after the current has ceased to flow. If the current is strong enough, the core saturates such that no further increase in magnetic flux is observed with further increases in current. A reverse current will reverse the magnetic flux, but not until it has reached a switching threshold, which is not the same for both current directions.

This property is called *hysteresis* and is shown in figure 9-3. Core materials are chosen for a square hysteresis loop. Once magnetized, the core is never returned to the original zero magnetic state (unless it is demagnetized). A 1 state is defined as one flux direction and a 0 state as the other flux direction. For data readout,

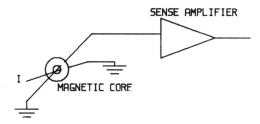

FIGURE 9-2 Core and sense amplifier

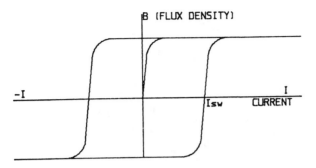

FIGURE 9–3 Hysteresis curve

another wire, called a *sense wire,* is passed through the core. The core acts as a miniature transformer.

When a reverse current is applied on the write wire and the flux is in a 1 direction, the flux will reverse, inducing a small current pulse in the sense wire that is proportional to the rate of the flux change. This is amplified with the sense amplifier.

When a reverse current is applied on the write wire and the flux is in a 0 direction, the flux will *not* reverse, thus failing to induce an output pulse. The presence of a read pulse is recognized as a 1; the absence of a read pulse is recognized as a 0.

The output pulse, only a few millivolts, amplified in the sense amplifier sets a flip flop of a memory output register. Only one output register, with the number of flip flops equal to the bits per word, is needed for the memory. The properties of the magnetic core memory differ from the TTL or CMOS version:

- When power is turned OFF, the flux remains; thus information is not lost. The memory is *not volatile*.
- When data is read with a flux reversal, the information is lost. Memory readout is destructive.

Unless the data just read is not needed any longer, the read data is rewritten into the same location, that is, it is restored.

9–3 MEMORY ADDRESSING AND DATA LINE INTERCONNECTIONS

We can choose the basic cell of figure 9–1(a) to construct a small memory of two addresses, each address containing a four-bit word. However, the cell of figure 9–1(a) is incomplete. There must be a way to select one set of cells over another set of cells; in other words, each set of cells must be specifically addressed. A modification of the circuit in figure 9–1(a), shown in figure 9–4(a), includes an address

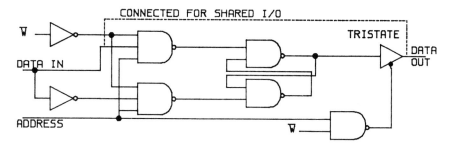

FIGURE 9–4(a) Addressable memory cell

input and a tristate output gate that accepts new data in write mode and addressed and sends out data if in read mode and addressed.

Figure 9–4(a) can be drawn as a basic block as in figure 9–4(b). The memory cell has

>one data input line
>one data output line
>one address select line
>one read-write mode line

We are now ready to draw the logic for the 2 × 4 memory in figure 9–5, meaning two addresses and four bits per word.

The output gates of each cell are either open collector or tristate gates, allowing the dot-ORing of all like-outputs bits. One could obviously expand the system of figure 9–5 to as many bits per word and addresses as needed. As the circuit is drawn, bit expansion would be in the row direction. Each additional bit has its own data input and output line, but the entire row shares the same address line. This is so because all bits in one row reside at the same address. Address expansion of figure 9–5 is done by adding more rows. Each row must receive its own address line, but like bits are shared with all other rows (or addresses). This is reasonable, since data bits can be sent to, or received from, the entire memory, but only the addressed row will respond.

In a system, as in figure 9–5, there are as many rows and also as many address lines as there are addresses. As long as the system is small, this is acceptable. For large memory systems the number of required address lines becomes overwhelming. The address scheme in figure 9–5 is one-dimensional, with one dedicated address line for each location.

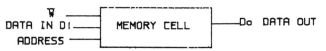

FIGURE 9–4(b) One-dimensional address memory cell

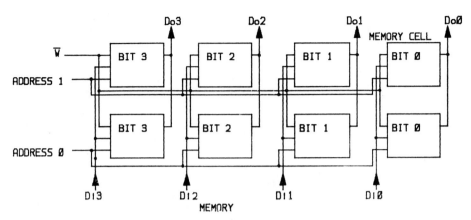

FIGURE 9-5 Memory with two-location, four-bit, one-dimensional addressing

The number of address lines can be greatly reduced by using a two-dimensional address scheme. The memory of figure 9-4(b) is modified as in figure 9-6 to receive two address lines: a horizontal address line and a vertical address line. Only if both lines are simultaneously true will the memory cell respond.

Let us choose a memory with only one bit per word (the reason for that choice will soon become apparent) and N addresses, consisting of X number of rows and Y number of columns. Each row has a horizontal address line connected to each cell in a row. Each column has a vertical address line connected to each cell in the column. Thus each cell receives two address lines, one horizontal and one vertical line. Only one horizontal and one vertical address line are true at any one time. The location where the two true lines meet (or intersect) is the selected location. The number of intersections, or addresses, is equal to the product of $X \times Y = N$. Memories are usually organized with equal numbers of rows and columns, that is, $X = Y$. Thus $X^2 = N$ or $X = \sqrt{N}$. We need X row address lines and also X column address lines or a total of $2X = 2\sqrt{N}$ address lines. Thus the number of required address lines has decreased from N to $2\sqrt{N}$. This constitutes a considerable saving in address lines. A 256-location memory needs only 16 horizontal and 16 vertical lines rather than 256 in one dimension.

We can illustrate the concept by showing the architecture of a 4×4 memory matrix, that is, 16 addresses but only one bit per word.

Consider the memory of figure 9-7 to be a memory with N addressable locations where $\sqrt{N}$ wires (in this case, $\sqrt{16} = 4$) are the horizontal address and $\sqrt{N}$ are the vertical address wires. Such a matrix is a memory plane for one bit and is called a bit plane. Several bit planes can be stacked on top of each other to extend the one bit per word memory to n bits per word. The stacking of the planes forms a third dimension. That last added dimension is the *data dimension*. The address lines are fed in parallel to all bit planes, selecting the very same location or address on each plane. Each plane, however, receives its own data input/output lines. There are as many bit planes in the third dimension as there are bits per word. We have a three-dimensional memory with two dimensions for addressing and one dimension for data bits.

Memory

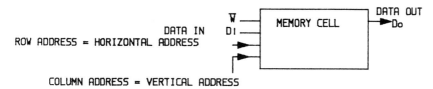

FIGURE 9–6 Two-dimensional address memory cell

The number of address lines can be reduced further by decoding the address lines for each of the two dimensions. If we consider the example of figure 9–7 with two sets of four address lines of which only one line is true for each set, then two sets of two address lines can select one out of four in each dimension. We thus distinguish between the internal address lines, of which only one is true in each dimension coming from the address decoder, and the external address lines in binary code, which are the input to the address decoder. The internal address lines and the address decoder are part of the memory chip. The user presents the memory with an address in binary code.

The memory of figure 9–7 has 16 addresses. Since $2^4 = 16$, we need only

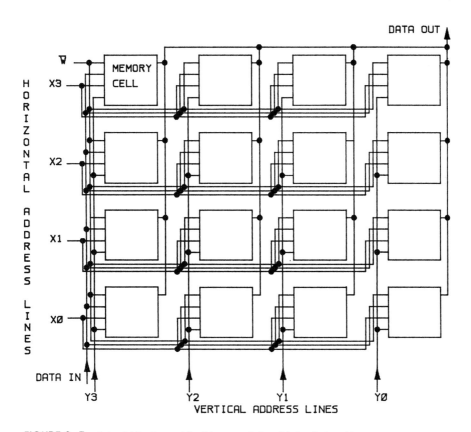

FIGURE 9–7 4 × 4 bit plane, 16 addresses, 1 data bit in, 1 data bit out

four external address inputs. These can be named A_0, A_1, A_2, and A_3, two of which form the horizontal address and two of which form the vertical address. The decoders are again internal to the memory chip, and the user need not worry which is horizontal and which is vertical. In other words, the internal memory chip structure is transparent to the user who presents four address bits, and one of 16 locations is selected. The memory bit plane of figure 9–7 is now extended to include four bit planes and the address decoder as in figure 9–8.

The memory of figure 9–8 can, of course, be expanded to provide more memory capacity as well as more bits per word. The address decoder expands accordingly. If the capacity is as large as 64K, then 16 address bits are required to be decoded to the two sets of 256 horizontal and vertical internal address lines, respectively. The number of addressable locations is 2^n, where n is the number of external address bits.

We cannot expect to have every conceivable memory size commercially available. A variety of different memory chips exist in modular form, and we use as many chips as are needed to construct a memory that satisfies our requirement. If a memory needs expansion, it is necessary to select one memory chip while deselecting others. The memory structure is therefore not complete unless a chip select

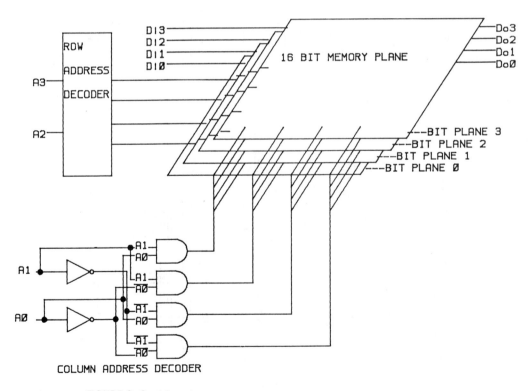

FIGURE 9–8 16×4 memory

Memory

line is included. When a chip is not active, it will neither respond to nor interfere with other chips in the system.

In addition to the above discussed input/output lines required to operate a memory chip there are several options.

Data lines need not provide separate input and output lines. Each data bit input and output line can be shared, saving I/O pins. When data is written to the memory, the output line need not be read; conversely, when data is read, the data input line is not needed. Read and write operations are mutually exclusive. In many cases in which input and output have a different source and destination it is convenient to have separate I/O lines, and chips with that feature are available. The shared I/O line is better adapted for the typical bus architecture found in digital systems.

Some chips have a special output enable line, which is useful for read multiplexing to a common destination.

Even though one chip select line is adequate, some chips include two chip select lines to facilitate chip select decoding.

To review the above discussion, a memory chip has

- n address inputs for 2^n locations and as many data lines as there are bits per word
  or twice as many lines as there are bits/word if I/O lines are not shared
- A read-write mode line
- One or more chip select lines
- An optional output enable line
- A power supply and ground

There is a choice in organizing a chip of given capacity by compromising between address space and word length. There is no optimum or recommended way to organize a memory, except for the fact that the number of pins required is a function of organization. It is desirable to minimize the number of pins to obtain the smallest possible chip for the largest possible bit capacity. For example, a chip has 1024 memory cells. These can be arranged to be in

- 1024 addresses but only one bit/word
- 512 addresses but two bits/word
- 256 addresses but four bits/word
- 128 addresses but eight bits/word
- 64 addresses but 16 bits/word, etc.

In table 9-1 we count the number of required pins for each of the above cases.

It becomes clear that smaller word length is more favorable. It is preferable to build a 1024 addresses $\times$ 8 bit word memory with eight 1024 $\times$ 1 chips than eight 128 $\times$ 8 chips. Yet each is available, the HM6518 by Harris and CDP1823 by RCA, respectively.

TABLE 9–1 Memory organization and required I/O pins

	1024	512	256	128	64
Address pins	10	9	8	7	6
Data pins (shared I/O lines)	1	2	4	8	16
Read-write mode pin	1	1	1	1	1
Chip select pins	1	1	1	1	1
Power pins	2	2	2	2	2
Total pins	15	15	16	19	26

Larger-capacity memory will require more pins and consequently larger packages. A 22-pin package, for example, may be organized as 64K × 1 like the IDT7187:

Address bits	16
Data in	1
Data out	1
Chip select	1
Read-write mode	1
Power	2
	22 pins

Dual in-line packages are the most popular. They come in the narrow 0.3″ wide style up to 22 pins. Beyond that the package is usually 0.6″ wide and occupies more "real estate."

We are now at a point at which we can design a memory system with given chips. Let us approach the subject one step at a time.

The chip has a fixed number of address bits and a fixed word length. To design a particular memory system, we might want to expand the memory to have

- More bits per address
- More addresses
- Both of the above

Suppose we work out simple examples for all three possibilities.

EXAMPLE 9–1

We have 1K × 1 memory chips and wish to design a 1K × 2 memory. Since both bits share the same memory address, it is logical that all address lines are shared. The data bits are separate, since they represent different bits, even if of the same two-bit word. The read-write command input $\overline{W}$ is shared, since both chips operate together either in the read or write mode. The chip select inputs $\overline{CS}$ may be tied to a true level = 0. (See figure 9–9.) ∎

―――――――――――――――――――――――――――――――――――― EXAMPLE 9–2

We have the same 1K × 1 memory chip and wish to design a 2K × 1 memory. In this case the input bit as well as the output bit are shared and therefore tied together. It is of course assumed that the output drivers are tristate such that only one selected chip uses the output line. All address lines are also connected together. The chip select line is used to select one chip while deselecting the other. Thus only one chip is active at a given time. This is decoded with a single inverter. When one chip select line is true, the other must be false.

Since we have a 2K memory, we really need 11 rather than 10 bits to address all locations. The chip has ten address lines, and the chip select becomes the eleventh address line. It is actually the most significant address bit of the two chip system. (See figure 9–10.)

The read-write inputs $\overline{W}$ are tied together because the deselected chip will not respond to either a read or write command. ∎

For simplicity, groups of address and data lines that are respectively connected are shown as wire busses rather than individual lines. That is often the practice to facilitate schematic reading.

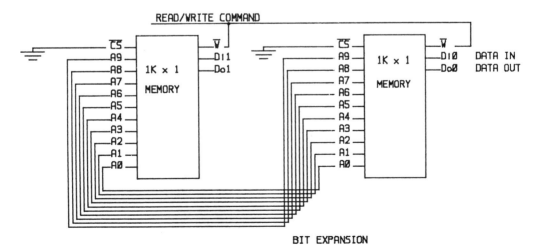

BIT EXPANSION

FIGURE 9–9 1K × 2 memory

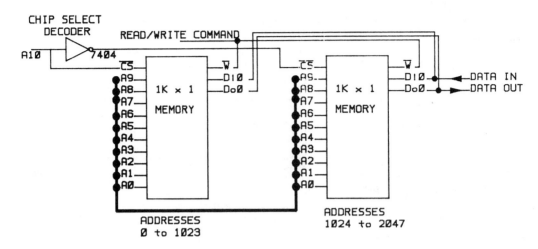

FIGURE 9–10 Address expansion 2K × 1 memory

EXAMPLE 9–3

We have 1K × 1 memory chips and wish to design a 2K × 2 memory. Now we need four chips, two chips to double the word length and two more to obtain double address space.

Looking at the last two memory implementations, we can establish a few observed rules:

1. Two output lines are shared as data bit 0, and the other two are shared as data bit 1. The same is true for the data input lines.

2. All address lines are connected together regardless of whether the system is expanded in the address or data fields.

3. The chip select of the chips that belong to the same address set are connected together. The different chip select sets are decoded such that if one chip select set is true we address locations 0 to 1023 and if the other chip select set is true we address locations 1024–2047.

4. The read-write command input is shared. This is independent of memory organization.

These rules, as we shall see, apply to larger systems as well. Figure 9–11 shows the required system interconnections. ■

EXAMPLE 9–4

We choose a small memory chip like the RCA CDP1822 (see the appendix to this chapter), which is 256 addresses deep and 4 bits wide, that is, a 256 × 4 memory. We wish to design a memory that is 1K addresses deep and 8 bits wide, that is, a

Memory

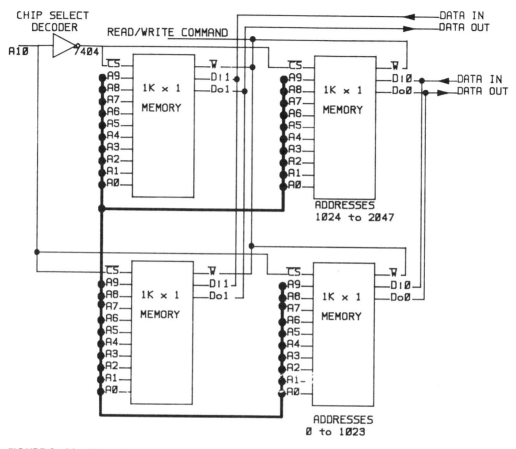

FIGURE 9-11 2K × 2 memory

1K × 8 memory system. (See figure 9–12.) Here two chip select lines are available: CS_2 active positive and $\overline{CS_1}$ active negative. The availability of two chip select lines simplifies the chip select decode logic. Follow the logic of A_9 and A_8 and convince yourself that one set of two chips is selected for each of the four possible A_9/A_8 combinations. ∎

EXAMPLE 9-5

Figure 9–13 expands an IDT 6116, a 2K × 8 memory chip to a 16K × 8 system. Each chip has 11 address lines, A_0 to A_{10}; one chip select line, $\overline{CS}$; eight data I/O lines, I/O_0 to I/O_7; read-write mode, $\overline{W}$; and an output enable line, $\overline{OE}$. Including power supply and ground, we have a 24-pin chip. We need eight chips for the system and a chip select decoder to select one of eight chips. A 74138 is a 3-line to 8-line decoder, thus expanding the address capability from 11 to 14 address lines. Fourteen address lines uniquely locate one of 16,384 locations within the memory, since $2^{14} = 16,384$. As in previous examples,

278   Chapter 9

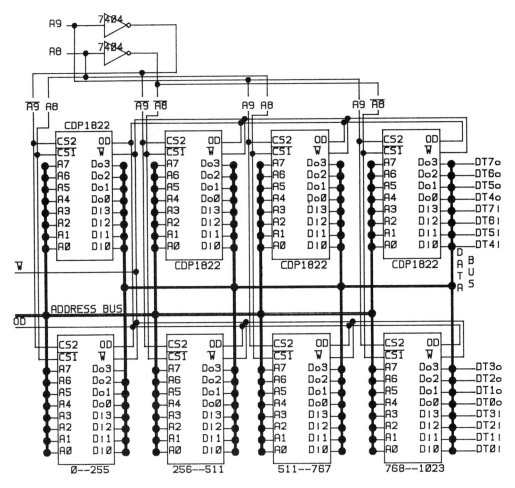

FIGURE 9–12 1K × 8 memory; two chip select inputs allow simpler decoding

1. All like address lines of the memory chips are tied together; they are shared.
2. All read-write mode lines are shared.
3. All like I/O lines are shared.
4. The chip select lines are decoded. ∎

EXAMPLE 9–6

The previous example can be repeated with an IDT6168 4K × 4 memory rather than the IDT6116 2K × 8 memory. The capacity of the chip is the same, namely 16K, but the memory organization is different. The chip has more address lines but four fewer data lines, and the output enable line is omitted. Thus the chip is reduced to 20 pins rather than 24 pins as in Example 9–5. The chip select decoder

Memory

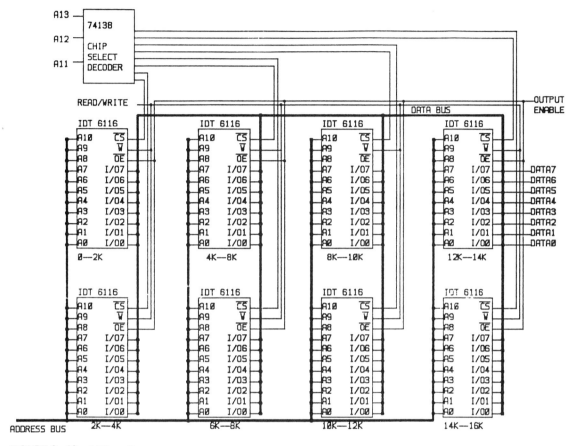

FIGURE 9–13 16K × 8 memory

is also different. Here we need a 2-line to 4-line decoder, since the address expansion is from 12 to 14 bits. A 74139 has two sections of 2-line to 4-line decode. We need only one section. The same interconnect rules as before apply. Figure 9–14 shows the arrangement. ■

9–4 MEMORY TIMING

Attention must be paid to memory timing. There is a specified sequence of events that, if not adhered to, will cause the write or read operation to malfunction, or worse, to destroy the contents of the memory.

Timing diagrams like figures 9–15 and 9–16 are always included in the memory chip data sheets (see the appendix to this chapter). The double set of lines indicates address or data changes, not a fixed pattern.

T_{rc} is the read cycle time and refers to the minimum time required to hold the address true to obtain an output reading.

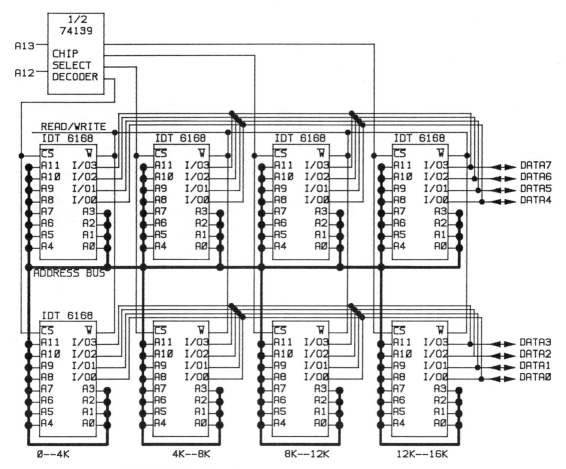

FIGURE 9–14 16K × 8 memory

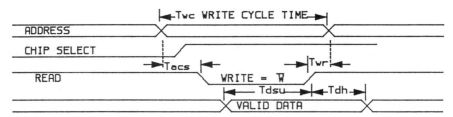

Tacs ADDRESS TO WRITE SET-UP TIME
Tdsu DATA TO WRITE SET-UP TIME
Tdh DATA HOLD FROM WRITE
Twr WRITE REASE TIME

FIGURE 9–15(a) Read cycle waveforms

Memory

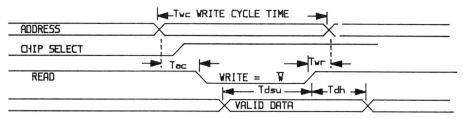

FIGURE 9-15(b) Write cycle waveforms

Chip select can but must not necessarily become true at the same time as the rest of the address. When both chip select and address are true, then T_{cx} refers to the delay before the tristate gates change to a low-impedance state. That does not mean that the memory has located the right address as yet; the output might be active, but the correct data is not yet available.

T_{co} refers to the delay before valid output data is available. During read mode the read line is typically high. In that mode the memory content cannot be changed; thus none of the timing is critical as far as the memory is concerned. The device receiving the information must wait for the maximum specified T_{co} time before accepting the output data.

In write mode, careful adherence to timing is required.

T_{wc} is the write cycle time and refers to the minimum time required to hold the address true to enter data into the memory.

T_{acs} refers to the address to chip select setup time. It is not critical as long as the read-write mode line $\overline{W}$ is still in read mode.

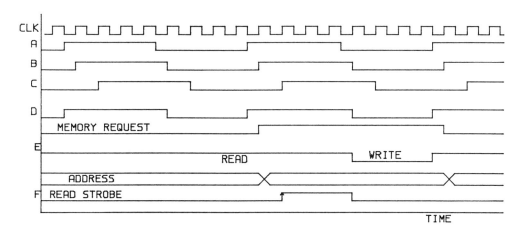

FIGURE 9-16 Read-write timing

TABLE 9–2 Memory timing specifications

	Read Mode		Write Mode
T_{rc}	160 nsec	T_{wc}	160 nsec
T_{cx}	100 nsec	T_{acs}	0 nsec
T_{co}	110 nsec	T_{wr}	40 nsec
		T_{dsu}	50 nsec
		T_{dh}	10 nsec

T_{wr} is the write release time and is *critical*. The write mode should be released before the address is changed; otherwise, it is possible to also write the data into some other location, destroying the existing content.

T_{dsu} refers to the time when input data must be valid. It is advisable to have the data ready as early as possible.

T_{dh} refers to the time required to hold the data beyond the time that the write mode is released. This ensures that the contents of the written address are not altered just before release.

Typical values for a high-speed CMOS memory chip are given in table 9–2.

It is, of course, important to observe all timing limitations in order to avoid faulty memory operation, but particular attention must be paid to two limitations. In read mode, data can be strobed out only after data is available, that is, after T_{co}. For the delay times quoted in table 9–2 this is 110 nsec. In write mode, data must be held 10 nsec and the address 40 nsec past write mode = active.

With these criteria in mind we can design a read-write timing control circuit for a particular example.

EXAMPLE 9–7

Assume that we want to access a memory address, read the content, and then write some other data into the same address. This is known as a read-modify-write mode. (See figure 9–17.)

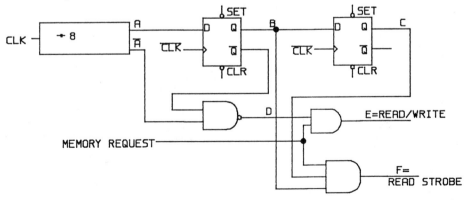

FIGURE 9–17 Read-write timing logic

Memory

If we have a 5 MHz clock in the system and the read-write cycle time is 1.6 μs, dividing the clock, which is 200 nsec per cycle, by eight will result in the correct 1.6 μs memory cycle time. Let's call it waveform A. (See figure 9–16.) This can be delayed by 100 nsec or half a clock cycle by clocking another flip flop with $\overline{\text{Clock}}$. We get waveform B. Address changes are controlled with B.

The OR function of $A + B = D$ results in a shortened write command pulse, allowing 100 nsec write release time, satisfying T_{wr}.

Wave form C is another 200 nsec delayed from B. The leading edge of C is used to strobe the valid read data onto a bus or register, satisfying T_{co}.

9–5 APPLICATION EXAMPLES

Memories are not only used for data storage and retrieval, but can also be an active subsystem of more sophisticated processors.

EXAMPLE 9–8 Ping-Pong Memory

High-speed processors are often faced with the problem of continually updating the memory content as well as continued readout. There is no time to do both simultaneously. Furthermore, the updating process is in synchronism with a processor or data acquisition system, while the readout must respond to timing requirements of some external device. The memory becomes a data buffer between two systems. That problem is solved by putting two memories in parallel. While one memory is in a processing mode, the other reads out the results of a previous process with an address scan counter. At chosen intervals the mode of the two memories is switched. This is called memory *ping-ponging*. Each, the processor and the scan counter, has access to both memories in a double-pole double-throw switch fashion. The output data bus must be similarly switched between the two memories. As usual, the switching is accomplished with tristate gates as in figure 9–18. ∎

EXAMPLE 9–9 Sectioned Memory

It is sometimes necessary to access a memory for read or write at a faster rate than the advertised cycle time permits. Such is the case when an image stored in memory is to be displayed at video speed. A television image is conveniently generated with 512 × 512 pixels. Each pixel is one point on the screen with an intensity determined by an eight-bit ''bit pattern.'' To avoid flicker on the screen, the image is repeated 30 times a second. The total number of pixels accessed per second is 512 × 512 × 30 = 7,884,320 pixels, or approximately 127 nsec per pixel. The required access time falls just within the limits of T_{co} in table 9–2. A higher-resolution image of 1024 × 1024 pixels, however, requires access times that are four times faster.

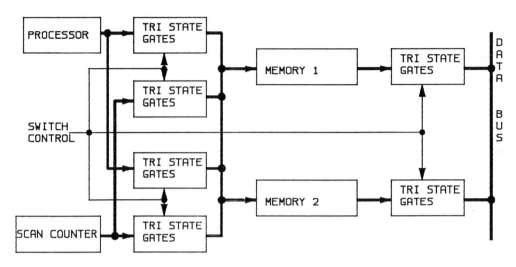

FIGURE 9-18 Ping-pong buffer memory

This problem can be overcome by separating the address bus and memory into four sections while keeping the output data bus common. Suppose we have a 32 MHz clock and derive 8 MHz thereof with 125 nsec per cycle. Each memory section is accessed in 125 nsec intervals, which is adequate for read access time, but in overlapping fashion as in figure 9-19. The output of each section is sampled in 31.25 nsec intervals, with the 32 MHz clock.

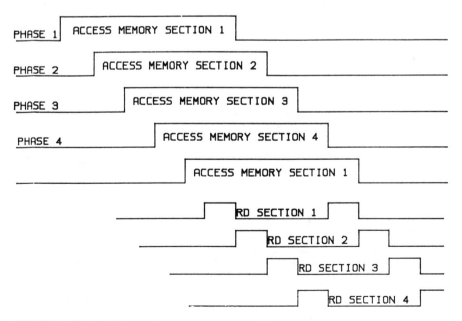

FIGURE 9-19(a) High-speed memory access method

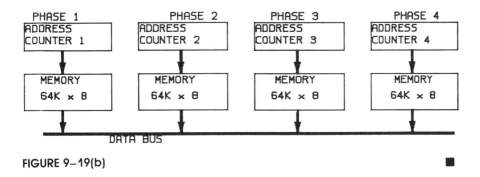

FIGURE 9–19(b)

──────────────── EXAMPLE 9–10 Histogram Generator

A frequent application in data acquisition systems is the accumulation of statistical patterns. Suppose we throw a die repeatedly, and with every throw one of six numbers comes up. A memory with only six locations is initially cleared. The first throw yields a number, say 3. Address 3 is accessed, and a 1 is entered, indicating that the number 3 has come up once. The second throw yields a 5. Address 5 is accessed, and again a 1 is entered. The third throw again comes up with a 3. Address 3 is accessed, and the content increased by one. Now address 3 contains a 2, indicating that the number 3 has been found twice. After n repetitions the six memory locations contain the frequency of occurrence for each number. The statistical distribution is obtained by scanning the memory sequentially from address 1 to address 6. (See figure 9–20.)

Such distributions, called histograms, can be generated with an accumulating memory. One application is found in nuclear instrumentation.

Suppose an instrument in outer space samples the type and energy content of free particles encountered. Every time a particle hits a detector, a pulse is generated with an amplitude proportional to energy. The pulse height is converted with an A/D converter to eight bits or 256 energy levels. Just as in the previous die experiment, one of 256 numbers comes up. That number becomes the memory address, and the content of that address is increased by one.

Suppose the expected rate of particle encounters is 4000 per second. We sample for 1 s, then look at the distribution. We must allow for a maximum accumulation of 4000 samples in one bin, an unlikely but possible situation. The memory must therefore be 12 bits wide ($2^{12} = 4096$) and 256 addresses long. We need a 256 × 12 memory.

The task at hand can be accomplished in three steps:

1. The incoming data determines the address; the memory is in read mode.
2. The contents of the present address is loaded into an incrementer and updated.
3. The mode is changed to write, and the incremented content is rewritten into the same address.

This is a read-modify-write operation.

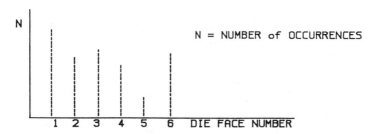

FIGURE 9–20 Statistical distribution (histogram)

One way to increment the data obtained in the read mode is to load the data into a counter (in parallel mode) then count up one. (See figure 9–21.) Another is to feed the data into an adder and load the adder output into a register. The counter (or register) output must be isolated from the memory I/O line with tristate gates. This is so because in read mode the memory data lines are in an output mode and

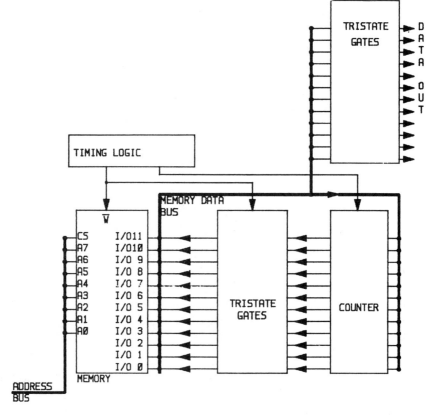

FIGURE 9–21 The address is multiplexed between an event counter for writing into the memory or microprocessor for data readout.

Memory

cannot compete with the counter output. An output shall not feed an output; this constitutes a memory data bus contention. As soon as the memory is in write mode, though, the memory I/O lines are in an input mode and will accept the counter output.

At the end of the accumulation period the memory address bus ceases to receive data from the event processor and is switched to a scan counter. The memory is now in read mode only and sends data out. At the same time the output data tristate gates are enabled, and the distribution data are fed to an appropriate output device such as a display or microprocessor system. ■

EXAMPLE 9–11   First-in–First-out Memory (FIFO)

There are special memory applications in which it is not necessary to access every location upon demand, but rather information is to be stored for later removal in the order in which it was entered. The first word entered is also the first word removed. Such a memory is called a *first-in–first-out memory* (FIFO). At first glance, such a scheme is nothing but a shift register n bits wide (for n bits per word) and with as many stages as there are words to be stored. The difficulty is that data arrives from a processor at one rate and is removed by some other system at some other rate. The two system clocks might not even be synchronized. If the removal rate is consistently slower than the arrival rate, the shift register memory will eventually fill up and be unable to accommodate further storage. If the removal rate is consistently faster than the data arrival rate, the shift register memory will fill only one location or none and is not needed in the first place. The usual application is when both average arrival and removal rates are approximately equal, but there are times when the removal system is not quite ready at the same time as the processor, and time buffering is required. Data can be entered only at the top and can be removed only at the bottom of the memory. The following characteristics of a FIFO emerge:

1. Data entered into an empty memory shall fall to the bottom as fast as possible and wait there until removed.
2. The next set of data shall follow but stop one location before the bottom (not destroy, or write over, the data previously entered).
3. As long as no data is removed, the new data just stacks from the bottom up.
4. As soon as one data word is removed, the entire remaining data stack moves one position down.
5. When data arrival and data removal are simultaneous, the data stack moves down, as above, and the new data follows as fast as possible to the top of the stack.

The concept is simple, but the design of the timing requirements is more complicated and falls into the category of sequential logic design. Figure 9–22 shows the system concept with a timing control block. The detailed timing logic for a fast and slower version will be derived in the chapter on sequential logic (Chapter 11). ■

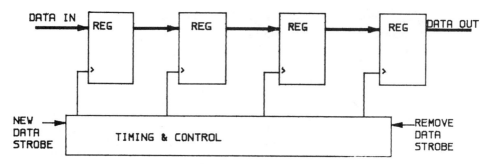

FIGURE 9-22 First-in-first out (FIFO) memory

EXAMPLE 9-12 Last-in-First-out Memory (LIFO)

Last-in-first-out memory is known as LIFO. While the FIFO found application as a buffer, the LIFO is used as a *stack* in microprocessors. When a program is interrupted by an interrupt and shifted to a service routine, or by a call to a subroutine, a return address must be stored for later reference, and interim data must be saved for later use. If several such interrupts occur, a corresponding number of return addresses and save data are *pushed* onto the stack. The system returns to the main program in the reverse order in which the interrupts occurred. The last data resides at the top of the stack. When recalled, the last data *pops* out of the top location. In microprocessor language this is known as a *push and pop operation*.

One could, of course, design a shift register memory similar to the FIFO with the capability of shifting left as well as right. A push instruction would enter new data at the top, while shifting all other data one position down the stack. A pop instruction would remove data from the top of the stack and move all other data one position up the stack. The LIFO is usually not accessed at high speed, nor does it operate between two nonsynchronized systems as does the FIFO. It is therefore simpler to assign a RAM section of the system memory to serve as a stack and address successive locations, rather than physically moving data from one memory location to the next. A special register, called the *stack pointer*, holds the first or starting address of the stack and is incremented and decremented in accordance with push and pop instructions. ■

EXAMPLE 9-13 Table Look-up Memory

Mathematical tables are a frequent application. Rather than computing trigonometric values, for example, it is much faster to address a memory location and read the desired value. Such tables can be stored as a permanent reference. At other times a programmer can compute a special table, load RAM with the result, and then use that section of RAM for later execution.

Another application may be encoding and decoding. The data at hand becomes the address to the memory; the contents of that address is the desired encoded data. Or a programmer might wish to display a plot of values spanning decades. A logarithmic plot is called for. A set of logarithmic values is stored in RAM. Incoming data are the address to the RAM. The output is the transformed data. Such tables are often called *transfer characteristic tables*. A wide range of translation or encoding problems can be handled in this fashion. ∎

9–6 READ-ONLY MEMORIES

Many functions within a computer system (or a special instrument) require repeated execution of a given algorithm or a program that is a permanent, unalterable part of a system. It is most convenient to store the permanent instructions in a set of locations, point to the program or algorithm when needed, execute the instructions sequentially, and exit when done. The content of such a memory need never be changed. Not only is the content of that memory permanent, but it must not be lost when power to the system is turned off. We talk of a *read-only memory* or ROM.

The internal cell structure of such a memory is therefore different from the ordinary RAM. It is usually a switch-controlled matrix with the row lines as the input and column lines as outputs.

The simplest matrix we can think of is a 2 × 2 matrix, shown in Example 9–14 and figure 9–23 with two address locations.

EXAMPLE 9–14

Suppose the desired content of the ROM is

    01 for address 0

    11 for address 1

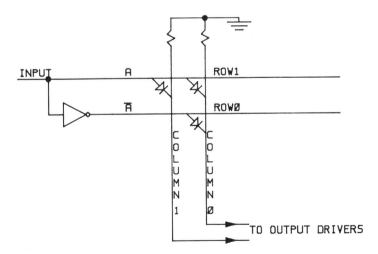

FIGURE 9–23 2 × 2 ROM

One address bit decoded with an inverter will select either A or $\overline{A}$ to be true. If address 0 is selected, $\overline{A}$ is true, let's say high; then row 0 is high, and so is column 0 because of the closed connection at the intersection of row 0 and column 0. Column 1 will remain at 0 because row 0 and column 1 are not connected. The columns read 01.

If address 1 is selected, A is high, that is, row 1 is high. Row 1 is connected to both columns 0 and 1; therefore both columns 0 and 1 are high, and the columns read 11.

The number of bits per output word is equal to the number of columns, and the number of memory addresses is equal to the number of rows. This is, of course, a one-dimensional address scheme and practical only for a small address set. ∎

A second example extends the concept to an excess 3 encoder.

EXAMPLE 9–15

The encoder of figure 9–24 is easily understood. If a 5 is to be encoded to excess 3, an 8 is expected. In other words,

Input = 0101

Output = 1000

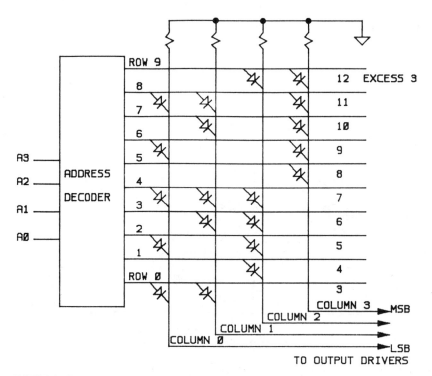

FIGURE 9–24 Read-only memory excess 3 encoder

Memory

The input ($= 5$) goes directly to the address decoder, which in turn sets row 5 high. Row 5 is connected to column 4 only. Thus column 4 is pulled high, while columns 1, 2, and 3 remain low. Reading the four output columns, we see 1000. The row column connections simply reflect the desired output pattern for a given row. The entire pattern for inputs 0 to 9 is thus "imprinted" in the matrix interconnection pattern in Figure 9–24. ∎

For larger memories, two-dimensional addressing as in RAMs is appropriate. An example will be shown later.

There are three types of read-only memories:

ROM: Read-only memory

PROM: Programmable read-only memory

EPROM: Erasable programmable read-only memory

They differ from each other by the kind of contact between row and column.

The ROM is *mask-programmable,* that is, during the final steps of manufacturing a mask determines which of the row-column contacts are to be left in place or etched away, leaving the desired interconnect pattern. The ROM program is included in the manufacturing process.

The PROM uses a fusible link material for row-column contacts. The contact can later be opened, allowing the data pattern to be determined by the user after the device has been manufactured. Figure 9–25(a) shows a typical fuse cell. A heavier than normal current applied to a selected row while the columns are in a particular data pattern will blow the fuses in selected locations. Once programmed, the chip cannot be changed.

The EPROM has the additional advantage of being erasable. The contact element is a floating gate, avalanche injection, charge storage device. The floating gate is electrically isolated, then charged by avalanche injection to create a conductive channel. Control is maintained by a second transistor in series as shown in figure 9–25(b). The trapped charges can maintain the conductive channel for years.

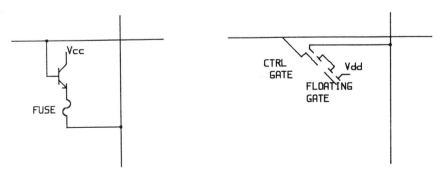

FIGURE 9–25(a)   Typical fuse cell   **FIGURE 9–25(b)**   Erasable contact

When irradiated with ultraviolet light, however, the trapped charges dissipate, opening the connection, and the chip can be reprogrammed.

Recent developments provide an electric path for charge removal. This type is called electrically erasable programmable read-only memory (EEPROM). It is easier to work with, since errors in programming are quickly corrected.

Two-dimensional addressing is done by selecting both rows and columns and allowing only one bit per memory plane. There are as many bits per word as there are bit planes. The rows are selected in the same fashion as the one-dimensional address scheme in figure 9–23. Columns are selected by enabling an output gate associated with the column of interest. Figure 9–26 shows the arrangement.

A commonly used EPROM is the $2K \times 8$ 2716 by Intel, Hitachi, National, TI, Thomson and OKI. The memory cells are interconnected to form a split 128×128 cell matrix as shown in figure 9–27. Each section is further divided into four subsections of 16×128 cells each. Each of these subsections is connected to a column decoder, which selects one of 16 columns. A sense amplifier and an output buffer, which is associated with the particular bit, send the data out. The block diagram for a single-bit data flow is shown in figure 9–28.

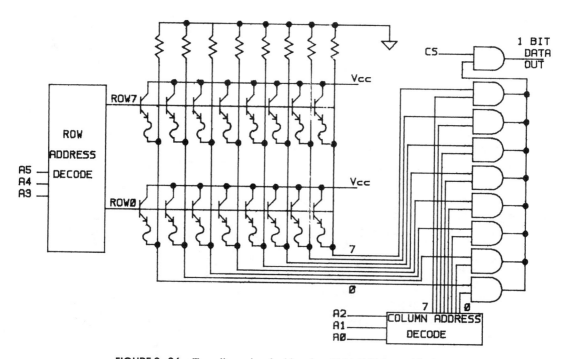

FIGURE 9–26 Two-dimensional addressing 64-bit ROM, one bit plane (Courtesy of Intel Corporation)

Memory

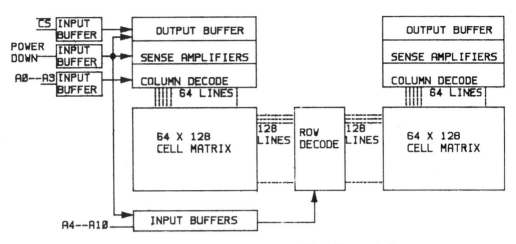

FIGURE 9-27 2716 Intel ROM block diagram (Courtesy of Intel Corporation)

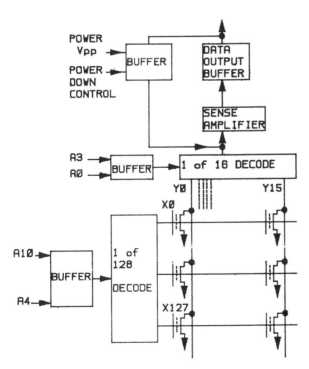

FIGURE 9-28 2716 ROM SINB (single-bit data flow) (Courtesy of Intel Corporation)

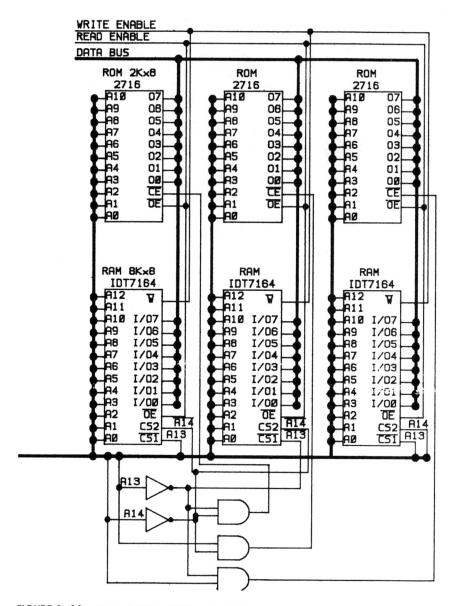

FIGURE 9–29 6K × 8 ROM, 24K × 8 RAM

Memory

Programming an EPROM

This is done with an EPROM programmer. The desired data pattern can be entered from a computer or manually. A start control activates the timing chain to generate the required address, data setup, and hold times. In manual operation a hex address is selected, the corresponding data is entered, and the programmer advances to the next location. For verification a verify mode will step through all the addresses, allowing for visual verification of the memory contents. The programmer also has a duplicating mode. This is a most convenient feature. In entering data it is most probable that some locations will contain errors. A second EPROM chip is inserted in the programmer, and the program from the initially programmed latched chip is copied up to but not including the error location. The proper data is manually entered, and the copy process can continue to the next error location.

Erasing Data

The recommended erasure procedure for the 2716 is exposure to shortwave ultraviolet light that has a wavelength of 2537 Å. The erasure time is approximately 20 minutes using a 12,000 $\mu W/cm^2$ ultraviolet lamp.

Constant exposure to room-level fluorescent lighting will erase the data in approximately 3.5 years. One month of direct exposure to sunlight will do the same. It is therefore a good idea to cover the erase window with an opaque tape to protect the data.

In many applications the EPROM is used only to download a program from EPROM to RAM. After a system is thus initialized, there is no further need for the EPROM. The EPROM includes a power-down mode to save system power consumption.

A typical application in figure 9–29 combines EPROM and RAM on a computer address, data, and control bus.

9–7 DYNAMIC MEMORIES

The RAM memory applications discussed use *static* memory cells, that is, once a cell is in a 1 or 0 state, it remains there until changed by rewriting. *Dynamic* memory cells store the information in a capacitor associated with a gate. However, the charge will leak out. To preserve the entered information, all cells of the memory system must be refreshed approximately every 2 ms. This is inconvenient, since all memory read-write operations must be periodically interrupted to accommodate a refresh cycle. On the other hand, dynamic memories use less power and less real estate, thus packing more memory capacity at a lower cost into each chip. Static memory capacity and power consumption per chip are improving with every year. As the performance gap between static and dynamic memories narrows, the former finds increased favor with the logic designer.

Dynamic memories up to 256K bits of storage are available. This is accomplished with only a 16-pin chip. The saving of pins is done by entering the address in two steps. Row address and column address are separately latched into the device for either a read or write operation. Thus only 9 rather than 18 address pins are needed. A row address strobe ($\overline{RAS}$) and a column address strobe ($\overline{CAS}$) latch the complete address into an address buffer. In spite of the additional step of accessing the memory, the access time is as fast as 120 nsec.

Dynamic memories come also in small capacities of 256×5 such as the TM4256 or intermediate capacity such as Intel's 2116. Let us look at internal structure of the 2116 as shown in figure 9–30.

The address is entered in two steps into the two seven-bit latches. There are two sets of 128 columns, and each column has an associated sense amplifier. There are two sets of 64 rows. Either the upper or lower set of 64 rows is selected with row bit 6. The intersection of the selected row and column determines which of the sense amplifiers to read in read mode or which cell to write into in write mode.

The sense amplifier compares the level of an addressed cell to a reference cell. The level stored in this reference is less than the minimum allowable stored high level and greater than the maximum allowable stored low level. During a sensing operation the degraded level of a cell is restored to its original full or no charge. This is also what is done during refresh.

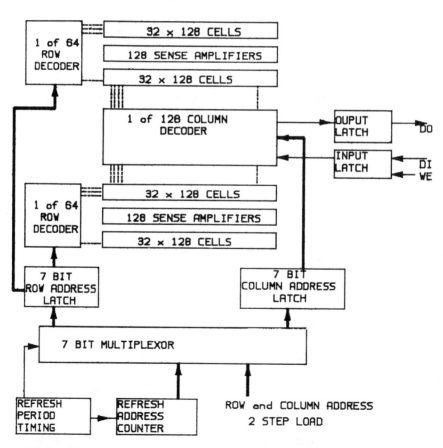

FIGURE 9–30 Intel 2116 dynamic memory (Courtesy of Intel Corporation)

Each of the 128 columns is activated every 2 ms to restore leaked-off charge. Each refresh cycle may take as much as 500 nsec. The total time lost for refresh is 64 μs or 3.2% of the available memory time. The loss of memory time is acceptable in most cases, but the system operation interruption is often difficult to accommodate. The logic for memory refresh requires an address multiplexor, which switches at the appointed time from system to the refresh address generator.

9–8 CORE MEMORIES

Until recently, core memories dominated the market for fast mass storage of information. They are still much in use but find decreased application in new designs. We shall nevertheless describe the basic architecture and operation. Core memories are arranged in three-dimensional fashion much like solid-state memories, that is, in bit planes as in figures 9–7 and 9–8.

Each plane is a matrix of cores rather than solid-state latches. There are as many planes as there are bits in a word. The number of addressable locations (or words) is equal to the number of bits on a plane. Cores are as small as 10 mils in outer diameter. Thus a 3″ × 3″ square can accommodate a matrix of 256 × 256 cores or 65,536 addresses. A small sample shown in figure 9–31 is a 4 × 4 matrix, a 16-bit plane.

It is evident from figures 9–2 and 9–3 that the magnetic flux changes direction if and only if enough current is applied to overcome the hysteresis. I_{sw} is the switching current. If less than the switching current is applied, the flux remains unchanged. This property allows two-dimensional coincident selection. One wire enters each core from the X direction, the other from the Y direction. If a core is to be switched, that is, a flux reversal is to take place, then the sum of the current in both wires must exceed the hysteresis threshold. Each energized wire carries only half the required switching current. As in the solid-state case, when an address is selected, one wire in each dimension is energized. The two wires intersect only in one core. That core receives enough current to react. All other cores in the affected row and column receive only half as much current and will not react. It can be considered a current AND function. Thus with a given current direction the core will flip to a known flux direction, which can be looked at as a 1. The reverse current under the same address condition will flip the core back to a 0 state. So far we have accomplished only the writing of a 1 into a core with a subsequent reset. Merely coincident-selecting a core accomplished the writing of a 1. To write a 0, a third wire through the core is needed. When a core is coincident-selected and at the same time a reverse $I_{sw}/2$ current is applied to the third wire, the total current is insufficient for switching. Thus the core remains in a 0 state. That third current is called an *inhibit current* and the third wire an *inhibit wire*.

To repeat

- If a core is selected and an inhibit current is not applied, a 1 is written.
- If a core is selected and an inhibit current is applied, a 0 is written.

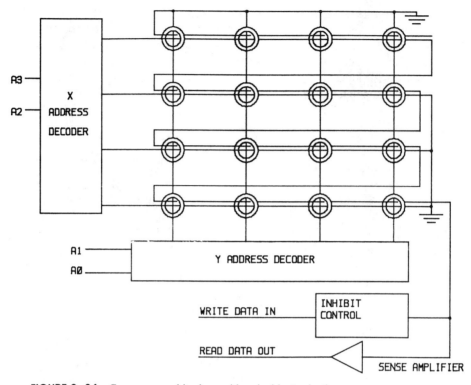

FIGURE 9-31 Core memory bit plane with coincident selection

$$\text{Write 1:} \quad \frac{1}{2}I + \frac{1}{2}I + 0*I > I_{sw}$$

$$\text{Write 0:} \quad \frac{1}{2}I + \frac{1}{2}I - \frac{1}{2}I < I_{sw}$$

The above discussion assumed that the core in question was in a 0 state before we attempted to write a bit. This, of course, might not be the case. It is therefore necessary to reset all addressed cores (one addressed core in each bit plane) to 0 by first applying a reverse current to the address lines. This sets all cores of that address to 0. The *write cycle* must consist of two operations: clear followed by write.

The read operation is done by selecting a core with a reverse current in both the *X* and *Y* dimensions. As was stated before, all addressed cores will reset to 0. The third wire or inhibit wire is now used as a *sense wire*. The third wire is therefore called *sense inhibit wire*. If a core was in a 1 state and flips to a 0 state as a result of the reverse current, a current pulse is induced in the sense wire. This pulse is amplified with a sense amplifier and sets a flip flop to 1. If a core was in a 0 state, no pulse is induced, and a 0 is set in the flip flop.

Memory

Notice that all addressed cores were inherently reset to 0, and the information is lost. This is called destructive readout. The content of that address is actually transferred to the output flip flops, more commonly known as the information register. The content of that address is usually restored by rewriting the very same cores.

For ordinary nondestructive readout the *read cycle* must consist of a read followed by a restore.

The read and clear operations are really identical, and so are the write and restore operations. The difference between a read and a write cycle lies only in the use of the sense inhibit wire. In a read cycle the sense inhibit wire measures the flux change and allows the data to be entered into the information register. In a write cycle the sense inhibit wire is activated in accordance with the data to be written from the information register.

The write sequence of events is as follows:

1. The address is applied to the address decoder. As in the solid-state case the address is decoded from 2^N to N X-lines and N Y-lines internal to the memory system.
2. The write data must be available in the information register.
3. The write mode must be established.
4. A cycle initiate pulse is sent to start the process.
5. The memory control logic clears the addressed location, then writes the data.
6. A busy status is sent to the user register.

The read sequence of events is as follows:

1. The address is applied to the address decoder.
2. The read mode must be established.
3. A cycle initiate pulse is sent to start the process.
4. The data is read and strobed into the information register with a data available status.
5. The data is restored (unless different data is to be written or the user wishes to clear the memory).
6. A busy status is sent to the user register.

A typical read-write cycle takes 800 nsec. In the read mode the information is available to the user before the complete cycle is done, as indicated by the data available status.

The memory is also often used in a read-modify-write mode when different data than just read is being restored. This saves time if the old information is of no further use.

An example of a small core memory is shown in figure 9–32. Each plane consists of nine 16 × 8 matrices and there are five bit planes. The total number of bits is 5,760. Notice the IDT 2K × 8 static CMOS RAM in the same picture.

FIGURE 9–32 Comparison of a magnetic core memory and a CMOS RAM

9–9 BUBBLE MEMORY

Magnetic storage does not necessarily require an individual core for each bit. A small magnetic domain can be induced locally on a larger magnetic surface. This is done by applying an external magnetic field to a thin film of ferromagnetic crystal. The induced magnetic domain is a *bubble*. It is separated from an adjacent bubble by what is called a *Bloch wall*. The ferromagnetic sheet lies between two permanent magnets, which give it an initial magnetic polarization. When local magnetization is reversed, it is designated as a 1. The absence of such a reversal is designated as a 0. Two sets of orthogonal coils (arranged at right angles to each other) induce a rotating magnetic field when excited by 90 degree phase-shifted currents. This rotating field drags the bubbles with it. The bubbles thus move in shift register fashion. (See figure 9–33.)

Of course, the bubble memory is not random access. The information is serially available similar to a disk. Data is serially arranged in single bit streams and propagate at a modest rate of 100 KHz. The bubble memory is a contender for the rotating disk. Mechanical structures such as disks are not optimum in a marine or airborne environment. Instead of physically moving the storage medium, the bubble memory remains mechanically at rest while the information moves in rotating motion.

Figure 9–34 shows one quadrant of an Intel 7110. The memory is composed of the following elements:

- 80 storage loops of 4096 bits each. This is 327,680 bits. The excess storage above 256,000 provides redundancy and extra storage for error correction.

Memory

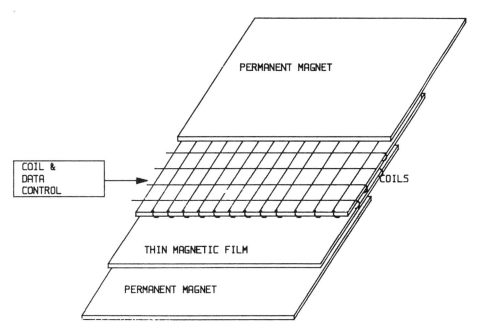

FIGURE 9–33 Magnetic bubble memory construction (Courtesy of Intel Corporation)

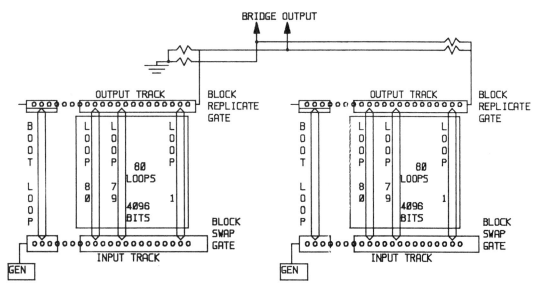

FIGURE 9–34 Intel 7110 bubble memory loop architecture (Courtesy of Intel Corporation)

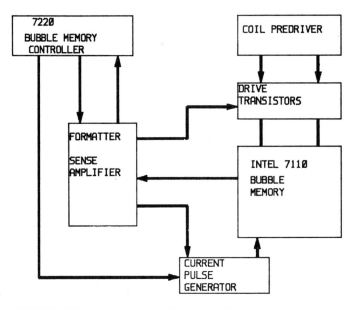

FIGURE 9–35 Block diagram of single-bubble memory system, 128K bytes

- A replicating generator that either writes new information in or recirculates old data.
- The input track and swap gates, which transfer the bubble from the input track to the respective loop.
- The output track and replicate gate, which transfer the bubble from the loop to the output track and the output detector.
- The boot loop, which returns information for recirculation.
- The detector, which senses the bubble and produces an output voltage on the order of a few millivolts, which are sent to the sense amplifier.

The memory requires several support chips as shown in figure 9–35.

9–10 ASSOCIATIVE MEMORIES

The storage systems discussed in this chapter dealt with information stored in predetermined locations. Either the programmer or the computer under control of a memory management program chooses the memory address. When information is to be retrieved, the computer is directed to that specific address. The associative memory stores information without keeping track of the specific location. Information is subsequently retrieved by investigating the content. The memory is content addressed.

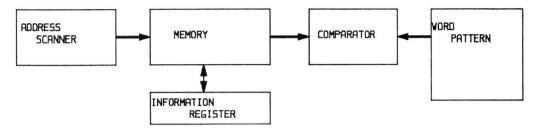

FIGURE 9-36 Associative memory organization

This is probably not too different from the way in which we store information in our own brains. We never know where a particular set of information is located within our brain cell structure. We search our brain until we associate the information found with what we are looking for, hence the term *associative memory*. Of course, we do not know how we do our memory search. A method must be found to do that search in an electronic memory. Suppose we have a particular word pattern and we want to find its exact or closest match within the memory. One way is to load a descriptor into a register and proceed to scan the entire memory, word by word, until the match to the descriptor is located. This is too time consuming. Another way to approach the problem is to scan the memory for a match in the most significant position only. Some addresses will match that requirement; others will not. Those that do not will not be searched again. Those that do are next scanned and compared to the next lesser significant bit. In that manner the memory is scanned with decreasing sets of interest, for a "one-bit compare" as often as there are bits in a word. With every bit compare, a set of nonmatching locations is eliminated until the last scan yields either one exact match or several close matches. What is considered a close match depends on the problem. Moreover, the descriptor might point only to specific bits within a word. Many schemes to approach this problem have been proposed, depending on the application and the amount of complexity one can afford.

Such memory structures (see figure 9-36) find application in learning computers or self-organizing systems that imitate or study biological nervous systems. Associative computers are computers that use associative memories. Such computers can learn from their own set of experiences, learn to recognize patterns, and carry out functions as a result of their own internal, rather than externally applied, programming.

9-11 CCD MEMORY

Shift registers can be looked at as memory or as a digital delay line. Such devices are also needed in the analog world. Of course one can always build an analog delay line with appropriate L-C networks. The idea here, however, is to shift analog information down a line in discrete steps similar to a clocked digital shift register.

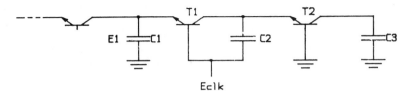

FIGURE 9–37 Charge deficit transfer in a bucket brigade

The circuit of figure 9–37 stores analog information while shifting in discrete steps. The voltage stored in capacitor C_1 is transferred via the current source of a transistor to capacitor C_2 within half a clock cycle and further transferred to C_3 within the next half clock cycle. The two-stage operation is needed in order to have alternate empty capacitors for further charge transfer. (In a digital shift register, bits are transferred out of a position and a previous stage into that position simultaneously.)

The amount of charge transfer can be explained by the following set of statements:

1. The charge on a capacitor C_1 is $Q_{c1} = E_1 \times C_1$ when T_1 is turned off and where E_1 is the input voltage.
2. The charge on capacitor C_1 changes to $Q_{c1} = E_{clk} \times C_1$ when T_1 is turned on and where E_{clk} is the clock voltage.
3. The change of charge of capacitor C_1 is therefore

$$dQ = C_1(E_{clk} - E_1)$$

4. The current for that change of charge is delivered by the collector current of T_1, which means that an equivalent charge dQ has been removed from C_2.
5. The original charge of capacitor C_2 was

$$Q = E_{clk} \times C_2$$

with the collector at zero potential.
6. The final charge of capacitor C_2 is the original charge minus the change of charge:

$$Q_{c2} = E_{clk} \times C_2 - C_1(E_{clk} - E_1) = E_1 \times C_1 \quad \text{if } C_1 = C_2$$

This process is called *charge deficit transfer* and the circuit a *bucket brigade*, reminiscent of fire control before the days of fire engines.

This concept has been implemented in monolithic structures. The devices are called charge-coupled devices or CCDs. The capacitors can also be charged directly if a photosensitive dielectric is included so that the capacitors assume a charge proportional to incident light. In that case an entire array of capacitors can receive an image that is shifted out after an appropriate integration time (several milliseconds). The array is arranged by rows and columns; rows are shifted out in

Memory

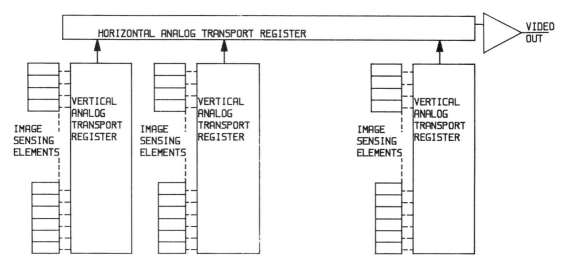

FIGURE 9–38 CCD202 100 × 100 area image sensor (Courtesy of Fairchild CCD Imaging)

sequential manner as in a television image display. Such *image memories* are widely in use for image data processing work in conjunction with computer analysis.

The Fairchild obsolete CCD202, for example, was a 100 × 100 pixel array, capable of clocking data at a 6 MHz rate. It was arranged as in figure 9–38 and was designed for use in videocameras that require low power consumption.

REVIEW QUESTIONS

1. What is a *D* type latch? How does it differ from a *D* type master/slave flip flop?
2. How does the CMOS latch work?
3. How is information stored in a magnetic core?
4. Which of the basic memory cells are volatile?
5. What is meant by hysteresis?
6. Discuss one-dimensional versus two-dimensional addressing.
7. What is a bit plane?
8. Discuss an address decoder.
9. Discuss chip architecture versus chip pinout requirements (in terms of word length and address space).
10. Why can all data outputs of all locations on one bit plane be tied together, that is, share one output line?
11. Why can all data inputs to each location on a bit plane be tied together, that is, share one input line?

12. If addressing is two-dimensional, why are these memories called three-dimensional?
13. What function does the chip select line serve?
14. Discuss the read-write mode line.
15. Discuss read-write cycle timing.
16. Why is it critical to take the memory out of write mode before allowing the address to change?
17. What is a histogram?
18. Discuss a read-modify-write mode.
19. The memory increment scheme of figure 9–21 incorporates tristate gates. Why is the isolation needed?
20. Discuss the high-speed access method of figure 9–19.
21. What is meant by memory ping-ponging?
22. What is a FIFO?
23. What is a LIFO?
24. What is a table look-up memory and how is it used?
25. Discuss ROM and PROM.
26. Can you think of an application in which the use of EPROMs is not recommended?
27. How does one protect an EPROM from losing its content?
28. What are the pros and cons of dynamic memories?
29. How is information stored in a dynamic memory?
30. How does one overcome the capacitor discharge problem in a dynamic memory?
31. Why is a core memory read destructive?
32. In what way are read and write cycles in a core memory similar?
33. Why are address lines in a core memory common to all bit planes while the sense inhibit line is unique to each bit plane?
34. What is a bubble memory?
35. How do bubbles propagate?
36. How does an associative memory differ from a conventional memory?
37. How does analog information propagate in a CCD memory?

PROBLEMS

1. Draw a schematic for a 4 × 4 one-dimensional memory.
2. Draw one bit plane for a 16-bit two-dimensional array.

Memory

3. How many input address lines are needed for a
   (a) 1K × 1 memory
   (b) 1K × 4 memory
   (c) 4K × 8 memory
   (d) 16K × n memory
   (e) 512K × n memory

4. How many horizontal and vertical address lines are there internal to the memories of Problem 3?

5. How many bit planes are there in a
   (a) 1K × 8 memory?
   (b) 512K × 8 memory?

6. Include an address decoder for Problem 2. Design the decoder on a gate level.

7. A given 1K × 1 memory chip has one chip select line, $\overline{CS}$. Design a chip select decode network for
   (a) 16K × 1 memory
   (b) 16K × 8 memory
   (c) 64K × 8 memory

8. A given 4K × 2 memory has two chip select lines, CS_1 and $\overline{CS_2}$. Design a chip select decode network for
   (a) 16K × 4 memory
   (b) 1024K × 8 memory

9. Design a 4K × 2 memory using a Texas Instrument MS 4027 (without read-write timing control) having the following features: 4K × 1 organization; one chip select line; shared I/O line.

10. Include read-write control logic in Problem 9.

11. Design a 64K × 16 memory using an IDT 6116 chip.

12. Design a 2K × 4 memory using an RCA MWS 5101.

13. Design the control logic for a sectioned memory of figure 9–19.

14. Design the address bus input multiplexor for figure 9–21. The address input switches between a processor and a scan counter.

15. How deep must the FIFO of figure 9–22 be to function without data loss if the average data in and data out rate is the same, data arrives at intervals of 1–2 μs, and data is removed at a rate varying from 0.5 to 16 μs. What function would a FIFO full or empty status play?

16. Write the binary pattern for a 16 × 16 address table look-up memory that transforms incoming data into:
    (a) Excess 3 code
    (b) Gray code
    (c) Binary to BCD
    (d) BCD to binary
    (e) Logarithmic
    (f) Degrees to radians

17. Draw a one-dimensional address ROM as in figure 9–23 that will encode 0 to 15 (0000 to 1111) into a Gray code:

    0000 to 0000
    0001   0001
    0010   0011
    0011   0010
    0100   0110
    0101   0111
    0110   0101
    0111   0100
    1000   1100
    1001   1101
    1010   1111
    1011   1110
    1100   1010
    1101   1011
    1110   1001
    1111   1000

18. Draw a one-dimensional address ROM as in figure 9–24 that will contain the word ELECTRONICS in ASCII code:

    E 1000101
    L 1001100
    E 1000101
    C 1000011
    T 1010100
    R 1010010
    O 1001111
    N 1001110
    I 1001001
    C 1000011
    S 1010011

19. Draw a two-dimensional address PROM as in figure 9–26 that will encode binary to BCD. Indicate the blown fuses.

    0000 to 0 0000
    0001   0 0001
    0010   0 0010
    0011   0 0011

0100	0 0100
0101	0 0101
0110	0 0110
0111	0 0111
1000	0 1000
1001	0 1001
1010	1 0000
1011	1 0001
1100	1 0010
1101	1 0011
1110	1 0100
1111	1 0101

20. Draw a memory system for a 64 × 16 RAM and a 2K × 16 EPROM. Show the data line interconnect, address line interconnect, chip select logic, and read-write mode interconnect. Use bus lines as in figure 9–29 where appropriate.

CHAPTER 9 APPENDIX: DATA SHEETS

RCA CMOS LSI Products
CDP1822, CDP1822C

256-Word by 4-Bit LSI Static Random-Access Memory

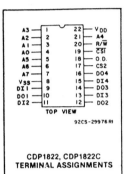

CDP1822, CDP1822C
TERMINAL ASSIGNMENTS

Features:
- Low operating current — 8 mA at V_{DD} = 5 V and cycle time = 1 μs
- Industry standard pinout
- Two Chip-Select inputs — simple memory expansion
- Memory retention for standby battery voltage of 2 V min.
- Output-Disable for common I/O systems
- 3-State data output for bus-oriented systems
- Separate data inputs and outputs

The RCA-CDP1822 and CDP1822C are 256-word by 4-bit static random-access memories designed for use in memory systems where high speed, low operating current, and simplicity in use are desirable. The CDP1822 features high speed and a wide operating voltage range. Both types have separate data inputs and outputs and utilize single power supplies of 4 to 6.5 volts for the CDP1822C and 4 to 10.5 volts for the CDP1822.

Two Chip-Select inputs are provided to simplify system expansion. An Output Disable control provides Wire-OR capability and is also useful in common Input/Output systems. The Output Disable input allows these RAMs to be used in common data Input/Output systems by forcing the output into a high-impedance state during a write operation independent of the Chip-Select input condition. The output assumes a high-impedance state when the Output Disable is at high level or when the chip is deselected by $\overline{CS1}$ and/or CS2.

The high noise immunity of the CMOS technology is preserved in this design. For TTL interfacing at 5-V operation, excellent system noise margin is preserved by using an external pull-up resistor at each input.

The CDP1822 and CDP1822C types are supplied in 22-lead hermetic dual-in-line side-brazed ceramic packages (D suffix), in 22-lead dual-in-line plastic packages (E suffix). The CDP1822C is also available in chip form (suffix H).

OPERATIONAL MODES

MODE	INPUTS				OUTPUT
	Chip Select 1 $\overline{CS_1}$	Chip Select 2 CS_2	Output Disable OD	Read/ Write $R/\overline{W}$	
Read	0	1	0	1	Read
Write	0	1	0	0	Data In
Write	0	1	1	0	High Impedance
Standby	1	X	X	X	High Impedance
Standby	X	0	X	X	High Impedance
Output Disable	X	X	1	X	High Impedance

Logic 1 = High   Logic 0 = Low   X = Don't Care

(Courtesy of GE Solid State)

Memory

1800-Series Memories
CDP1822, CDP1822C

DYNAMIC ELECTRICAL CHARACTERISTICS at $T_A = -40$ to $+85°C$, $V_{DD} \pm 5\%$,
Input $t_r, t_f = 20$ ns, $V_{IH} = 0.7\ V_{DD}$, $V_{IL} = 0.3\ V_{DD}$, $C_L = 100$ pF

CHARACTERISTIC		V_{DD} (V)	CDP1822 Min.†	CDP1822 Typ.*	CDP1822 Max.	CDP1822C Min.†	CDP1822C Typ.*	CDP1822C Max.	UNITS
Write Cycle Times (Fig. 2)									
Write Cycle	t_{WC}	5	500	—	—	500	—	—	
		10	300	—	—	—	—	—	
Address Setup	t_{AS}	5	200	—	—	200	—	—	
		10	110	—	—	—	—	—	
Write Recovery	t_{WR}	5	50	—	—	50	—	—	
		10	40	—	—	—	—	—	
Write Width	t_{WRW}	5	250	—	—	250	—	—	
		10	150	—	—	—	—	—	
Input Data Setup Time	t_{DS}	5	250	—	—	250	—	—	
		10	150	—	—	—	—	—	ns
Data In Hold	t_{DH}	5	50	—	—	50	—	—	
		10	40	—	—	—	—	—	
Chip-Select 1 Setup	$t_{\overline{CS1}S}$	5	200	—	—	200	—	—	
		10	110	—	—	—	—	—	
Chip-Select 2 Setup	t_{CS2S}	5	200	—	—	200	—	—	
		10	110	—	—	—	—	—	
Chip-Select 1 Hold	$t_{\overline{CS1}H}$	5	0	—	—	0	—	—	
		10	0	—	—	0	—	—	
Chip-Select 2 Hold	t_{CS2H}	5	0	—	—	0	—	—	
		10	0	—	—	0	—	—	
Output Disable Setup	t_{ODS}	5	200	—	—	200	—	—	
		10	110	—	—	—	—	—	

†Time required by a limit device to allow for the indicated function.
*Typical values are for $T_A = 25°C$ and nominal V_{DD}.

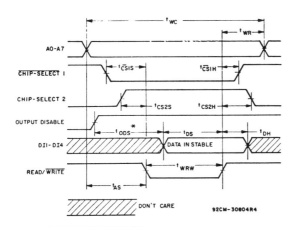

* t_{ODS} IS REQUIRED FOR COMMON I/O OPERATION ONLY; FOR SEPARATE I/O OPERATIONS, OUTPUT DISABLE IS DON'T CARE.

(Courtesy of GE Solid State)

RCA CMOS LSI Products
CDP1822, CDP1822C

DYNAMIC ELECTRICAL CHARACTERISTICS at $T_A = -40$ to $+85°C$, $V_{DD} \pm 5\%$, Input $t_r, t_f = 20$ ns, $V_{IH} = 0.7\,V_{DD}$, $V_{IL} = 0.3\,V_{DD}$, $C_L = 100$ pF

CHARACTERISTIC		V_{DD} (V)	CDP1822			CDP1822C			UNITS
			Min.†	Typ.*	Max.	Min.†	Typ.*	Max.	
Read Cycle Times (Fig. 1)									
Read Cycle	t_{RC}	5	450	—	—	450	—	—	
		10	250	—	—	—	—	—	
Access from Address	t_{AA}	5	—	250	450	—	250	450	
		10	—	150	250	—	—	—	
Output Valid from Chip-Select 1	t_{DOA1}	5	—	250	450	—	250	450	
		10	—	150	250	—	—	—	
Output Valid from Chip-Select 2	t_{DOA2}	5	—	250	450	—	250	450	ns
		10	—	150	250	—	—	—	
Output Active from Output Disable	t_{DOA3}	5	—	—	200	—	—	200	
		10	—	—	110	—	—	—	
Output Hold from Chip-Select 1	t_{DOH1}	5	20	—	—	20	—	—	
		10	20	—	—	—	—	—	
Output Hold from Chip-Select 2	t_{DOH2}	5	20	—	—	20	—	—	
		10	20	—	—	—	—	—	
Output Hold from Output Disable	t_{DOH3}	5	20	—	—	20	—	—	
		10	20	—	—	—	—	—	

†Time required by a limit device to allow for indicated function.
*Typical values are for $T_A = 25°C$ and nominal V_{DD}.

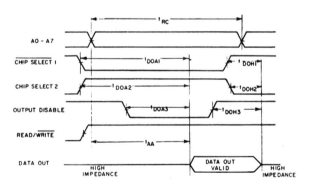

Read cycle timing waveforms.

(Courtesy of GE Solid State)

RCA CMOS LSI Products
MWS5101DL2, MWS5101DL3, MWS5101EL2, MWS5101EL3

256-Word by 4-Bit LSI Static Random-Access Memory

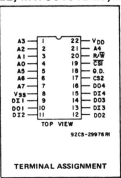

TERMINAL ASSIGNMENT

Features:
- Industry standard pinout
- Very low operating current — 8 mA at V_{DD} = 5 V and cycle time = 1 μs
- Two Chip-Select inputs — simple memory expansion
- Memory retention for standby battery voltage of 2 V min.
- Output-Disable for common I/O systems
- 3-State data output for bus-oriented systems
- Separate data inputs and outputs

The RCA-MWS5101 is a 256-word by 4-bit static random-access memory designed for use in memory systems where high speed, very low operating current, and simplicity in use are desirable. It has separate data inputs and outputs and utilizes a single power supply of 4 to 6.5 volts.

Two Chip-Select inputs are provided to simplify system expansion. An Output Disable control provides Wire-OR capability and is also useful in common Input/Output systems. The Output Disable input allows these RAMs to be used in common data Input/Output systems by forcing the output into a high-impedance state during a write operation independent of the Chip-Select input condition. The output assumes a high-impedance state when the Output Disable is at high level or when the chip is deselected by $\overline{CS1}$ and/or CS2.

The high noise immunity of the CMOS technology is preserved in this design. For TTL interfacing at 5-V operation, excellent system noise margin is preserved by using an external pull-up resistor at each input.

For applications requiring wider temperature and operating voltage ranges, the mechanically and functionally equivalent static RAM, RCA-CDP1822, may be used.

The MWS5101 types are supplied in 22-lead hermetic dual-in-line side-brazed ceramic packages (D suffix), in 22-lead dual-in-line plastic packages (E suffix), and in chip form (H suffix).

OPERATIONAL MODES

MODE	Chip Select 1 $\overline{CS}_1$	Chip Select 2 CS_2	Output Disable OD	Read/ Write $R/\overline{W}$	OUTPUT
READ	0	1	0	1	Read
WRITE	0	1	0	0	Data In
WRITE	0	1	1	0	High Impedance
STANDBY	1	X	X	X	High Impedance
STANDBY	X	0	X	X	High Impedance
OUTPUT DISABLE	X	X	1	X	High Impedance

Logic 1 = High   Logic 0 = Low   X = Don't Care

(Courtesy of GE Solid State)

Chapter 9

General-Purposes Memories

MWS5101DL2, MWS5101DL3, MWS5101EL2, MWS5101EL3

MAXIMUM RATINGS, *Absolute-Maximum Values:*

DC SUPPLY-VOLTAGE RANGE (V_{DD})
   (All voltage referenced to V_{SS} terminal) .... −0.5 to −7 V
INPUT VOLTAGE RANGE, ALL INPUTS .... −0.5 to V_{DD} + 0.5 V
DC INPUT CURRENT, ANY ONE INPUT .... ±10 mA
POWER DISSIPATION PER PACKAGE (P_D):
   For T_A = −40 to +60°C (PACKAGE TYPE E) .... 500 mW
   For T_A = +60 to +85°C (PACKAGE TYPE E) .... Derate Linearly at 12 mW/°C to 200 mW
   For T_A = −55 to +100°C (PACKAGE TYPE D) .... 500 mW
   For T_A = +100 to +125°C (PACKAGE TYPE D) .... Derate Linearly at 12 mW/°C to 200 mW
DEVICE DISSIPATION PER OUTPUT TRANSISTOR
   FOR T_A = FULL PACKAGE-TEMPERATURE RANGE (All Package Types) .... 100 mW
OPERATING-TEMPERATURE RANGE (T_A):
   PACKAGE TYPE D .... −55 to +125°C
   PACKAGE TYPE E .... −40 to +85°C
STORAGE TEMPERATURE RANGE (T_{stg}) .... −65 to +150°C
LEAD TEMPERATURE (DURING SOLDERING):
   At distance 1/16 ± 1/32 inch (1.59 ± 0.79 mm) from case for 10 s max. .... +265°C

OPERATING CONDITIONS at T_A = Full Package-Temperature Range
For maximum reliability, operating conditions should be selected so that operation is always within the following ranges:

CHARACTERISTIC	LIMITS ALL TYPES		UNITS
	Min.	Max.	
DC Operating-Voltage Range	4	6.5	V
Input Voltage Range	V_{SS}	V_{DD}	

STATIC ELECTRICAL CHARACTERISTICS at T_A = 0 to 70°C, V_{DD} = 5 V ±5%.

CHARACTERISTIC		TEST CONDITIONS		LIMITS MWS5101D MWS5101E			UNITS
		V_O (V)	V_{IN} (V)	Min.	Typ.•	Max.	
Quiescent Device Current, I_{DD}	L2 Types	−	0.5	−	25	50	μA
	L3 Types	−	0.5	−	100	200	
Output Voltage: Low-Level, V_{OL}		−	0.5	−	0	0.1	V
High-Level, V_{OH}		−	0.5	4.9	5	−	
Input Low Voltage, V_{IL}		−	−	−	−	1.5	
Input High Voltage, V_{IH}		−	−	3.5	−	−	
Output Low (Sink) Current, I_{OL}		0.4	0.5	2	4	−	mA
Output High (Source) Current, I_{OH}		4.6	0.5	−1	−2	−	
Input Current,▲ I_{IN}		−	0.5	−	−	±5	
3-State Output Leakage Current,* I_{OUT}	L2 Types	0.5	0.5	−	−	±5	μA
	L3 Types	0.5	0.5	−	−	±5	
Operating Current, I_{DD1}#		−	0.5	−	4	8	mA
Input Capacitance, C_{IN}		−	−	−	5	7.5	pF
Output Capacitance, C_{OUT}		−	−	−	10	15	

• Typical values are for T_A = 25°C and nominal V_{DD}.
▲ All inputs in parallel.
* All outputs in parallel.
# Outputs open-circuited; cycle time=1 μs.

(Courtesy of GE Solid State)

RCA CMOS LSI Products
MWS5101DL2, MWS5101DL3, MWS5101EL2, MWS5101EL3

DYNAMIC ELECTRICAL CHARACTERISTICS at T_A = 0 to 70°C, V_{DD} = 5 V ±5%, t_r, t_f = 20 ns, V_{IH} = 0.7 V_{DD}, V_{IL} = 0.3 V_{DD}, C_L = 100 pF

CHARACTERISTIC		LIMITS MWS5101D, MWS5101E						UNITS
		L2 Types			L3 Types			
		Min.†	Typ.●	Max.	Min.†	Typ.●	Max.	
Read Cycle Times (Fig. 1)								
Read Cycle	t_{RC}	250	–	–	350	–	–	
Access from Address	t_{AA}	–	150	250	–	200	350	
Output Valid from Chip-Select 1	t_{DOA1}	–	150	250	–	200	350	
Output Valid from Chip-Select 2	t_{DOA2}	–	150	250	–	200	350	
Output Active from Output Disable	t_{DOA3}	–	–	110	–	–	150	ns
Output Hold from Chip-Select 1	t_{DOH1}	20	–	–	20	–	–	
Output Hold from Chip-Select 2	t_{DOH2}	20	–	–	20	–	–	
Output Hold from Output Disable	t_{DOH3}	20	–	–	20	–	–	

† Time required by a limit device to allow for the indicated function.
● Typical values are for T_A = 25°C and nominal V_{DD}.

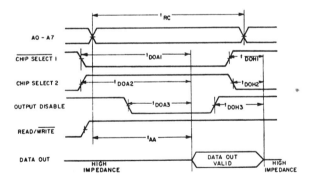

Read cycle timing waveforms.

(Courtesy of GE Solid State)

1800-Series Memories
CDP1822, CDP1822C

DYNAMIC ELECTRICAL CHARACTERISTICS at $T_A = -40$ to $+85°C$, $V_{DD} \pm 5\%$,
Input $t_r, t_f = 20$ ns, $V_{IH} = 0.7\ V_{DD}$, $V_{IL} = 0.3\ V_{DD}$, $C_L = 100$ pF

CHARACTERISTIC		V_{DD} (V)	CDP1822 Min.†	Typ.*	Max.	CDP1822C Min.†	Typ.*	Max.	UNITS
Write Cycle Times (Fig. 2)									
Write Cycle	t_{WC}	5	500	—	—	500	—	—	
		10	300	—	—	—	—	—	
Address Setup	t_{AS}	5	200	—	—	200	—	—	
		10	110	—	—	—	—	—	
Write Recovery	t_{WR}	5	50	—	—	50	—	—	
		10	40	—	—	—	—	—	
Write Width	t_{WRW}	5	250	—	—	250	—	—	
		10	150	—	—	—	—	—	
Input Data Setup Time	t_{DS}	5	250	—	—	250	—	—	ns
		10	150	—	—	—	—	—	
Data In Hold	t_{DH}	5	50	—	—	50	—	—	
		10	40	—	—	—	—	—	
Chip-Select 1 Setup	$t_{\overline{CS}1S}$	5	200	—	—	200	—	—	
		10	110	—	—	—	—	—	
Chip-Select 2 Setup	t_{CS2S}	5	200	—	—	200	—	—	
		10	110	—	—	—	—	—	
Chip-Select 1 Hold	$t_{\overline{CS}1H}$	5	0	—	—	0	—	—	
		10	0	—	—	0	—	—	
Chip-Select 2 Hold	t_{CS2H}	5	0	—	—	0	—	—	
		10	0	—	—	0	—	—	
Output Disable Setup	t_{ODS}	5	200	—	—	200	—	—	
		10	110	—	—	—	—	—	

†Time required by a limit device to allow for the indicated function.
*Typical values are for $T_A = 25°C$ and nominal V_{DD}.

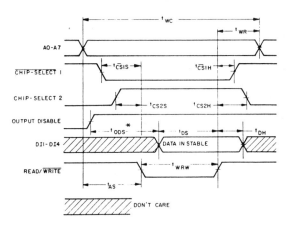

Write cycle timing waveforms.

(Courtesy of GE Solid State)

Memory

CMOS STATIC RAMS
16K (2K x 8-BIT)

IDT6116SA
IDT6116LA

FEATURES:
- High-speed
  - Military — 35/45/55/70/90/120/150ns (max.)
  - Commercial — 30/35/45/55/70/90ns (max.)
- Low-power operation
  - IDT6116SA
    Active: 180mW (typ.)
    Standby: 100µW (typ.)
  - IDT6116LA
    Active: 160mW (typ.)
    Standby: 20µW (typ.)
- Battery backup operation — 2V data retention voltage (LA version only)
- Produced with advanced CEMOS™ high-performance technology
- CEMOS process virtually eliminates alpha particle soft-error rates (with no organic die coatings)
- Single 5V (±10%) power supply
- Input and output directly TTL-compatible
- Static operation: no clocks or refresh required
- Standard 24-pin DIP, 24-pin THINDIP or plastic DIP, 28- and 32-pin LCC, or 24-Lead Flatpack
- Military product available 100% screened to MIL-STD-883, Class B

DESCRIPTION:
The IDT6116SA/LA is a 16,384-bit high-speed static RAM organized as 2K x 8. It is fabricated using IDT's high-performance, high-reliability technology — CEMOS. This state-of-the-art technology, combined with innovative circuit design techniques, provides a cost-effective alternative to bipolar and fast NMOS memories.

Access times as fast as 30ns are available with maximum power consumption of only 495mW. The circuit also offers a reduced power standby mode. When $\overline{CS}$ goes high, the circuit will automatically go to, and remain in, a standby power mode as long as $\overline{CS}$ remains high. In the standby mode, the low-power device consumes less than 20µW typically. This capability provides significant system level power and cooling savings. The low-power (L) version also offers a battery backup data retention capability where the circuit typically consumes only 1µW to 4µW operating off of a 2V battery.

All inputs and outputs of the IDT6116SA/LA are TTL-compatible and operation is from a single 5V supply, simplifying system designs. Fully static asynchronous circuitry is used, requiring no clocks or refreshing for operation, providing equal access and cycle times for ease of use.

The IDT6116SA/LA is packaged in either a 24-pin, 600 and 300 mil DIPs or 32- and 28-pin leadless chip carriers, providing high board-level packing densities.

The IDT6116SA/LA Military RAM is 100% processed in compliance to the test methods of MIL-STD-883, Method 5004, making it ideally suited to military temperature applications demanding the highest level of performance and reliability.

PIN CONFIGURATIONS

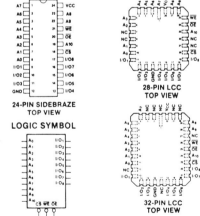

A_0-A_{10}	ADDRESS	$\overline{WE}$	WRITE ENABLE
I/O1-I/O8	DATA INPUT/OUTPUT	$\overline{OE}$	OUTPUT ENABLE
$\overline{CS}$	CHIP SELECT	GND	GROUND
V_{CC}	POWER		

FUNCTIONAL BLOCK DIAGRAM

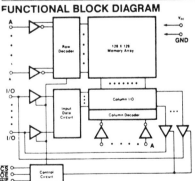

(Courtesy of IDT Corporation)

IDT6116SA/IDT6116LA CMOS STATIC RAMS 16K (2K x 8-BIT) MILITARY AND COMMERCIAL TEMPERATURE RANGES

AC ELECTRICAL CHARACTERISTICS (V_{CC} = 5V ± 10%, All Temperature Ranges)

SYMBOL	PARAMETER	6116SA30[1] 6116LA30[1] MIN.	MAX.	6116SA35/45[4] 6116LA35/45[4] MIN.	MAX.	6116SA55 6116LA55 MIN.	MAX.	6116SA70 6116LA70 MIN.	MAX.	6116SA90 6116LA90 MIN.	MAX.	6116SA120[2] 6116LA120[2] MIN.	MAX.	UNIT
READ CYCLE														
t_{RC}	Read Cycle Time	30	—	35/45	—	55	—	70	—	90	—	120	—	ns
t_{AA}	Address Access Time	—	30	—	35/45	—	55	—	70	—	90	—	120	ns
t_{ACS}	Chip Select Access Time	—	30	—	35/45	—	50	—	65	—	90	—	120	ns
t_{CLZ}	Chip Select to Output in Low Z[3]	5	—	5	—	5	—	5	—	5	—	5	—	ns
t_{OE}	Output Enable to Output Valid	—	18	—	20/25	—	40	—	50	—	65	—	80	ns
t_{OLZ}	Output Enable to Output in Low Z[3]	5	—	5	—	5	—	5	—	5	—	5	—	ns
t_{CHZ}	Chip Deselect to Output in High Z[3]	—	18	—	20/25	—	30	—	35	—	40	—	40	ns
t_{OHZ}	Output Disable to Output in High Z[3]	—	18	—	20/25	—	30	—	35	—	40	—	40	ns
t_{OH}	Output Hold from Address Change	5	—	5	—	5	—	5	—	5	—	5	—	ns
WRITE CYCLE														
t_{WC}	Write Cycle Time	30	—	35/45	—	55	—	70	—	90	—	120	—	ns
t_{CW}	Chip Select to End of Write	20	—	25/30	—	40	—	40	—	55	—	70	—	ns
t_{AW}	Address Valid to End of Write	20	—	25/30	—	45	—	65	—	80	—	105	—	ns
t_{AS}	Address Setup Time	0	—	0	—	5	—	15	—	15	—	20	—	ns
t_{WP}	Write Pulse Width	15	—	20/25	—	40	—	40	—	55	—	70	—	ns
t_{WR}	Write Recovery Time	0	—	0	—	5	—	5	—	5	—	5	—	ns
t_{OHZ}	Output Disable to Output in High Z[3]	—	18	—	20/25	—	30	—	35	—	40	—	40	ns
t_{WHZ}	Write to Output in High Z[3]	—	18	—	20/25	—	30	—	35	—	40	—	40	ns
t_{DW}	Data to Write Time Overlap	15	—	15/20	—	25	—	30	—	30	—	35	—	ns
t_{DH}	Data Hold from Write Time	0	—	0	—	5	—	5	—	5	—	5	—	ns
t_{OW}	Output Active from End of Write[3]	0	—	0	—	0	—	0	—	0	—	0	—	ns

NOTES:
1. 0°C to +70°C temperature range only.
2. -55°C to +125°C temperature range only.
3. This parameter guaranteed but not tested.
4. Data is preliminary for military devices only.

(Courtesy of IDT Corporation)

Memory

IDT6116SA/IDT6116LA CMOS STATIC RAMS 16K (2K x 8-BIT) MILITARY AND COMMERCIAL TEMPERATURE RANGES

TIMING WAVEFORMS OF READ CYCLE NO. 1[(1)]

READ CYCLE 2[(1,2,4)]

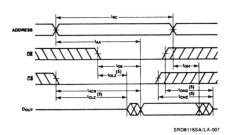

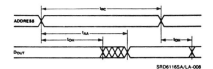

READ CYCLE 3[(1,3,4)]

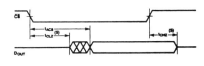

NOTES:
1. $\overline{WE}$ is High for Read Cycle.
2. Device is continuously selected, $\overline{CS} = V_{IL}$.
3. Address valid prior to or coincident with $\overline{CS}$ transition low.
4. $\overline{OE} = V_{IL}$.
5. Transition is measured ±500mV from steady state with 5pF load (including scope and jig). This parameter is sampled and not 100% tested.

TIMING WAVEFORMS OF WRITE CYCLE 1[(1)] ($\overline{WE}$ CONTROLLED)

TIMING WAVEFORMS OF WRITE CYCLE 2[(1)] ($\overline{CS}$ CONTROLLED)

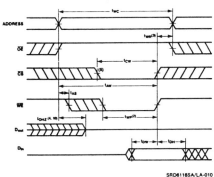

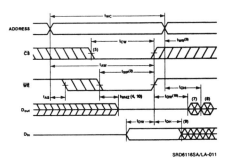

NOTES:
1. $\overline{WE}$ must be high during all address transitions.
2. A write occurs during the overlap (t_{WP}) of a low $\overline{CS}$ and a low $\overline{WE}$.
3. t_{WR} is measured from the earlier of $\overline{CS}$ or $\overline{WE}$ going high to the end of write cycle.
4. During this period, I/O pins are in the output state so that the input signals of opposite phase to the outputs must not be applied.
5. If the $\overline{CS}$ low transition occurs simultaneously with the $\overline{WE}$ low transitions or after the $\overline{WE}$ transition, outputs remain in a high impedance state.
6. $\overline{OE}$ is continuously low ($\overline{OE} = V_{IL}$).
7. D_{OUT} is the same phase of write data of this write cycle.
8. D_{OUT} is the read data of next address.
9. If $\overline{CS}$ is low during this period, I/O pins are in the output state. Then the data input signals of opposite phase to the outputs must not be applied to them.
10. Transition is measured ±500mV from steady state with a 5pF load (including scope and jig). This parameter is sampled and not 100% tested.

(Courtesy of IDT Corporation)

CMOS STATIC RAMS
16K (4K × 4 BIT)

IDT6168SA
IDT6168LA

FEATURES:
- High-speed (equal access and cycle time)
  —Military: 25/35/45/55/70/85/100ns (max.)
  —Commercial: 20/25/35/45/55ns (max.)
- Low-power consumption
  —IDT6168SA
     Active: 225mW (typ.)
     Standby: 100µW (typ.)
  —IDT6168LA
     Active: 225mW (typ.)
     Standby: 10µW (typ.)
- Battery backup operation—2V data retention voltage (IDT6168LA only)
- Available in high-density 20-pin DIP, 20-pin plastic DIP and 20-pin leadless chip carriers
- Produced with advanced CEMOS™ high-performance technology
- CEMOS process virtually eliminates alpha particle soft-error rates (with no organic die coatings)
- Bidirectional data input and output
- Single 5V (±10%) power supply
- Input and output directly TTL-compatible
- Three-state output
- Static operation: no clocks or refresh required
- Military product available 100% screened to MIL-STD-883, Class B

DESCRIPTION:
The IDT6168 is a 16,384-bit high-speed static RAM organized as 4K × 4. It is fabricated using IDT's high-performance, high-reliability technology—CEMOS. This state-of-the-art technology, combined with innovative circuit design techniques, provides a cost effective alternative to bipolar and fast NMOS memories.

Access times as fast as 20ns are available with maximum power consumption of only 550mW. The circuit also offers a reduced power standby mode. When $\overline{CS}$ goes high, the circuit will automatically go to, and remain in, a standby mode as long as $\overline{CS}$ remains high. In the standby mode, the device consumes less than 100µW, typically. This capability provides significant system-level power and cooling savings. The low-power (LA) version also offers a battery backup data retention capability where the circuit typically consumes only 1µW operating off of a 2V battery.

All inputs and outputs of the IDT6168 are TTL-compatible and operate from a single 5V supply, thus simplifying system designs. Fully static asynchronous circuitry is used, which requires no clocks or refreshing for operation, and provides equal access and cycle times for ease of use.

The IDT6168 is packaged in either a space-saving 20-pin, 300 mil DIP or 20-pin leadless chip carrier, providing high board-level packing densities.

The IDT6168 Military RAM is 100% processed in compliance to the test methods of MIL-STD-883, Method 5004, making it ideally suited to military temperature applications demanding the highest level of performance and reliability.

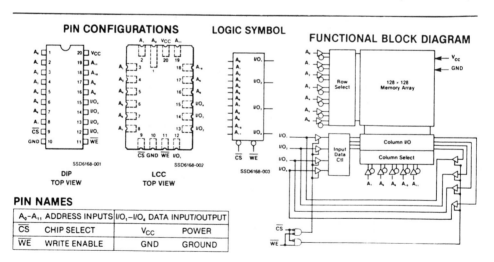

PIN NAMES

A_0–A_{11}	ADDRESS INPUTS	I/O_1–I/O_4	DATA INPUT/OUTPUT
$\overline{CS}$	CHIP SELECT	V_{CC}	POWER
$\overline{WE}$	WRITE ENABLE	GND	GROUND

(Courtesy of IDT Corporation)

Memory

IDT6168SA/IDT6168LA CMOS STATIC RAM 16K (4K x 4-BIT)
MILITARY AND COMMERCIAL TEMPERATURE RANGES

AC ELECTRICAL CHARACTERISTICS (V_{CC} = 5V ± 10%, All Temperature Ranges)

SYMBOL	PARAMETER	6168SA20[1] 6168LA20[1] MIN.	MAX.	6168SA25 6168LA25 MIN.	MAX.	6168SA35 6168LA35 MIN.	MAX.	6168SA45 6168LA45 MIN.	MAX.	6168SA55 6168LA55 MIN.	MAX.	6168SA70[2] 6168LA70[2] MIN.	MAX.	UNIT
READ CYCLE														
t_{RC}	Read Cycle Time	20	—	25	—	35	—	45	—	55	—	70	—	ns
t_{AA}	Address Access Time	—	20	—	25	—	35	—	45	—	55	—	70	ns
t_{ACS}	Chip Select Access Time	—	20	—	25	—	35	—	45	—	55	—	70	ns
t_{OH}	Output Hold from Address Change	5	—	5	—	5	—	5	—	5	—	5	—	ns
t_{LZ}	Chip Selection to Output in Low Z[3]	5	—	5	—	5	—	5	—	5	—	5	—	ns
t_{HZ}	Chip Deselect to Output in High Z[3]	—	10	—	10	—	15	—	15	—	25	—	30	ns
t_{PU}	Chip Select to Power Up Time[3]	0	—	0	—	0	—	0	—	0	—	0	—	ns
t_{PD}	Chip Select to Power Down Time[3]	—	20	—	25	—	35	—	40	—	50	—	60	ns
t_{RCS}	Read Command Set-Up Time	-5	—	-5	—	-5	—	-5	—	-5	—	-5	—	ns
t_{RCH}	Read Command Hold Time	-5	—	-5	—	-5	—	-5	—	-5	—	-5	—	ns
WRITE CYCLE														
t_{WC}	Write Cycle Time	20	—	20	—	30	—	40	—	50	—	60	—	ns
t_{CW}	Chip Select to End of Write	20	—	20	—	30	—	40	—	50	—	60	—	ns
t_{AW}	Address Valid to End of Write	20	—	20	—	30	—	40	—	50	—	60	—	ns
t_{AS}	Address Setup Time	0	—	0	—	0	—	0	—	0	—	0	—	ns
t_{WP}	Write Pulse Width	20	—	20	—	30	—	40	—	50	—	60	—	ns
t_{WR}	Write Recovery Time	0	—	0	—	0	—	0	—	0	—	0	—	ns
t_{DW}	Data Valid to End of Write	13	—	13	—	17	—	20	—	20	—	25	—	ns
t_{DH}	Data Hold Time	3	—	3	—	3	—	3	—	3	—	3	—	ns
t_{WZ}	Write Enable to Output in High Z[3]	—	7	—	7	—	13	—	20	—	25	—	30	ns
t_{OW}	Output Active from End of Write[3]	0	—	0	—	0	—	0	—	0	—	0	—	ns

NOTES:
1. Available over 0°C to +70°C temperature range only.
2. Available over -55°C to +125°C temperature range only. Also available 85 and 100ns military devices.
3. This parameter is guaranteed but not tested.

(Courtesy of IDT Corporation)

IDT6168SA/IDT6168LA CMOS STATIC RAM 16K (4K x 4-BIT) — MILITARY AND COMMERCIAL TEMPERATURE RANGES

TIMING WAVEFORM OF READ CYCLE NO. 1[1,2]

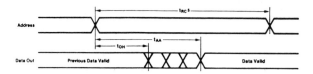

TIMING WAVEFORM OF READ CYCLE NO. 2[1,3]

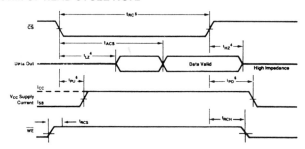

NOTES:
1. $\overline{WE}$ is high for READ cycle.
2. $\overline{CS}$ is low for READ cycle.
3. Address valid prior to or coincident with $\overline{CS}$ transition low.
4. Transition is measured ±200mV from steady state voltage with specified loading in Figure 2. This parameter is sampled and not 100% tested.
5. All READ cycle timings are referenced the last valid address to the first transitioning address.
6. This parameter is sampled and not 100% tested.

TIMING WAVEFORM OF WRITE CYCLE NO. 1 ($\overline{WE}$ CONTROLLED)[1]

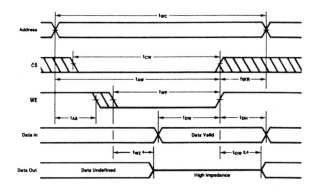

(Courtesy of IDT Corporation)

chapter 10

synchronous sequential logic

10-1 INTRODUCTION

Chapters 1–7 were strictly devoted to familiarize the student with combinatorial logic as well as the operating characteristics of the various available chips. Boolean algebra and Karnaugh mapping are powerful tools to minimize the number of required gates for a given design requirement. However, the discussion dealt only with a given set of input conditions and the desired output at one point in time. The output was not dependent on the history of previous events.

In many instances the logic designer faces the problem of designing a network that responds not only to a set of present events but also to a set of previous events. The algorithm for handling such design problems falls under the title of "sequential logic." Since previous events are involved, it becomes immediately obvious that memory elements are needed. After all, the circuits cannot deal with inputs that are forgotten. The most convenient memory element is the flip flop. Any of the four common flip flops can be used: R/S, D, T, or J/K. Which one to choose becomes a problem in logic minimization.

As we know, flip flops have control and clock inputs. The control inputs are prepared during a given clock cycle and then transferred to the flip flop output at either the positive or the negative clock edge. The algorithm deals mainly with the design of the proper gating networks for the flip flop control inputs and has therefore been called "flip flop programming." It becomes clear that all output state changes occur at clock time; therefore the logic becomes synchronous.

Sequential logic is applied to the design of synchronous counters as well as a wide variety of control and timing networks.

Before delving into the subject it is necessary to have a second look at flip flops and their dynamic behavior.

FIGURE 10-1 D flip flop

10-2 FLIP FLOP EXCITATION TABLES

From previous discussions we are familiar with the flip flop truth table, in which the output Q is evaluated in terms of a given set of input conditions. Let us look at the D flip flop truth table:

D	Q
0	0
1	1

This means, of course, that the output Q follows the input D at clock-time. Another way to look at the device is to ask the question "In order for Q to change from 0 to 1 at clock time, what must D be? The answer is obviously 1. Let us call the state of Q now Q_n and the desired state of Q after the next clock edge Q_{n+1}. (See figure 10-1.) Four possible transitions—Q changes from 0 to 0, from 0 to 1, from 1 to 0, or from 1 to 1—and the corresponding control input excitation to cause the desired output transition are shown in the D excitation table:

D	Q_n	Q_{n+1}
0	0	0
1	0	1
0	1	0
1	1	1

A similar table is given for the T flip flop (figure 10-2):

T	Q_n	Q_{n+1}
0	0	0
1	0	1
1	1	0
0	1	1

The first and the last transition mean no change at all. This is acceptable because one needs to control the flip flop either to change state or not to change state.

There is no new information in the excitation table that was not in the truth

Synchronous Sequential Logic

FIGURE 10-2 *T* flip flop

FIGURE 10-3 *J/K* flip flop

table. It is just a more convenient way to present the behavior of the device in designing sequential networks. The *J/K* flip flop is somewhat more complicated. For Q to transfer from a 0 state to a 0 state, J must be 0, and K could be either 0 or 1. For Q to transfer from a 0 state to a 1 state, J must be 1, and K could be either 0 or 1. For Q to transfer from a 1 state to a 0 state, J could be either 0 or 1, but K must be 1. For Q to transfer from a 1 state to a 1 state, J could be either 0 or 1, but K must be 0.

For two of the four possible transitions, K is a "don't care" condition, and for the other two transitions the same holds true for J. "Don't care" conditions are entered as X on the table. Thus for the *J/K* flip flop (figure 10-3) we get the following table:

J	K	Q_n	Q_{n+1}
0	X	0	0
1	X	0	1
X	1	1	0
X	0	1	1

For the *R/S* flip flop (figure 10-4) we get the following table:

S	R	Q_n	Q_{n+1}
0	X	0	0
1	0	0	1
0	1	1	0
X	0	1	1

FIGURE 10-4 *R/S* flip flop

10-3 SYNCHRONOUS COUNTERS

Synchronous counters have been discussed, and some examples were shown, but only to the extent of what they look like and how they differ from ripple counters. Here we will design counters of any sequence and any modulo. The design of a synchronous counter is facilitated by dividing the task into several steps.

- Step 1. Start with a statement of the desired count sequence.
- Step 2. Determine the number of flip flop stages.
- Step 3. State the desired transitions in a *now-next table* (or transition table).
- Step 4. List the required flip flop control input conditions in a Karnaugh map (a control map).
- Step 5. Reduce the map entries to Boolean equations.
- Step 6. Use the Boolean equations to implement the input logic to the chosen flip flops.

This procedure is best explained and justified by examples.

EXAMPLE 10-1 3 Bit Up Counter

The simplest example to start with is a three-bit counter, counting from 0 to 7. The variables chosen are A, B, and C. It is best to let A be the least significant bit; thus if the count length is to be extended, D, E, and so on can be appended in the more significant direction.

- Step 1. Count sequence: 0-1-2-3-4-5-6-7-repeat.
- Step 2. Eight states (from 1) require three variables, C, B, and A.
- Step 3. The Now column in table 10-1 should always (regardless of the desired

TABLE 10-1 Transition table

State	Now			Next		
	C	B	A	C	B	A
0	0	0	0	0	0	1
1	0	0	1	0	1	0
2	0	1	0	0	1	1
3	0	1	1	1	0	0
4	1	0	0	1	0	1
5	1	0	1	1	1	0
6	1	1	0	1	1	1
7	1	1	1	0	0	0

Synchronous Sequential Logic

count sequence) be listed in simple ascending order. Note that the Now column does not say anything about the desired count sequence; it merely lists all possible states. The Next column lists the states to which the flip flops must change. Each row entry into the Next column states the expected change. Thus the expected transitions are read *horizontally* in each row to the adjacent column. Notice that reading the Next column from top to bottom will not say anything about the count sequence.

- Step 4. Before attempting to transfer the information from the transition table to a Karnaugh map, we want to make a few comments about the map. Each square in the map represents a Now state. For example, the row 00 and column 00 mean $DCBA = 0000 = 0$; the row 11 and column 10 mean $DCBA = 1110 = 14$.

The entire map of figure 10–5 can be filled with the now states; a weighted map is thus created. This map can be used as a reference to quickly and correctly identify a square into which we must enter the flip flop control condition.

To avoid confusion and remain consistent from one problem to the next, it is advisable to establish a convention for the Karnaugh map labeling, as follows: Label D, C, B, A as shown in figure 10–5 and agree that A is the least significant bit. This keeps the weighted map consistent for the entire discussion. Choosing A as the least significant bit allows word expansion above D to E, F, and so on in an orderly fashion.

We repeat: The numbers in the map represent the now state into which we enter the necessary 1, 0, or X condition for the control input, so that the particular flip flop considered will go to the assigned "next state" after the next clock pulse.

For the example at hand we need only the upper half of the weighted map in figure 10–6 (three-variable map). We need one map each for the C, B, and A flip flops. Let us choose J/K flip flops for this implementation. $J_c K_c$, $J_b K_b$, and $J_a K_a$ are the control inputs for the three flip flops. C, B, and A are the three variables that are actually the three flip flop outputs Q_c, Q_b, and Q_a.

From this point on it is convenient to call the flip flop outputs just by their variable names. Instead of Q_a or Q_b, we call them A and B. The inputs are called J_a, K_a, J_b, and so on. The entries in the maps allow us to find a logic network that

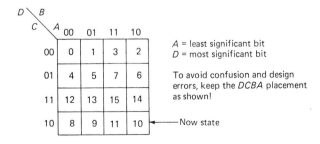

FIGURE 10–5 Weighted map

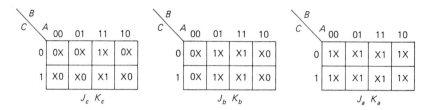

FIGURE 10–6 Control maps

forces the given flip flop to change state or not to change state at any given clock cycle as required by the stated count sequence. The entries are obtained directly from the Now-Next columns and the appropriate flip flop excitation table. Look, for example, at the C columns in table 10–1.

For the 0 state, Q_c is to change from 0 to 0. Thus according to the J/K excitation table where $Q_n = 0$ changes to $Q_{n+1} = 0$, we find a 0X requirement for the J/K control inputs. Thus the entry into the weight = 0 box ($C = 0, B = 0, A = 0$) of the $J_c K_c$ map, which is the upper left-hand corner box, is 0X. This means that the counter is just starting from count 0 ($Q_c, Q_b,$ and $Q_a = 0$) and the C flip flop is to stay in a 0 state when clocked. So it will for the 1 and the 2 states, and the same 0X entries hold. When the counter reaches the 3 state, the requirement changes. Now the control inputs must be prepared to change state from 0 to 1 with the next clock cycle. Looking at the J/K excitation table where $Q_n = 0$ changes to $Q_{n+1} = 1$, we find a 1X requirement for the J/K control inputs. Thus 1X is entered into the $J_c K_c$ map where $C = 0, B = 1,$ and $A = 1$ (that is, the 3 state). This is a lengthy explanation for a simple process. It can be repeated in a step-by-step form:

- Step 1. Look at the first row of the Now-Next column of the flip flop of interest and find the required change for that state.
- Step 2. Go to the appropriate excitation table and find the required control input condition for the required change.
- Step 3. Find the Now box in the map for the flip flop of interest and enter the requirement found in Step 2.
- Step 4. Proceed until the map is filled.
- Step 5. The next difficulty that arises is the fact that the map has double entries. How do we circle the 1's, 0's and X's to get a minimum solution? First it is to be realized that for every entry the first letter pertains to J and the second to K. We can simply split each map into two maps, one for J and one for K. For the C flip flop this becomes the maps in figure 10–7.

Now it is obvious that the $BA = 11$ column can be circled for both the J and K maps. We get

$$J_c = BA \quad \text{and} \quad K_c = BA$$

Synchronous Sequential Logic

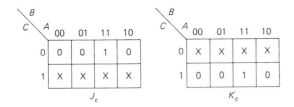

FIGURE 10–7 Control maps

Note that for the J map the X in the 7 box was chosen to be 1, for the K map the X in the 3 box was chosen to be 1, and the other X's were chosen to be 0.

Splitting the maps into two maps is too laborious. With a little experience it is quite simple to forego the circle drawing. Look at the combined J/K map in figure 10–6. Mentally eliminate the right-hand entries pertaining to K, draw the circles mentally around the 1's and X's pertaining to J, and write down the corresponding Boolean term for J. Then mentally eliminate the left-hand entries pertaining to J, draw the circles mentally around the 1's and X's pertaining to K, and write down the corresponding Boolean term for K. From the C flip flop it is simple to see that $J_c = BA$ and $K_a = BA$. From the B flip flop it is easy to see that $J_b = A$ and $K_a = A$. From the A flip flop it is simple to see that $J_a = 1$ and $K_a = 1$.

There is a second solution for J_a and K_a. Namely, let the X's for J_a in the $BA = 01$ and $BA = 11$ columns be 0. Then $J_a = \overline{A}$. Let the X's for K_a in the $BA = 00$ and $BA = 10$ columns be 0. Then $K_a = A$.

At first it seems that the second solution is not as well minimized as the first. After all, it was previously stated that the best optimization is obtained by circling as many variables as possible. Yet an implementation of the first solution required both J_aK_a inputs to be tied high. It is not good practice to tie TTL logic inputs directly to V_{cc} without a current-limiting resistor. (The practice is to tie ten logic inputs through one 1K resistor to V_{cc}.)

An implementation of the second solution requires only two wires from J_a to $\overline{A}$ and K_a to A. From a design practice viewpoint the second solution is simpler.

Generally, the simplest solution for a given Karnaugh map is the maximum circling of variables; that is, we try to enclose as many squares on the map as possible, using as many X's as possible. In the J_aK_a map in figure 10–6, that would yield $J_a = 1$ and $K_a = 1$, which is not optimum from an implementation viewpoint. The point is that in some cases a less than maximum circling is chosen. No fixed rule exists as to when to use a less than maximum circling. It is up to the ingenuity of the designer to recognize the advantages.

- Step 6. The last step is to implement the circuit in figure 10–8 using the Boolean expressions established in Step 5:

$$J_c = BA \qquad J_b = A \qquad J_a = \overline{A}$$
$$K_c = BA \qquad K_b = A \qquad K_a = A$$

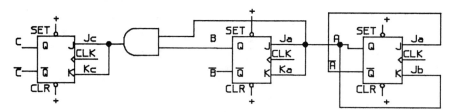

FIGURE 10–8 Synchronous counter with J/K flip flops

Note that the circuit is unconventionally drawn from right to left. This was done so that the most significant bit of the counter would appear on the left. (Otherwise, the binary value of the output would appear mirror imaged.) ■

EXAMPLE 10–2

To illustrate how to work with four variables, we can repeat Example 10–1 for a count from 0 to 15.

- Step 1. Count sequence: 0–1–2–3–4–5–6–7–8–9–10–11–12–13–14–15–repeat.
- Step 2. 16 states from 0 to 15 require four variables, $D, C, B,$ and A.
- Step 3. See figure 10–9(a).
- Step 4. See figure 10–9(b).

Now				Next			
D	B	C	A	D	C	B	A
0	0	0	0	0	0	0	1
0	0	0	1	0	0	1	0
0	0	1	0	0	0	1	1
0	0	1	1	0	1	0	0
0	1	0	0	0	1	0	1
0	1	0	1	0	1	1	0
0	1	1	0	0	1	1	1
0	1	1	1	1	0	0	0
1	0	0	0	1	0	0	1
1	0	0	1	1	0	1	0
1	0	1	0	1	0	1	1
1	0	1	1	1	1	0	0
1	1	0	0	1	1	0	1
1	1	0	1	1	1	1	0
1	1	1	0	1	1	1	1
1	1	1	1	0	0	0	0

FIGURE 10–9(a) Transition table for Example 10–2

Synchronous Sequential Logic

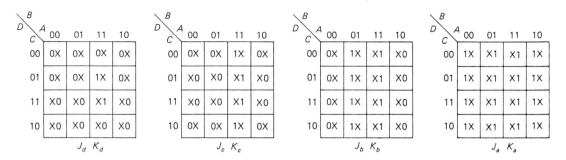

FIGURE 10-9(b) Control maps for Example 10-2

- Step 5.

$$J_d = CBA \quad J_c = BA \quad J_b = A \quad J_a = \overline{A}$$
$$K_d = CBA \quad K_c = BA \quad K_b = A \quad K_a = A$$

You will notice in figure 10-9 that the input to the second flip flop $= A$, the input to the third flip flop $= BA$, and the input to the fourth flip flop $= CBA$. It becomes evident that the input to the fifth flip flop would be $DCBA$, to the sixth flip flop it would be $EDCBA$, and so on. It is often not necessary for the designer to extend the design to more variables if a longer sequence is required. The extended Boolean pattern becomes obvious.

We can also look at the four-bit up counter with an additional requirement in mind. Suppose this four-bit counter is a chip with expansion capability to more bits. In that case a carry output $= DCBA$ to the next set must be provided. There must also be an input terminal to receive such a carry from a previous less significant counter. Figure 10-10(a) can be modified to figure 10-10(b). The input inverter allows the input extension to be tied to ground when not being used. Since the expected input carry is inverted, the output carry is generated with a NAND rather than an AND.

- Step 6. See figures 10-10(a) and 10-10(b). ∎

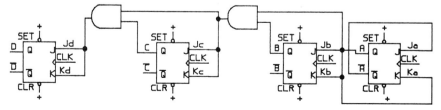

FIGURE 10-10(a) Synchronous counter with J/K flip flop

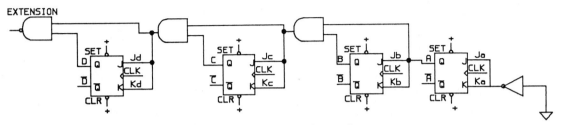

FIGURE 10–10(b) Synchronous counter with J/K flip flop and extension gates

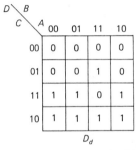

1. $D_d = D\bar{C} + D\bar{B} + D\bar{A} + \bar{D}CBA$
2. $D_d = D(\bar{C} + \bar{B} + \bar{A}) + \bar{D}CBA$
3. $D_d = D(\overline{BCA}) + \bar{D}(CBA)$
4. $D_d = D \oplus \overline{CBA}$

1. $D_c = C\bar{B} + C\bar{A} + \bar{C}BA$
2. $D_c = C(\bar{B} + \bar{A}) + \bar{C}BA$
3. $D_c = C(\overline{BA}) + \bar{C}(BA)$
4. $D_c = C \oplus BA = \overline{C \oplus \overline{BA}}$

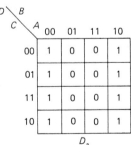

1. $D_b = B\bar{A} + \bar{B}A$
2. $D_b = B \oplus A$
3. $D_b = \overline{B \oplus \bar{A}}$

$D_a = \bar{A}$

FIGURE 10–11 Control maps for Example 10–3

Synchronous Sequential Logic

EXAMPLE 10–3 D Flip Flop Up Counter

Of course, the same function can be implemented with D flip flops. Making use of the D excitation table and using the Now-Next columns of the previous example, we proceed directly to Steps 4 and 5 in figure 10–11. This is the function that is implemented in a 74163 synchronous binary counter (see figure 10–12) (See the TI TTL data book, Second Edition, 1981, p. 7-193) if one ignores the Load, Clear, and Enable lines. The functions in line 1 (for D_d, D_c, and D_b) are read directly off the map. Even if they are minimum functions, they are not necessarily in the most convenient form for implementation. The designer often looks for a repetitive pattern, which makes the fabrication simpler, sometimes even at the cost of some additional gates. Rearrangement of the equations for D_d, D_c, and D_b simplifies the terms into a nice Exclusive OR pattern. Note that if we had a fifth stage, it would be easily recognized to be

$$D_e = \overline{E \oplus \overline{DCBA}} \quad etc.$$

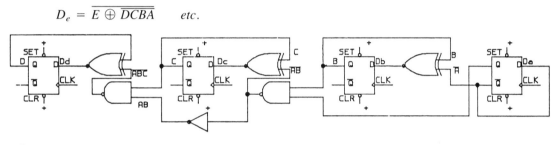

FIGURE 10–12 Synchronous counter with D flip flops

EXAMPLE 10–4 Decade Counter

Let us design a decade counter with T flip flops that makes use of "don't care" states; that is, states from 10 to 15 will not occur, and thus the Next column will have X's from the respective Now states 10 to 15.

- Step 1. Count sequence: 0–1–2–3–4–5–6–7–8–9–0, etc.
- Step 2. Ten states require four flip flops, D, B, C, and A.
- Step 3. See figure 10–13(a).
- Step 4. See figure 10–13(b).
- Step 5.

$$T_d = CBA + DA \quad T_c = BA \quad T_b = \overline{D}A \quad T_a = 1$$

- Step 6. See figure 10–14.

The circuit of figure 10–14 is the 74192 decade up counter section (see TI TTL Data Book, 1985, p. 3-756) if the down count gating, Clear, and Load are ignored.

Now				Next			
D	*C*	*B*	*A*	*D*	*C*	*B*	*A*
0	0	0	0	0	0	0	1
0	0	0	1	0	0	1	0
0	0	1	0	0	0	1	1
0	0	1	1	0	1	0	0
0	1	0	0	0	1	0	1
0	1	0	1	0	1	1	0
0	1	1	0	0	1	1	1
0	1	1	1	1	0	0	0
1	0	0	0	1	0	0	1
1	0	0	1	0	0	0	0
1	0	1	0	X	X	X	X
1	0	1	1	X	X	X	X
1	1	0	0	X	X	X	X
1	1	0	1	X	X	X	X
1	1	1	0	X	X	X	X
1	1	1	1	X	X	X	X

FIGURE 10–13(a) Transition table for Example 10–4

T_b

DC \ BA	00	01	11	10
00	0	1	1	0
01	0	1	1	0
11	X	X	X	X
10	0	0	X	X

T_a

DC \ BA	00	01	11	10
00	1	1	1	1
01	1	1	1	1
11	X	X	X	X
10	1	1	X	X

T_d

DC \ BA	00	01	11	10
00	0	0	0	0
01	0	0	1	0
11	X	X	X	X
10	0	1	X	X

T_c

DC \ BA	00	01	11	10
00	0	0	1	0
01	0	0	1	0
11	X	X	X	X
10	0	0	X	X

FIGURE 10–13(b) Control maps for Example 10–4

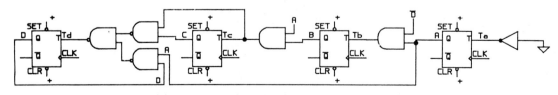

FIGURE 10–14 Decade counter with *T* flip flops ■

Synchronous Sequential Logic

EXAMPLE 10–5 Two-Mode (Up-Down) Counter

The previous examples dealt with counters that repeat a sequence in cyclical manner, hence the name *ring counters*. At no point in the sequence was the circuit called upon to make a decision. The counters were input independent. The simplest decision a counter can make is to count in two modes as a function of some control input setting. The most obvious would be an up-down counter that follows the sequence determined by a logic level or switch setting.

A four bit up-down counter becomes a five-variable problem in which the switch setting S is the fifth variable. We will cover one example of a five-variable problem later in the chapter. For now the method of two-mode counting is illustrated with one fewer variable, that is, flip flops C, B, and A and switch S. (See figure 10–15(a). A recognizable control equation pattern will emerge, from which we can expand the circuit to four flip flops.

- Step 1. Count sequence $S = 0$: 0–1–2–3–4–5–6–7–0, etc. Count sequence $S = 1$: 7–6–5–4–3–2–1–0–7, etc.
- Step 2. There are four variables, S, C, B, and A, but only three flip flops. Let S be the most significant bit. This is convenient, since the lower half of the map is then dedicated to the $S = 1$ sequence.
- Step 3. See figure 10–15(a). We need two Next columns, one for $S = 0$ and one for $S = 1$.
- Step 4. See figure 10–15(b). Note that in this case the $S = 1$ column starts in the weight = 8 box.
- Step 5.
    $$T_c = \overline{S}BA + S\overline{B}\overline{A} \qquad T_b = \overline{S}A + S\overline{A} \qquad T_a = 1$$
- Step 6. Implement the equations in figure 10–16(a). Notice the control equation pattern:
    $$T_a = 1 \qquad T_b = \overline{S}A + S\overline{A} \qquad T_c = \overline{S}BA + S\overline{B}\overline{A}$$

Now C B A	S = 0 Next C B A	S = 1 Next C B A
0 0 0	0 0 1	1 1 1
0 0 1	0 1 0	0 0 0
0 1 0	0 1 1	0 0 1
0 1 1	1 0 0	0 1 0
1 0 0	1 0 1	0 1 1
1 0 1	1 1 0	1 0 0
1 1 0	1 1 1	1 0 1
1 1 1	0 0 0	1 1 0

FIGURE 10–15(a) Transition table for Example 10–5

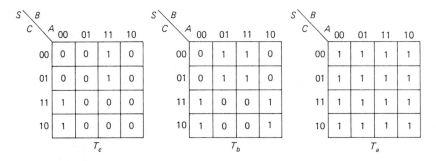

FIGURE 10–15(b) Control maps for Example 10–5

It follows by inspection that

$$T_d = \overline{S}CBA + S\overline{C}\,\overline{B}\,\overline{A}$$

A slightly different implementation in figure 10–16(b) can be chosen to take advantage of the repeated AND pattern. ∎

EXAMPLE 10–6 Counter with Chosen Sequence

Suppose we want a counter of some sequence other than straight up or down. The method remains the same. This time, let one sequence be a Gray up count in one mode and a modulo 5 counter in the other mode. A Gray counter has the advantage of changing only one variable with each clock pulse. Thus the detection of a particular count state within the sequence with gates makes sure that no glitches occur due to race conditions. The mod 5 counter will have three "don't care" states. All we might be interested in is a divide by 5 without caring what the particular sequence is. The three "don't care" states may then be placed anywhere. It turns out that a judicious choice of "don't care" locations within the map might reduce the gate count. This kind of state assignment is called *secondary assignment*. (See figure 10–17.)

- Step 1. The count sequence for Gray count is $S = 0$: 0–1–3–2–6–7–5–4–0. The count sequence for mod 5 count is $S = 1$: 0–1–2–7–4–0, etc. This choice of secondary assignments lines up the 1's and X so as to allow better minimization than a straight 0 to 4 count.
- Step 2. There are four variables: S, C, B, and A. Choose J/K flip flops.
- Step 3. See figure 10–17(a).
- Step 4. See figure 10–17(b).

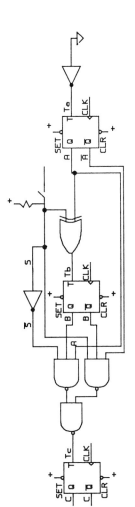

FIGURE 10–16(a) Three-bit up-down counter

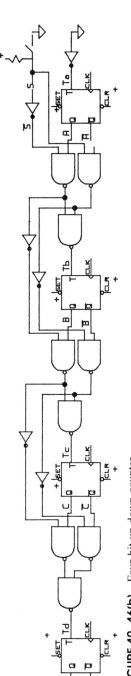

FIGURE 10–16(b) Four-bit up-down counter

Now C B A	S = 0 Next C B A	S = 1 Next C B A
0 0 0	0 0 1	0 0 1
0 0 1	0 1 1	0 1 0
0 1 0	1 1 0	1 1 1
0 1 1	0 1 0	X X X
1 0 0	0 0 0	0 0 0
1 0 1	1 0 0	X X X
1 1 0	1 1 1	X X X
1 1 1	1 0 1	1 0 0

FIGURE 10–17(a) Transition table for Example 10–6

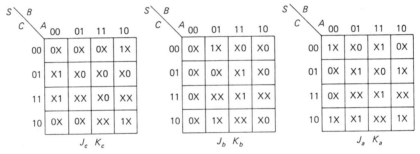

FIGURE 10–17(b) Control maps for Example 10–6

- Step 5.

$$J_c = B\overline{A} \qquad J_b = \overline{C}A \qquad J_a = S\overline{C} + BC + \overline{B}\,\overline{C}$$
$$K_c = \overline{B}\,\overline{A} \qquad K_b = CA \qquad K_a = S + \overline{C}B + C\overline{B}$$

- Step 6. Implement the equations in figure 10–18.

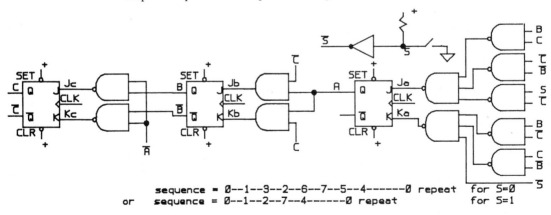

FIGURE 10–18 Two-mode counter

Synchronous Sequential Logic

The above procedure for two-mode counters can be extended to multimode counters by writing that many more Next columns. The number of variables to be handled becomes too cumbersome. Karnaugh mapping is acceptable up to six variables. Beyond six variables, numerical methods based on the Quine McClusky method are preferred that lend themselves to computer program solutions. A large number of problems either do not involve more than six variables or involve a recognizable repetitive pattern that reduces the problem to manageable proportions.

―――――――――――――――――――――――――――――――― **EXAMPLE 10–7** Shift Register

Let us design the well-known shift register, as in figure 10–20, with D flip flops for three stages.

- Step 1. There is no particular sequence. Whatever state the flip flops happen to be in, including the input to stage A, are shifted left with each clock pulse. (The pattern is input dependent.)
- Step 2. We decided on three stages and three variables. The input, however, becomes a fourth variable $= I$, which should be located below the least significant bit, A. We start with Step 3, since no required input sequence is specified.
- Step 3. See figure 10–19(a).
- Step 4. See figure 10–19(b).
- Step 5.

$$D_c = B \qquad D_b = A \qquad D_a = I$$

- Step 6. Implement the equations in figure 10–20.

Now	Next
C B A I	C B A I
0 0 0 0	0 0 0 X
0 0 0 1	0 0 1 X
0 0 1 0	0 1 0 X
0 0 1 1	0 1 1 X
0 1 0 0	1 0 0 X
0 1 0 1	1 0 1 X
0 1 1 0	1 1 0 X
0 1 1 1	1 1 1 X
1 0 0 0	0 0 0 X
1 0 0 1	0 0 1 X
1 0 1 0	0 1 0 X
1 0 1 1	0 1 1 X
1 1 0 0	1 0 0 X
1 1 0 1	1 0 1 X
1 1 1 0	1 1 0 X
1 1 1 1	1 1 1 X

FIGURE 10–19(a) Transition table for Example 10–7

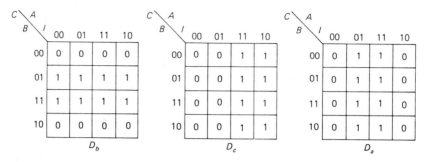

FIGURE 10–19(b) Control maps for Example 10–7

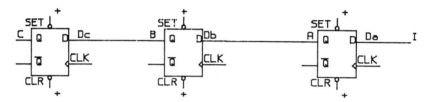

FIGURE 10–20 Shift register

This, of course, is a simple expected result. Of interest is the Now-Next table, which illustrates how all possible combinations can be listed. Notice that in each row the Next column is the Now column shifted left; what appears in the Next column under I (for the input) is not under the designer's control and therefore appears as X. ∎

EXAMPLE 10–8 A Self-Correcting Pointer

The shift register of figure 10–20 can of course be lengthened to any number of stages. A frequent application of shift registers is to feed the last-stage output back to the first-stage input and circulate a pattern indefinitely.

Suppose we want to circulate a single 1 in an eight-stage closed-loop shift register. The single 1 might, for example, control a data selector or multiplexor. The true output is a pointer similar to a rotating switch or commutator. We initially need only preset one of the stages to 1 and clear all other stages. This works quite well, except for the possibility of an error occurring at some later time. If for some reason there is a power supply glitch or some other noise condition, a second or more flip flops set to 1, or the existing 1 is lost, the single 1 pattern is destroyed with detrimental effects on the system.

To avoid that problem, the shift register can be designed to be self-correcting such that any pattern other than a single 1 will send the system to an all 0 followed

Synchronous Sequential Logic

by a 1 in position 0. The shift system can be designed for four flip flops and extended with a recognizable repeat pattern to more flip flops. Here again we start with Step 3, since the pattern is obvious from the problem statement.

Notice the following pattern in figure 10–21: All K inputs are returned to their own Q output. All J inputs receive the previous output just as in a shift register but also all the NOT outputs from all other flip flops except the preceding one. This means that any flip flop that is about to change to a 1 will not do so unless all other flip flops except the preceding one are in a 0 state. Hence if there is an error, an all 0 state will result.

Flip flop A, however, receives all NOT outputs including its own except the last stage (D). When all flip flops are in a 0 state, the last one must normally be in a 1 state. This is when A flips to a 1. If after an error there is an all 0 state, A will also flip into a 1 state, after which the system continues normally. (See figure 10–22.)

The extension pattern is obvious and is left as an exercise for the student.

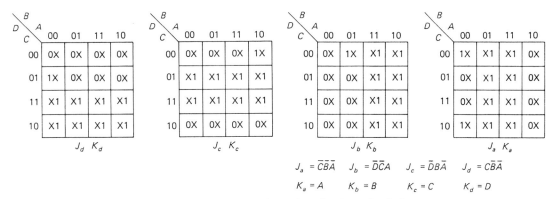

FIGURE 10–21 Transition table, control maps, and equations for Example 10–8

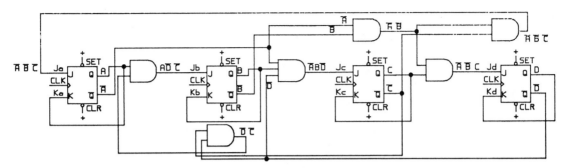

FIGURE 10-22 Self-correcting pointer

Finally, we implement the equations of figure 10-21, as shown in figure 10-22. ∎

Another interesting problem that falls into the category of designing a small section of a sequential circuit that can then be expanded by inspection is a successive approximation register. A well-known application of this register is the successive approximation A/D converter.

Such an A/D converter essentially consists of one analog comparator, a clocked successive approximation register with a D/A converter in the feedback loop. (See figure 10-23.)

The simplest version of an A/D converter uses a counter output converted to an analog voltage, which is then compared to the analog voltage to be measured. When the two are equal, the process is stopped. The last output of the counter must be a representation of the analog input. A ten-bit conversion can therefore take as many as 1024 clock cycles. This is too slow for many applications. A much faster way to obtain the answer is by successive approximation.

EXAMPLE 10-9 Successive Approximation Register

The idea here is to check whether the analog input is larger or smaller than half the maximum possible output. If it is larger, then the most significant bit of the

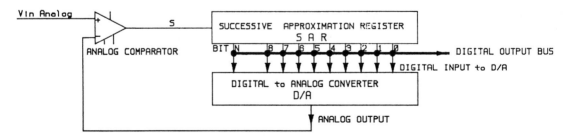

FIGURE 10-23 Successive approximation A/D converter

Synchronous Sequential Logic

digital output must be 1. If it is smaller, then the most significant bit of the digital output must be 0.

The above can be considered a test of the most significant bit. Once it is established, one proceeds to test the next bit.

If the MSB = 1, then the next test is to find whether or not the analog input is larger or smaller than ¾ of the maximum possible input. If the MSB = 0, then the next test is to find whether or not the analog input is larger or smaller than ¼ of the maximum possible input. In either of these cases the second bit is tested. If it is larger than the tested value, the second bit is 1; otherwise, it is 0.

The test method results in a tree as shown in figure 10-24 for a four-bit digital output. One starts with a test of half the maximum input, in this case 8. It always takes as many tests as there are bits to be established. Even if the input is 8 volts and it seems that only one bit needs to be tested, the system must nonetheless proceed to check that all other three 0's are indeed 0's.

Each vertical line in the tree represents one clock cycle and one test. (See figure 10-24.)

The successive approximation register has as many variables as there are bits to be tested.

Unlike the previous examples, in which the variable S remained constant for the duration of the sequence, here the control variable S may change after each test, setting the register (or the count sequence) into a different mode, that is, to the appropriate section of the tree.

The successive approximation register is preset to 8. Thus 8 V in analog form return to the comparator. If the input is larger than 8, then $S = 1$; otherwise, $S = 0$. After each test, S is either 0 or 1, depending on whether the D/A converted tested value is larger or smaller than the input. The test result S determines which branch of the tree is the Next state.

Another way to describe the process is as follows: The MSB is set to 1 and tested; that is, the D/A converter output is compared to the input. If the input is larger than the tested value (in this case 8), leave the MSB as is; otherwise, reset the MSB to 0 and at the same time set the next bit to 1, regardless of what decision was made on the MSB. Test the next bit. If the input is still larger, leave the second bit as is; otherwise, reset it to 0 and set the next bit to 1 regardless of what decision was made on the second bit. The process continues until all bits are tested.

The design of such a variable sequential counter proceeds in the same manner as in previous examples. To recognize a repetitive pattern, so that the design can easily be extended to more bits, four variables and the input variable S are needed.

- Step 1. The count sequence must be extracted from the tree.
- Step 2. As was stated above, the variables are S, D, C, B, and A.
- Step 3. With the procedure outlined above, the table in figure 10-25(a) results.
- Step 4. See figure 10-25(b).
- Step 5. More often than not, the Boolean equations obtained from the maps are not unique. Moreover, it is not always necessary to obtain a minimum set

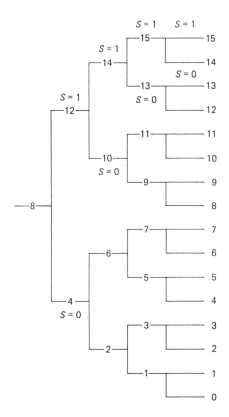

FIGURE 10-24 *A/D* converter tree

Now	$S = 0$ Next				$S = 1$ Next			
D C B A	D	C	B	A	D	C	B	A
0 0 0 0	X	X	X	X	X	X	X	X
0 0 0 1	0	0	0	0	0	0	0	1
0 0 1 0	0	0	0	1	0	0	1	1
0 0 1 1	0	0	1	0	0	0	1	1
0 1 0 0	0	0	1	0	0	1	1	0
0 1 0 1	0	1	0	0	0	1	0	1
0 1 1 0	0	1	0	1	0	1	1	1
0 1 1 1	0	1	1	0	0	1	1	1
1 0 0 0	0	1	0	0	1	1	0	0
1 0 0 1	1	0	0	0	1	0	0	1
1 0 1 0	1	0	0	1	1	0	1	1
1 0 1 1	1	0	1	0	1	0	1	1
1 1 0 0	1	0	1	0	1	1	1	0
1 1 0 1	1	1	0	0	1	1	0	1
1 1 1 0	1	1	0	1	1	1	1	1
1 1 1 1	1	1	1	0	1	1	1	1

FIGURE 10-25(a) Transition table for Example 10-9

Synchronous Sequential Logic

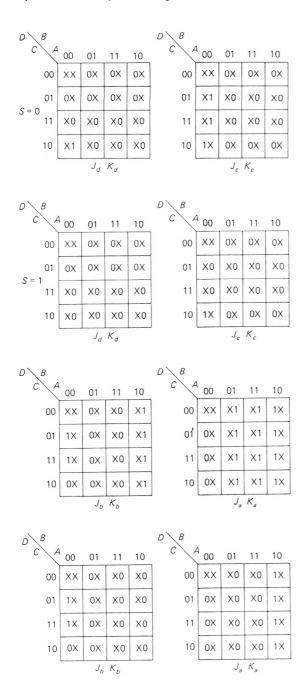

FIGURE 10–25(b) Control maps for Example 10–9

of terms. What we are looking for is an *optimum* set of terms. In this case the optimum term is a set of equations that show a repetitive pattern. This will allow expansion without further design. (See figure 10–26.)

$$J_d = 0 \quad \text{or}$$
$$J_d = E\overline{DCBA} \quad J_c = D\overline{CBA} \quad J_b = C\overline{BA} \quad J_a = B\overline{A}$$
$$K_d = \overline{SCBA} \quad K_c = \overline{SBA} \quad K_b = \overline{SA} \quad K_a = \overline{S}$$

It is not particularly difficult to reduce a five-variable map. Each J/K map consists of two maps, one for $S = 0$ and the other for $S = 1$. Look for a reducible pattern in either map and see whether the identical pattern exists in the other one. If so, then S is redundant; if not, then S must be included in the term, as either S or $\overline{S}$, depending on whether the circled pattern belongs to the S map or the $\overline{S}$ map.

- Step 6. By extending the pattern observed in J_a, J_b, and J_c it becomes obvious that $J_d = E\overline{DCBA}$. The E would be the fifth variable if we had designed the network to include such. Moreover, if E does not exist, then $J_d = 0$ is a good solution. Just ground the E input. (See figure 10–27.) This means that if we choose to extend the successive approximation register to eight bits, we can easily do so by feeding the least significant bit of the next set of four variables back to the J_d input of the first set of four variables. There must also be an input from a previous possible section.

Note that the K_a input is $\overline{S}$ and the last K output for the extended set is $\overline{SDCBA}$. If that output is ANDed with $\overline{S}$, the result is still $\overline{SDCBA}$. Thus an additional terminal provided at the K_a input network allows that extension. If not used, it is tied high. This is easily demonstrated in figure 10–26 by drawing the logic according to the above equations.

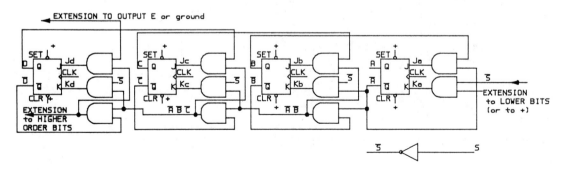

FIGURE 10–26 Four-bit successive approximation register

Synchronous Sequential Logic

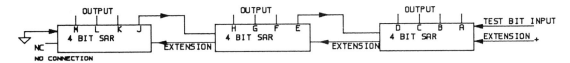

FIGURE 10-27 12-bit SAR

Another solution to the successive approximation register can be obtained by common sense reasoning. This implementation uses D flip flops rather than J/K flip flops. It requires twice as many flip flops but fewer control gates.

Ordinarily, a D implementation requires more gating than the J/K because one control input has fewer degrees of freedom than the two J/K inputs. One can, however, use the Set-Reset capability to advantage.

A start pulse sets the most significant bit to 1, while all other bits are left at 0, from a previous clear.

The next clock pulse sets Q_4 of the shift register to 1 and performs two tasks:

1. The most significant bit follows S as soon as the clock, somewhat delayed by the shift register flip flop and the AND gate, strobes the most significant D flip flop.
2. The next lesser significant flip flop C is unconditionally set to 1.

The next clock cycle shifts the clock Enable to Q_3 and output C to the value of S while flip flop B is unconditionally set to 1.

The process continues until all stages are tested and set.

The Set-Reset inputs are nonsynchronous. State changes occur whenever the Set or Reset is activated without being timed by the clock. The clock input to the flip flop under test is delayed and therefore not purely synchronous. The system in figure 10–28 becomes partially nonsynchronous but more economical in gate count. Gated and delayed clocking is not a recommended design practice. A careful look at the timing is necessary to make sure there are no possible race conditions.

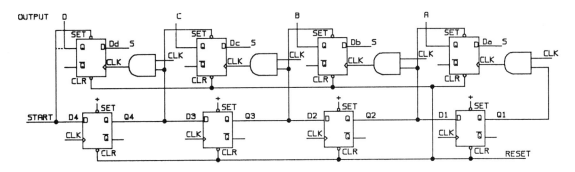

FIGURE 10-28 D flip flop SAR with shift register control

REVIEW QUESTIONS

1. What is the difference between a truth table and an excitation table?
2. Why is the Now table written in simple ascending order regardless of the desired sequence?
3. When you read a Next table from top to bottom, are you reading the desired sequence?
4. Figure 10–5 shows a weighted map. Why is it convenient to keep the same *DCBA* assignments for all problems?
5. In connection with Question 4, why is it convenient to let *A* be the least significant bit?
6. How do you determine the number of variables required for a given sequence?
7. How do you determine the number of required maps for a given problem?
8. Figure 10–6 shows three Karnaugh maps but four variables. Explain.
9. Why do mode control input variables not require separate maps?
10. Example 10–9 discusses an *A/D* converter with one control input *S*. In this case, *S* is not constant. What is it a function of?
11. Look at Example 10–9 and review how five-variable maps are handled.
12. Why is sequential analysis useful?

PROBLEMS

1. Expand the synchronous counter of figure 10–10(b) to eight bits.
2. Design a four-bit counter with *T* flip flops.
3. Design a decade counter with *D* flip flops.
4. Design a down counter 7 to 0. Use *J/K* flip flops.
5. Repeat Example 10–6 for *D* as well as *T* flip flops.
6. Design a three-bit Gray counter using *R/S* flip flops.
7. Design a modulo 12 counter 0 to 11 and return. Use *T* flip flops.
8. Design a counter with the sequence 1–3–5–7–6–4–2–0–repeat. Choose your own flip flop.
9. Design a divide-by-three counter.
10. Design a divide-by-five counter. The sequence is not important as long as the count repeats after five clock cycles. Try to find a solution that requires only three flip flops and no gates.
11. Design a modulo 6 counter 0 to 5 and return. Use *D* flip flops.
12. Design a two-mode counter with *J/K* flip flops.
    $S = 0$: Decade, 0 to 9 and return.
    $S = 1$: Modulo 6, 0 to 5 and return.

Synchronous Sequential Logic

13. Design a two-mode counter with T flip flops. When $S = 0$, 0–1–2–3–3–3–3 stays in 3. When $S = 1$, it continues to 4–5–6–7–0–0–0–0 and stays in 0 until $S = 0$.
14. Design a four-mode two-bit counter with J/K flip flops.
    $S_2 = 1, S_1 = 1$: Count 0–1–2–3 and stop.
    $S_2 = 1, S_1 = 0$: Count 0–1–2 and stop.
    $S_2 = 0, S_1 = 1$: Count 0–1 and stop.
    $S_1 = 0, S_2 = 0$: Count 0–0, that is, no count.
15. Design a four-mode counter using J/K flip flops.
    $S_2 S_1 = 00$: Sequence 0–1–2–3–4–5 and return.
    $S_2 S_1 = 01$: Sequence 5–4–3–2–1–0 and return.
    $S_2 S_1 = 10$: Sequence 0–1–3–2–6–7–5–4 and return.
    $S_2 S_1 = 11$: Sequence 0–1–2–3–4–5–6–7 and return.
16. Design a successive approximation register using D flip flops. Use the Now-Next columns from figure 10–25.
17. Expand the pointer shift register of figure 10–22 to eight bits.
18. Design a pointer that alternates outputs of flip flops A and B five times, then gives a single pulse output at flip flop C and repeats.
19. Design a three-bit shift-right, shift-left register.
20. Design a counter with D flip flops for the following sequence: 1–3–7–6–5–2–4–repeat. Could you use a shift register for this implementation? What happens if in state 0?
21. Redesign Problem 20 such that the zero state problem is overcome.

chapter 11

sequential logic continued

11-1 INTRODUCTION

The method that enables us to determine the required logic for sequential events was established in Chapter 10. In the following sections we shall broaden the subject by introducing state diagrams, sequential analysis, control pulse generation, and a few examples of event sequences of higher complexity.

11-2 STATE DIAGRAMS

The Now-Next transition table lists the required sequence of a set of events. It is a bookkeeping method that assures us that all possible states and transitions have been considered and that the resulting logic will not present us with unexpected surprises. In many cases the required sequence is sufficiently straightforward to be directly listed in the Now-Next transition table.

In more complex cases it is often helpful to precede the design or check the results of an analysis with a state diagram. This is nothing more than a pictorial representation of a sequence.

Take the simple counter of Example 10-1. Each state is represented with a circle labeled with a binary state and arrows to show how the counter advances. (See figure 11-1(a).)

A more complex state diagram also labels the arrows with the corresponding decision input variables. The up-down counter sequence of figure 10-16(a) is shown in figure 11-1(b). The external control input S condition is labeled alongside the arrows. The numbers in the circles represent the present state. The arrows indicate the transition with the next clock cycle. The input conditions S for which the desired transition occurs are stated alongside the arrows.

Sequential Logic Continued

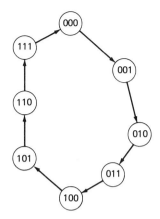

FIGURE 11–1(a) Three-bit counter

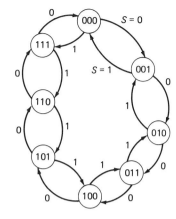

FIGURE 11–1(b) Three-bit up-down counter

EXAMPLE 11–1

Suppose a serial stream of three data bits arrives in any of eight possible combinations. If the arriving three-bit number is 101, a one-cycle pulse is generated. If the arriving three-bit number is 111, a two-cycle-long pulse is generated.

We do not know how many flip flop states are required, but we can assume that the system starts in a 0 state. Two possibilities exist out of state 0:

- A 0 arrives. This is not a valid beginning for 101 or 111. The system does not respond and stays in state 0.
- A 1 arrives. This is a good beginning. We go to state 1 and investigate further.

Two possibilities exist out of state 1:

- A 0 arrives. There is a possible 101. Go to state 2 and investigate further.
- A 1 arrives. There is a possible 111. Go to state 3 and investigate further. (Does this remind you of a poker game?)

Two possibilities exist out of state 2:

- A 0 arrives. Neither a 101 nor a 111 can be obtained. Go back to state 0.
- A 1 arrives. A 101 has been found. Go to state 4.

The system stays in state 4 for one clock cycle. Flip flop C will put out a pulse that is one clock cycle long.

Two possibilities exist out of state 4:

- A 0 arrives. Go back to state 0 and wait for a new sequence.
- A 1 arrives. This is possibly another good sequence. Go back to state 1 for further investigation.

Two possibilities exist out of state 1:

- A 0 arrives. This is a possible 101. Go to state 2.
- A 1 arrives. This is a possible 111. Go to state 3.

Two possibilities exist out of state 3:

- A 0 arrives. Neither a 101 nor a 111 can be obtained. Go back to state 0.
- A 1 arrives. A 111 has been found. Go to state 6.

Two possibilities exist out of state 6:

- A 0 or a 1 arrives. Go to state 7 in either case to achieve a two-cycle-long pulse on output C.

Two possibilities exist out of state 7:

- A 0 arrives. Go back to state 0 and wait for a new sequence.
- A 1 arrives. This is possibly another good sequence. Go back to state 1 for further investigation.

We need three flip flops to satisfy seven states. State 5 is a "don't care" state. The state diagram in figure 11–2 reflects the above reasoning. Notice that in state

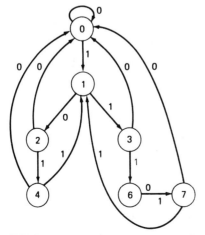

FIGURE 11–2 State diagram for recognition of a 101 or 111

Sequential Logic Continued

4 we have found a 101. Flip flop C is ON for one cycle. In states 6 and 7 we have found a 111. Flip flop C is ON for two cycles.

From the state diagram we can generate the Now-Next table as shown in figure 11–3. Implement the equations of figure 11–3 as shown in figure 11–4.

Now CBA	$S = 0$ Next CBA	$S = 1$ Next CBA
0 0 0	0 0 0	0 0 1
0 0 1	0 1 0	0 1 1
0 1 0	0 0 0	1 0 0
0 1 1	0 0 0	1 1 0
1 0 0	0 0 0	0 0 1
1 0 1	XXX	XXX
1 1 0	1 1 1	1 1 1
1 1 1	0 0 0	0 0 1

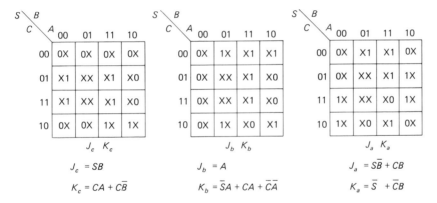

FIGURE 11–3 Transition table, control maps, and equations for Example 11–1

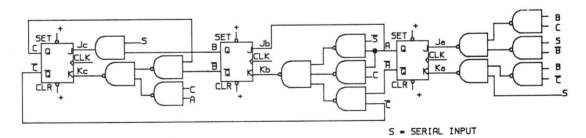

S = SERIAL INPUT

FIGURE 11–4 Recognition of a 101 or a 111

11–3 SEQUENTIAL ANALYSIS

The examples up to this point have demonstrated the design procedure. It is also necessary to have some means whereby one can draw a timing diagram to check that the circuit will indeed carry out the desired function. Unfortunately, there is no easy method to predict the nth step with a closed form mathematical expression. It is not difficult to construct a table with an arbitrary starting point and proceed with a step-by-step analysis from the beginning to the end of the sequence. One is often faced with the problem of looking at someone else's schematic and figuring out the intended operation of a flip flop feedback network. Just looking at the circuit will not tell us what waveforms to expect when displayed on the oscilloscope. It becomes beneficial to draw the expected wavetrain and compare it with the observed wavetrain to detect any malfunction.

- Step 1. From the given logic network, either your own or someone else's, determine the control input equations to all the flip flops.
- Step 2. Draw a sequence table as shown in figure 11–5.
- Step 3. Label the ordinate with all the control inputs and their corresponding Boolean equations as well as the flip flop outputs.
- Step 4. Number the abscissa in clock cycles.
- Step 5. Choose a starting value for the flip flop outputs. (All 0's is usually a good beginning.)
- Step 6. Determine the 1 or 0 value for all the control inputs at the chosen starting point and determine what each flip flop will do after the next clock cycle. Enter these new values next to the starting output value on the table.
- Step 7. Redetermine the new 1's and 0's for all the flip flop control inputs and repeat Step 6.

Suppose we have a circuit that looks like the one we obtained in Example 10–5 [figure 10–16(a)]. Without prior knowledge of the control equation we need only look at the control input circuit to establish the Boolean expressions.

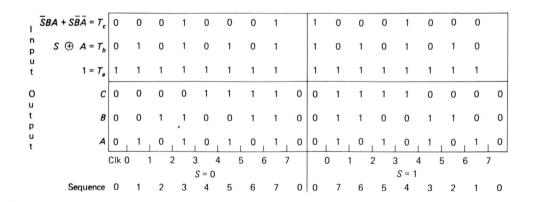

FIGURE 11–5 Sequence table for Example 11–2

Sequential Logic Continued

EXAMPLE 11-2

- Step 1. From figure 10–16(a) we write the control equations.
- Steps 2 through 7 involve the creation of the sequence table.

The sequence is exactly as specified in Example 10–5. Therefore the circuit was properly designed. ∎

EXAMPLE 11-3

Suppose we have a circuit as shown in figure 11–6 and we want to establish what it does.

First, find the control equations:

$$J_c = B\overline{A} \qquad J_b = CA + \overline{C}\overline{A} = \overline{C \oplus A} \qquad J_a = C$$
$$K_c = BA \qquad K_b = CA + \overline{C}\overline{A} = \overline{C \oplus A} \qquad K_a = CB$$

The sequence found in figure 11–7 is 0–2–4–5–7–repeat. States 1, 3, and 6 do not occur. It is necessary to find out what happens if by some chance the circuit finds itself in one of those states that do not normally occur. There is no way to know what the initial state of the circuit is after power turn-on. Without the help of a Reset pulse, will the circuit find its way into the proper sequence or not? The three questionable states are investigated in the sequence table in figure 11–7 by starting in the state of question. We find that the circuit will recover from state 6 into state 7 and then resume the normal required count sequence. States 1 and 3, however, are "stuck" states. There is no way ever to exit from them. There is no judgment as to whether this is good or bad. Perhaps this is exactly what the designer wants. If not, the circuit needs to be redesigned to make sure this does not happen. When a sequence with "don't care" states is designed, we use these to get better gate minimization. After the design is completed, it is necessary to investigate the circuit with the above procedure to establish proper operation and to make sure that "don't care" states are not "stuck" states.

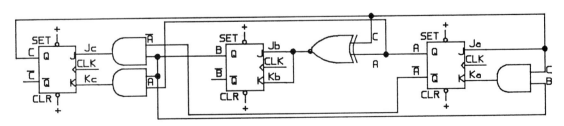

FIGURE 11-6 A counter with "stuck" states

		0	1	0	0	0	0	0	1
Input	$B\bar{A} = J_c$								
	$BA = K_c$	0	0	0	0	1	0	1	0
	$\overline{C \oplus A} = J_b$	1	1	0	1	1	0	0	0
	$C \oplus A = K_b$	1	1	0	1	1	0	0	0
	$C = J_a$	0	0	1	1	1	0	0	1
	$CB = K_a$	0	0	0	0	1	0	0	1
Output	C	0	0	1	1	1	0	0	0	0	1	1
	B	0	1	0	0	1	0	0	1	1	1	1
	A	0	0	0	1	1	0	1	1	1	0	1
	Clk	0	1	2	3	4	0	0	0			
	Sequence	0	2	4	5	7	0	1	1	3	3	6	7

FIGURE 11–7 Sequence table for Example 11–3

EXAMPLE 11–4

Let the input to a four-bit A/D converter be 13.2 V. The starting value of the successive approximation register is $1000 = 8$. J_d was grounded; therefore all entries for J_d in figure 11–8 are 0. In the first column, $S = 1$ because $13.2 > 8$.

$$K_d = \overline{S}\overline{C}\overline{B}\overline{A}. \text{ Since } S = 1 \text{ and } \bar{S} = 0, K_d \text{ must } = 0.$$
$$J_c = D\overline{C}\overline{B}\overline{A} \text{ and } D = 1, \overline{C} = 1, \overline{B} = 1, \overline{A} = 1. J_c \text{ must } = 1, \text{ etc.}$$

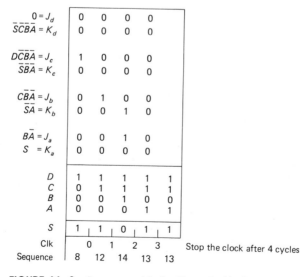

FIGURE 11–8 Sequence table for Example 11–4

Sequential Logic Continued

After every clock cycle you must determine the new value of S.

If the input $> DCBA$, $S = 1$.
If the input $< DCBA$, $S = 0$. ■

11-4 CONTROL PULSE GENERATION

An important application of the sequential design procedure is the design of control and timing networks. Timing and control must always fit the particular system considered. Starting sequences, handshake control between two subsystems, repetitive algorithms in arithmetic networks, memory timing control, and many other applications require the generation of pulses at given clock times. Unwanted glitches can have disastrous consequences for the system. Proper synchronous design ensures trouble-free operation.

Several examples will illustrate how to use what we have learned to generate pulses at desired time slots.

EXAMPLE 11-5

Suppose we want to generate a single pulse, one clock cycle wide, when a manual push-button is depressed. Switch bounce must not be seen by the system. (Switches are mechanical contacts. When the snap closes, the contact bounces like a ping pong ball. This causes the voltage output across the switch to flip back and forth between 1 and 0 until the switch has settled. Furthermore, as the switch contact surfaces approach in the process of closing, there is a transition time when closure is not well defined. What is contact? Zero distance between surfaces or perhaps 1 μin. or 20 μin.? What does constitute contact? In practice the two surfaces make and break contact several times before finally making permanent contact. The same happens during opening of a switch. As a result the output across a switch exhibits a string of pulses rather than a clean single step when closed or opened. This is called switch bounce. Such behavior can confuse a logic system and must be eliminated.)

One way to eliminate switch bounce is to set a flip flop. Since all bounces repeatedly set the flip flop, only one step is seen at the flip flop output. However, the step is not the pulse we want. It can serve to enable another flip flop circuit that generates the required pulse. The flip flop that was set by the switch can then be reset after an appropriate time delay; or simpler yet, we can leave it in the set state until some other switch in the system is depressed, which in turn resets the flip flop that was set earlier.

The single pulse circuit is similar to the two-mode counter in Example 10–5. Let S be the output of the flip flop set by the switch. If $S = 0$, nothing happens; if $S = 1$, then a flip flop will change from a 0 state to a 1 state, change back to a 0 state with the next clock cycle, and stay there as long as $S = 1$. There must be a second flip flop B that knows that a pulse has just been generated and keeps the system in a hold state until S goes to 0. Translated into a sequence statement, this becomes the task in figure 11–9.

State 3 does not occur; therefore we have the X's in figure 11–10. The equations of figure 11–10 are implemented in figure 11–11.

The analysis checks states 0, 1, and 2 for proper functioning but must also look at state 3 to make sure that the control logic will not get stuck or hang up. (See figure 11–12.) The timing diagram is shown in figure 11–13.

For $S = 0$ the circuit always returns to a 0 state, in either one or at most two clock cycles. When $S = 1$, the circuit proceeds properly from 0–1–2–2, etc. If a 3 state is reached by accident, it returns to state 0 for $S = 0$ and to state 2 for $S = 1$.

$S = 0$ Sequence: all states return to 0
$S = 1$ Sequence: 0-1-2-2-2-etc.

Now B A	$S=0$ Next B A	$S=1$ Next B A
0 0	0 0	0 1
0 1	1 0	1 0
1 0	0 0	1 0
1 1	X X	X X

FIGURE 11–9 Transition table for Example 11–5

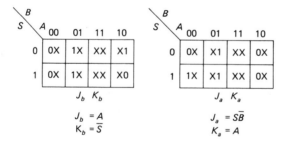

$J_b = A$
$K_b = \overline{S}$

$J_a = S\overline{B}$
$K_a = A$

FIGURE 11–10 Control maps for Example 11–5

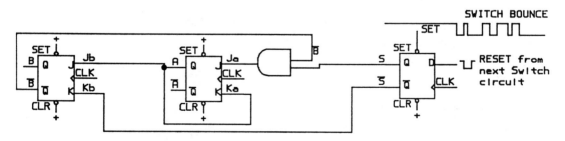

FIGURE 11–11 Single-pulse generator

Sequential Logic Continued

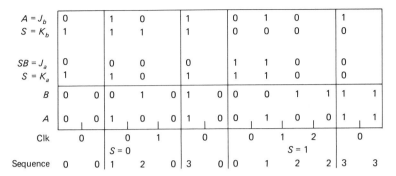

FIGURE 11-12 Sequence table for Example 11-5

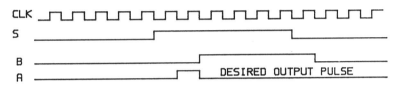

FIGURE 11-13 Single-pulse timing

EXAMPLE 11-6

Two pulses are to be generated on two lines following each other after the arrival of a control level S, but only once until S returns to 0. The two pulses follow each other immediately. The first pulse is one clock cycle wide, and the second is two clock cycles wide. It is useful to draw a timing diagram of the requirement as shown in figure 11-14. From the previous example it is clear that a hold state is needed until S returns to 0. This prevents the generation of further pulses while S is high. Two identical states within one sequence are not permitted, since it is not possible for the circuit to differentiate between them. (Switch bounce is not considered in this example.)

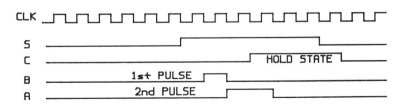

FIGURE 11-14 Two-pulse timing

Interpretation of the timing diagram tells us which states are required:

$S = 0$: Return to 0

$S = 1$: Sequence: 0–2–1–5–4–4–4, etc.

See figure 11–15. Figure 11–16 shows the implementation of the equations in figure 11–15. The analysis of the network in figure 11–16 is left as an exercise. "Don't care" states must be investigated.

| Now ||| S = 0 Next ||| S = 1 Next |||
C	B	A	C	B	A	C	B	A
0	0	0	0	0	0	0	1	0
0	0	1	X	X	X	1	0	1
0	1	0	X	X	X	0	0	1
0	1	1	X	X	X	X	X	X
1	0	0	0	0	0	1	0	0
1	0	1	X	X	X	1	0	0
1	1	0	X	X	X	X	X	X
1	1	1	X	X	X	X	X	X

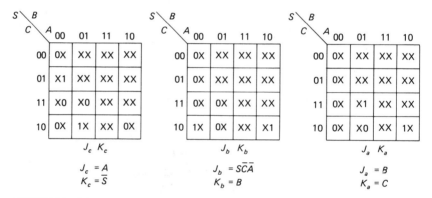

FIGURE 11–15 Transition table, control maps, and equations for Example 11–6

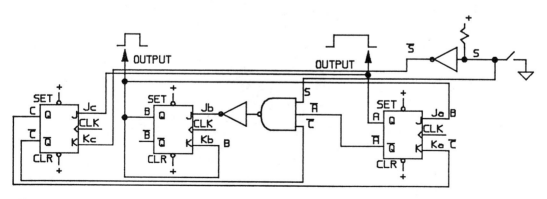

FIGURE 11–16 Two-pulse network

Sequential Logic Continued

EXAMPLE 11-7

An output pulse, one clock-cycle wide, is required if a pulse S_2 after a pulse S_1 has occurred. Here we are dealing with two external inputs. There must be four Next columns to account for all input combinations. As long as nothing happens, the system is in a 0 state. If an S_1 pulse arrives, the system goes into a "wait" state. When the S_2 pulse arrives, an output pulse is generated. S_1 and S_2 will never arrive simultaneously. Thus there must be a 0 state, a "wait" state, and an output state. Therefore two flip flops are needed. We could derive the Now-Next table from a timing diagram or from a state diagram, or we could bypass either step and proceed directly to the transition table in figure 11–17. The equations in figure 11–17 are implemented in figure 11–18.

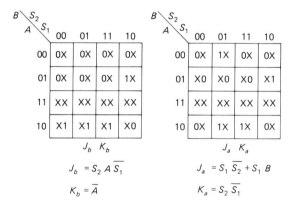

FIGURE 11–17 Transition table, control maps, and equations for Example 11-7

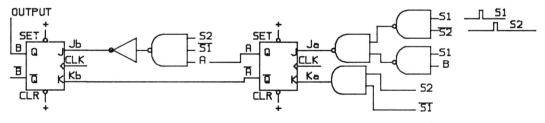

FIGURE 11–18 Output only if S_2 is later than S_1

EXAMPLE 11-8

Suppose we have a string of 0's and 1's arriving on a line. As long as 0's arrive, we want to transmit these as 0's. As soon as a 1 arrives, we want to transmit that 1 as a 1. All bits arriving thereafter shall be inverted. (This problem finds application in serial two's complement logic. Try this process: Choose an eight-bit number, write the two's complement, and compare it with the result of this method.) Let the control line be S and the arriving bits be N. Since the bits to be reversed are not to be delayed by one clock cycle, we will use a gate to obtain the final output. An associated output map can be made to show the required output for the corresponding flip flop state.

Only one flip flop is needed to indicate a new state within which bit reversal takes place.

Note that the Now-Next columns in figure 11–19 are essentially in Karnaugh map form. If a D flip flop is chosen, we can read the map directly to obtain the D input control and output equations.

The "don't care" states arise because if $S = 0$, the flip flop would never get into the 1 state. (See figure 11–20.)

	Flip flop states				Associated output			
	S N 0 0	S N 0 1	S N 1 1	S N 1 0	S N 0 0	S N 0 1	S N 1 1	S N 1 0
Now A	Next A	Next A	Next A	Next A	A	A	A	A
0 1	0 X	0 X	1 1	0 1	0 X	1 X	1 0	0 1

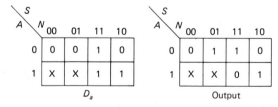

FIGURE 11–19 Transition table, output table, and control maps for Example 11–8

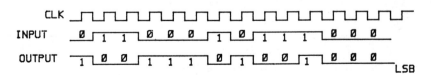

FIGURE 11–20 Serial two's complement timing diagram

$$D_a = SN + A \qquad \text{Output} = N\overline{A} + \overline{N}A = N \oplus A$$

The output Exclusive OR gate is a conditional inverter; therefore it appears as no surprise. (See figure 11–21.)

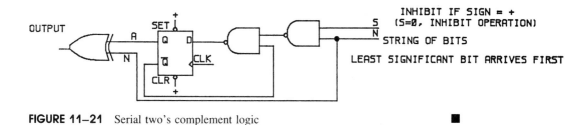

FIGURE 11–21 Serial two's complement logic ∎

EXAMPLE 11–9

Let S be a start pulse. The sequence counts to 8 and back to zero and will not repeat unless another start pulse arrives. (See figures 11–22 and 11–23.)

There will be an output as long as the counter is active, that is,

$$\text{Output} = CLK \cdot (C + B + A)$$

Now CBA	S = 0 Next CBA	S = 1 Next CBA
000	000	001
001	010	010
010	011	011
011	100	100
100	101	101
101	110	110
110	111	111
111	000	000

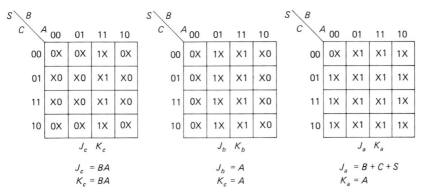

FIGURE 11–22 Transition table, control maps, and equations for Example 11–9

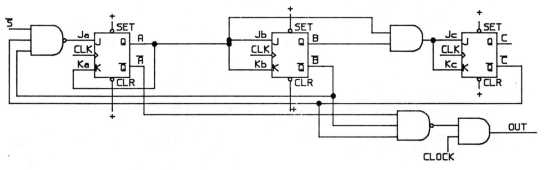

FIGURE 11-23 Eight clock cycles and stop

EXAMPLE 11-10

When two systems run on independent clocks, it often becomes necessary to hold off an operation until the way is cleared to proceed. This is a *handshake situation* between two systems.

Assume that pulse S_1 arrives but is to be transmitted only after pulse S_2 has come and gone. If S_1 arrives simultaneously with S_2, it must wait until S_2 occurs again. There are four possibilities to account for:

S_1 arrives before S_2.

S_2 arrives before S_1.

S_1 and S_2 arrive together.

Nothing arrives.

Let us furthermore assume that S_2 and S_1 are only one clock cycle wide and that S_1 will not arrive twice before an S_2 has occurred. Two flip flops are needed to handle the four cases.

The table of figure 11-24 needs some explanation:

- When *BA* are in 00 state and $S_2S_1 = 00$, nothing happens and the Next state is 00.
- When S_2 arrives alone, the circuit ignores the event.
- When S_1 arrives either alone or with S_2, *BA* goes to a 01 state, meaning that the circuit acknowledges the arrival of S_1.
- When *BA* are in a 01 state and an S_2 arrives, the circuit acknowledges the event by switching to a 10 state.

Sequential Logic Continued

	$S_2 S_1$ 00	$S_2 S_1$ 01	$S_2 S_1$ 11	$S_2 S_1$ 10
Now BA	Next BA	Next BA	Next BA	Next BA
0 0	0 0	0 1	0 1	0 0
0 1	0 1	X X	X X	1 0
1 0	1 1	X X	X X	X X
1 1	0 0	0 1	0 1	0 0

FIGURE 11–24 Transition table for Example 11–10

- After S_2 has disappeared, the circuit switches to a 11 state and sends S_1 out. Thus the 11 state is the output state, after which the circuit returns to a 00 state or to a 01 state if another S_1 has arrived.

Two maps are needed for B and A as shown in figure 11–25. There are four variables: S_2, S_1, B, and A. (In this case it is convenient to let S_1 be the least significant bit in order to retain a one-to-one correspondence to the table in figure 11–24.)

This is implemented in the network of figure 11–26.

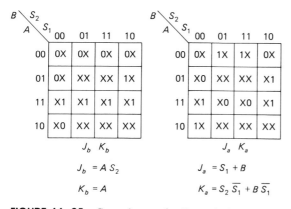

$J_b = A S_2$
$K_b = A$

$J_a = S_1 + B$
$K_a = S_2 \overline{S_1} + B \overline{S_1}$

FIGURE 11–25 Control maps for Example 11–10

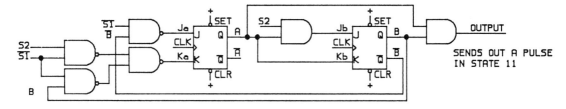

FIGURE 11–26 Synchronize two pulses

EXAMPLE 11-11

Differentiation is a process of measuring the rate of change of one variable with respect to another. One might be interested in measuring the rate of change of voltage with respect to time. In other words, measure the slope of a waveform. A particular example that is worthy of looking at is a digital peak detector. Assume that we have an analog pulse and want to determine the peak value as well as the precise time when the peak occurs. This is a sequential problem. We must continuously monitor at least two successive points along the curve in order to evaluate when a slope change occurs.

An analog pulse is sampled with a high-speed flash converter, and a succession of digital values is then processed. The sequence of events is as follows:

- Step 1. The digitized samples, let us say eight bits, are loaded in a two-stage shift register. At any one time the two registers hold the last two samples. The second stage holds the earlier value, and the first stage the last-entered value, that is, the later value.
- Step 2. Both shift register outputs are fed into a digital comparator. As long as the input pulse is rising, the later value must be larger than the previous one. The digital comparator gives a zero output, indicating a positive slope.
- Step 3. As the sampling process passes the peak, the slope reverses and the comparator has a 1 output, indicating a negative slope.
- Step 4. The last value in the second shift register must have been the peak or close to it. The comparator output change from 0 to 1 can be used as a strobe to remove the desired data from the second register.

The accuracy of the peak detection is proportional to the sampling frequency. By sampling often enough, one can come arbitrarily close to the actual peak. Ideally, we would want to detect a zero slope when two successive samples are equal. This is difficult to achieve in a practical situation for two reasons:

1. The peak might not be flat for a two-sample interval within the resolution of our eight bits.
2. The input pulse might have superimposed noise, which might indicate a false flat spot.

The same noise might also indicate a false peak when a negative slope is detected. It is therefore necessary to detect several sampling periods with a consistent negative slope before being reasonably sure that a true peak had occurred. It is assumed that a noise spike will be shorter in time than, say, three sampling periods.

The first detected peak value after a negative slope detection must therefore be stored in three additional shift registers while the slope continues to be monitored. If, during the monitoring period of three sampling periods, the comparator

Sequential Logic Continued

again shows a positive slope, then the suspected peak value is rejected, and the system continues to search for a true peak.

Consequently, we must design noise-monitoring control logic. The only information we have to work with is the comparator output. To repeat: We want to find a true peak as well as generate a pulse to indicate that a true peak has been found.

The sequence requires five states as shown in figure 11–27; three flip flops are needed. Three flip flops have eight states, and we can assign convenient states to given sequences of events.

The comparator output that is the control input to the three flip flops is S.

The three flip flop outputs are C, B, and A.

The output is taken at flip flop C in state 4.

The design criteria are as follows (for CBA):

- 000: The control logic starts in state 000. As long as nothing happens, it remains there. If a 1 arrives, the state changes to 01, indicating a slope change from positive to negative.
- 001: If a 0 follows (meaning $S = 0$) the previous $S = 1$, it must have been a false peak; return to state 000. If a 1 arrives, proceed to state 010.
- 010: If $S = 0$, return to state 000. It must have been a false peak. If $S = 1$, go to state 100. Flip flop C output $= 1$ means that a peak has been accepted. Use the flip flop C output $= 1$ to enable the output buffer in figure 11–30.
- 100: If $S = 0$, return to state 0. If $S = 1$, go to state 011. This is a "hold" state until the input has returned to 0.

Now		Next	
		Input = S	(slope)
State	CBA	0 CBA	1 CBA
0	000	000	001
1	001	000	010
2	010	000	100
3	011	000	011
4	100	000	011
5	101	XXX	XXX
6	110	XXX	XXX
7	111	XXX	XXX

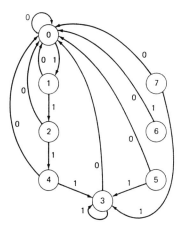

FIGURE 11–27 Transition table and state diagram for Example 11–11

- 011: If $S = 0$, return to state 000, ready for a new start. If $S = 1$, remain in state 011. Wait there until the input has returned to 0 (that is, the slope is 0 after the input is done).

States 101, 110, and 111 are "don't care" states.

The above reasoning translates into a Now-Next state table (transition table) as well as a state diagram in figure 11–27.

States 5, 6, and 7 are irrelevant. The desired transitions are mapped as shown in figure 11–28. The control logic implementation follows readily and is shown together with the needed shift register and output hold register in Figure 11–30.

The operation of the peak detector can be illustrated with a three-bit rather than an eight-bit word sequence. Suppose an incoming triangular input from the A/D converter uniformly increases from 000 to 110, then decrements back to 000. Notice that the peak occurred with sample 7. One sample time later, the slope $S = 1$,

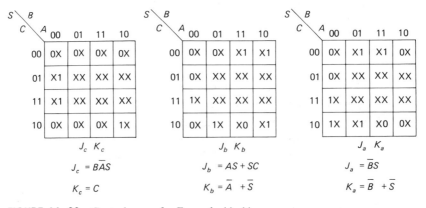

FIGURE 11–28 Control maps for Example 11–11

Sample	Reg 1	Reg 2	Reg 3	Reg 4	Reg 5	Buffer	Comparator output S	Flip flop output C	B	A
1	000	000	000	000	000	000	0	0	0	0
2	001	000	000	000	000	000	0	0	0	0
3	010	001	000	000	000	000	0	0	0	0
4	011	010	001	000	000	000	0	0	0	0
5	100	011	010	001	000	000	0	0	0	0
6	101	100	011	010	001	000	0	0	0	0
7	110	101	100	011	010	000	0	0	0	0
8	101	110	101	100	011	000	1	0	0	0
9	100	101	110	101	100	000	1	0	0	1
10	011	100	101	110	101	000	1	0	1	0
11	010	011	100	101	110	000	1	1	0	0
12	001	010	011	100	101	110	1	0	1	1
13	000	001	010	011	100	110	1	0	1	1
14	000	000	001	010	011	110	0	0	0	0

Enable buffer

FIGURE 11–29 Event sequence for a triangular pulse

Sequential Logic Continued 369

indicating a negative slope. At the end of sample 10 or the start of sample 11, the peak = 110 is loaded into the output buffer. The control circuit waits in state 011 until the slope = 0 for two sample periods, then returns to a 000 state.

The sequence of events for a triangular input with a peak = 110 is shown in figure 11–29. The implementation is shown in figure 11–30.

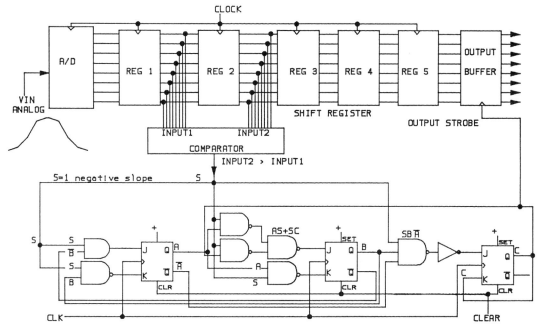

FIGURE 11–30   Digital peak detector

EXAMPLE 11–12

In Chapter 9 we discussed a FIFO and the control logic that determines how data is to proceed down the shift register until a busy register is encountered and then stop. Review the reasoning given in that application. In this example we want to design the required control logic for the FIFO.

Imagine an eight-bit-wide shift register that is partially filled. The filled positions must not be strobed lest we destroy the data in those locations. We must therefore know which locations contain data and develop a *busy map* of the system that is current for every clock cycle. Once the busy map is known, it is easy to decide which stages to strobe to advance new incoming data. The depth of the FIFO is, of course, a function of the system. All we need to do is to develop the timing concept for as short a memory as possible and recognize the repetitive pattern of control functions.

Such a pattern can be recognized from a memory that is only three positions deep. There are two external control signals

Data ready to be entered: S

Data requested for removal: T

The three memory stages are C, B, and A. The strobe functions to the registers are Z_c, Z_b, and Z_a.

As usual, we construct a Now-Next table (figure 11–31) followed by mapping (figure 11–32) and the desired strobe functions are read off the maps. Let the data enter at register A and be removed from register C. (See figure 11–33.)

The tables in figure 11–31 can be explained as follows.

For the busy table, assuming an empty memory and no data arrival, the busy map is obviously in a 000 state. If data arrives, $S = 1$, position A becomes busy, and the map moves to a 001 state. If T had requested data at the same time, it would not have made any difference because there was no data in position C to be removed. Thus in either case ($ST = 11$ or 10) the busy map goes to 001.

Take the case in which the busy table is in state 101, meaning that there is data in positions C and A. Suppose new data arrives and simultaneously data is removed, that is, $ST = 11$. The 1 from position C is replaced by the 0 in position B. The 1 from position A moves to position B, and the newly arriving data fills

State	Now CBA	ST 00 Next CBA	ST 01 Next CBA	ST 11 Next CBA	ST 10 Next CBA
0	000	000	000	001	001
1	001	010	010	011	011
2	010	100	100	101	101
3	011	110	110	111	111
4	100	100	000	001	101
5	101	110	010	011	111
6	110	110	100	101	111
7	111	111	110	111	111

Busy table

State	Now CBA	ST 00 $Z_c Z_b Z_a$	ST 01 $Z_c Z_b Z_a$	ST 11 $Z_c Z_b Z_a$	ST 10 $Z_c Z_b Z_a$
0	000	000	000	001	001
1	001	010	010	011	011
2	010	100	100	101	101
3	011	110	110	111	111
4	100	000	000	001	001
5	101	010	010	011	011
6	110	000	100	101	001
7	111	000	110	111	000

Output table

FIGURE 11–31 Busy table and output table

Sequential Logic Continued

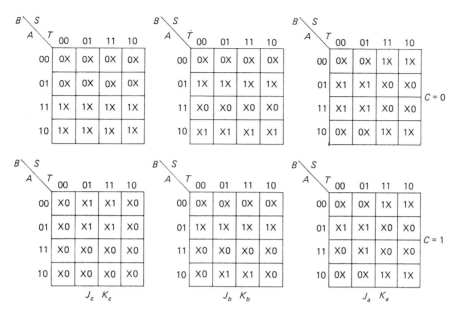

FIGURE 11–32(a) Output control maps

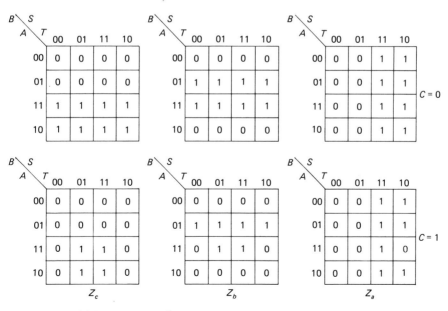

FIGURE 11–32(b) Busy control maps

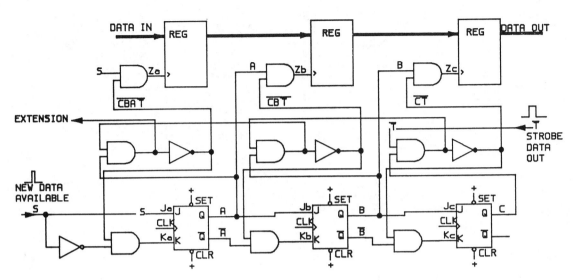

FIGURE 11-33 Fast FIFO memory with shift control

position A, setting busy A to 1. Thus 101 becomes 011. Similar reasoning fills the entire table.

The busy table indicates only which shift registers are to be strobed. The strobe logic must be implemented with the aid of the busy table. Z_c, Z_b, and Z_a are the strobe pulses applied to the respective shift registers.

For the output table, in state 0 when $CBA = 000$, the shift registers are empty. If no data arrives, $S = 0$, no strobes are applied; hence $Z_c Z_b Z_a = 000$. If data arrives, $S = 1$, only A need be strobed, so $Z_c Z_b Z_a = 001$.

In state 1 when $CBA = 001$, A is busy, and B and C are empty. If no data arrives, A is shifted to B. Thus B is strobed, and $Z_c Z_b Z_a = 010$. If new data arrives while in state 1, $ST = 11$ or 10; then A and B are strobed, and $Z_c Z_b Z_a = 011$.

In state 5, when $CBA = 101$, C and A are busy, and B is empty. If new data arrives but nothing is removed, $ST = 10$, and C will not be strobed, but A can move to B and the new data is entered into A; thus $Z_c Z_b Z_a = 011$. If data is removed while new data arrives, $ST = 11$, and $Z_c Z_b Z_a = 011$ as before because there was no data in B to be moved to C.

In state 7, when $CBA = 111$, all positions are busy. If new data arrives while data in C is not removed, $ST = 10$, no registers are strobed, and $Z_c Z_b Z_a = 000$; no data can be accepted.

Similar reasoning fills the entire output table. The busy control map and the output control map are shown in figures 11-32(a) and 11-32(b), where

$$J_c = B \qquad J_b = A \qquad J_a = S$$
$$K_c = \overline{B}(\overline{CT}) \qquad K_b = \overline{A}(\overline{BCT}) \qquad K_a = \overline{S}(\overline{CBAT})$$

Sequential Logic Continued

and where

$$Z_c = B(\overline{\overline{CT}}) \qquad Z_b = A(\overline{\overline{CBT}}) \qquad Z_a = S(\overline{\overline{CBAT}})$$

The extension to four stages is obvious. We get

$$J_d = C \qquad J_c = B \qquad J_b = A \qquad J_a = S$$
$$K_d = \overline{C}(\overline{\overline{DT}}) \qquad K_c = \overline{B}(\overline{\overline{DCT}}) \qquad K_b = \overline{A}(\overline{\overline{DCBT}}) \qquad K_a = \overline{S}(\overline{\overline{DCBAT}})$$
$$Z_d = C(\overline{\overline{DT}}) \qquad Z_c = B(\overline{\overline{DCT}}) \qquad Z_b = A(\overline{\overline{DCBT}}) \qquad Z_a = S(\overline{\overline{DCBAT}})$$

The functions inside the brackets can be used for K as well as Z. The circuit is readily implemented in figure 11–33. ∎

EXAMPLE 11–13

The control structure of the FIFO in figure 11–33 can be simplified if the timing requirements are relaxed. Let data be shifted into the next position only if that position is empty. In a true shift register, one can shift a block of data down one position with each clock cycle. Here we are shifting data only if the following position was emptied by a previous clock cycle. In other words, the data bubbles down until all positions near the bottom are filled. This implies that when data in a filled position are shifted, that location is left vacant for at least one cycle, and data residing one position above is shifted into the vacated position one clock cycle delayed. In this case the busy map becomes the output strobe map. This is so because a position empties for one cycle and then fills after a skipped cycle, allowing the busy signal to return to zero before becoming busy again. The leading edge of the busy signal thus becomes the strobe. The *busy-strobe map* changes to that shown in figure 11–34.

We can again explain a few of the map entries, after which all other entries are self-explanatory. If the FIFO is empty and data arrive, position A fills; thus 000 goes to 001. If nothing else happens, 001 goes to 010 and next to 100. Suppose A and B are filled and no new data arrives; then 011 goes to 101, then to 110. If

Now CBA	ST 00 Next CBA	ST 01 Next CBA	ST 11 Next CBA	ST 10 Next CBA
000	000	000	001	001
001	010	010	010	010
010	100	100	101	101
011	101	101	101	101
100	100	000	001	101
101	110	010	010	110
110	110	010	011	111
111	111	011	011	111

FIGURE 11–34 Slow shift FIFO transition table

more data arrives, 110 goes to 111. If T removes data from position C, stacked data proceeds to move down. Similar reasoning fills the entire table.

The mapping of the table in figure 11–34 is left as an exercise. The resulting control circuit is shown in figure 11–35.

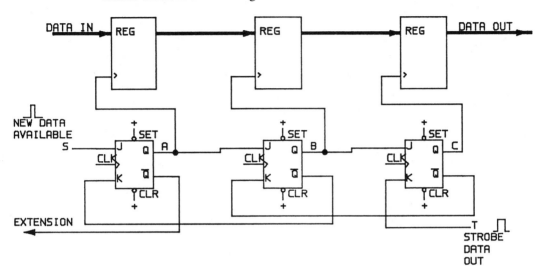

FIGURE 11–35 Slow FIFO memory with shift control

11–5 FLIP FLOP CONVERSION

During the early development of electronic circuits the flip flop was known as a bistable multivibrator. It is nothing more than a latch with Set and Reset inputs. It has two drawbacks: The output is not clocked and therefore is nonsynchronous, and Set and Reset applied simultaneously is a nonpermitted condition. A clocked output response is simply achieved with two additional gates. The truth tables and the circuit are shown in figures 11–36 and 11–37, respectively.

$RS = 11$ is not only an incompatible state but also a wasted state. Additional logic can make use of the $RS = 11$ control input to cause reliable toggling. Thus the R/S becomes a J/K flip flop.

S	R	Q
0	0	Q
0	1	0
1	0	1
1	1	?

Truth table

Q_n	Q_{n+1}	S	R
0	0	0	X
1	0	0	1
0	1	1	0
1	1	X	0

Excitation table

FIGURE 11–36 Truth tables for Example 11–14

Sequential Logic Continued

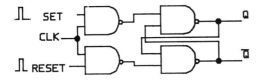

(NOTE : A SECOND SET DRIVEN FROM $\overline{CLK}$ WOULD MAKE IT A MASTER SLAVE RS FLIP FLOP)

FIGURE 11-37 Circuit for Example 11-14

EXAMPLE 11-14 R/S-to-J/K Conversion

With this goal in mind, let us construct a From-To table. In this case we want to transform an R/S to a J/K flip flop. We write the two excitation tables side by side and then read the required Boolean expressions for R and S in terms of J, K, and Q. (See figure 11–38.) The Q_{n+1} column serves only to define the SR and JK column. When S and R are read off the table in terms of J, K, and Q that exist now, Q_{n+1} in the future is ignored, since it is implied. "Don't cares" are ignored.

From		To			
S	R	J	K	Q_n	Q_{n+1}
0	X	0	X	0	0
1	0	1	X	0	1
0	1	X	1	1	0
X	0	X	0	1	1

From–To Table

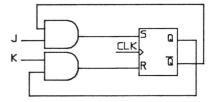

FIGURE 11-38 R/S to J/K conversion

There is a single 1 in the S column. It corresponds to J and Q: $S = J\overline{Q}$. There is a single 1 in the R column. It corresponds to K and Q: $R = KQ$. ∎

Similar From-To tables can be applied to convert any flip flop into another type, as shown in figure 11–39.

EXAMPLE 11-15 D-to-J/K Conversion

See the From-To table for D to J/K and figure 11–39. We get

$$D = J\overline{Q} + \overline{K}Q$$

D is true when Q changes from 0 to 1 *or* when Q stays in state 1. Again, input D is implemented in terms of J, K, and Q, ignoring the X. Note that Q_{n+1} does not enter into the Boolean equations. Q_{n+1} is in the table to get the proper D and JK transition conditions. D can be implemented only with the now conditions for J,

From	To			
D	J	K	Q_n	Q_{n+1}
0	0	X	0	0
1	1	X	0	1
0	X	1	1	0
1	X	0	1	1

From–To Table

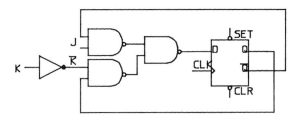

FIGURE 11-39 D to J/K conversion

K, and Q. When this is done, the indicated Q_{n+1} state will occur after the next clock pulse. ∎

REVIEW QUESTIONS

1. What is a good starting state for a sequential analysis problem?
2. If the chosen starting state in an analysis problem is a "stuck" state, what do you do?
3. In what sense is the generation of a control pulse a sequential problem?
4. In writing the Now-Next transition table for a control pulse problem, is it necessary to avoid repetition of identical states?
5. In connection with Question 4, if a repeated state is needed, what can be done to separate their identities?
6. How do we obtain the input control equations from a given logic schematic?
7. How can you determine that a given circuit hangs up in a "stuck" state?
8. Why is the A/D clock stopped after four cycles in Example 11-4?
9. Can the state sequence be read off the timing diagram in Example 11-6?
10. Can you think of applications in which it is necessary to generate a given number of clock pulses and stop?
11. Can you think of an application in which it is necessary to synchronize two random pulses?
12. In writing the From-To table for a flip flop conversion, the Q_{n+1} column appears. Since the entries in that column do not figure in the Boolean equation, why is that column needed?

PROBLEMS

1. Draw the state diagram for figure 10-10(a).
2. Draw the state diagram for figure 10-18.
3. Draw the state diagram for figure 11-18.

Sequential Logic Continued

4. Generate a state diagram for a serial bit stream recognition when:
   (a) The output pulse is one clock cycle wide for recognizing 011.
   (b) The output pulse is two cycles wide for recognizing 100.
5. Design the logic for the state diagram in Problem 4.
6. Analyze the circuit in figure 10–18.
7. Analyze the circuit in figure 10–22.
8. Analyze the circuit in figure 11–16. Will the circuit work if the control input S is a short pulse (one clock cycle wide)?
9. Analyze the successive approximation register as part of an *A/D* converter for a 9.5 V analog input.
10. Analyze the circuit in figure 11–40.
11. Analyze the circuit in figure 11–41 and draw a timing diagram.
12. Analyze the circuit in figure 11–42 and draw the timing diagram. Compare this problem to Problem 10 in Chapter 10.

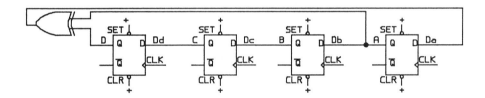

FIGURE 11–40

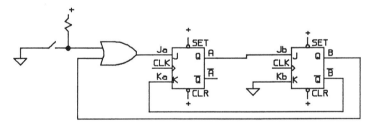

FIGURE 11–41

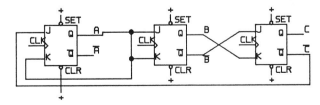

FIGURE 11–42

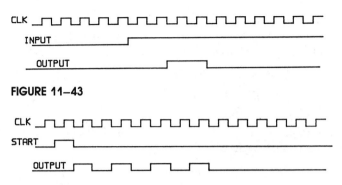

FIGURE 11-43

FIGURE 11-44

13. Design and analyze a circuit that will count the sequence 4-5-7-6-2-3-1-0-repeat.

14. Design a control pulse network such that a two clock cycle wide pulse is obtained but two cycles delayed from some initiating input gate, as shown in figure 11-43.

15. Design a control pulse network that puts out four consecutive pulses, each one clock cycle wide, after a start pulse has arrived. (See figure 11-44.) This sequence is not to be repeated before the next start pulse.

16. The serial two's complementer of figure 11-21 receives the following input: 1001000. Show the outputs of the two NAND gates, the Exclusive OR gate, and the flip flop for each clock cycle in a table. Draw the timing diagram.

17. The following set of numbers are measured samples of a Gaussian pulse. The peak detector of figure 11-30 detects and holds the peak value in the output buffer. Prepare a table similar to the table in figure 11-29 to show the sequence of events. The amplitude is given in a decimal equivalent of an eight-bit binary number

Amplitude	0	1	5	17	46	97	166	229	255	229	166	97	46	17	5	1	0	0
Time (μs)	0	0.2	0.4	0.6	0.8	1.0	1.2	1.4	1.6	1.8	2.0	2.2	2.4	2.6	2.8	3.0	3.2	3.4

18. Expand the three-position FIFO of figure 11-33 to six positions. Write the required equations and draw the circuit.

19. Write the Karnaugh maps for the slow FIFO in Example 11-13. Use the table in figure 11-34 for the map entries. Write the control equations.

20. Suppose you have four flip flops (figure 11-45) controlling the strobe inputs to four respective registers of a slow version FIFO. The inputs to the first register are data from some system. The fourth register has to hold the data until removed by some other system. Show by analyzing the four flip flops how the data proceeds through the register chain.

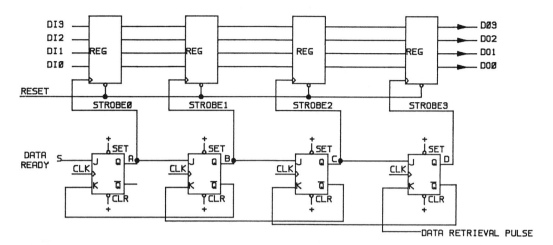

FIGURE 11-45

21. Convert a *T* flip flop to a *D* flip flop.
22. Convert a *J/K* flip flop to a *D* flip flop.
23. Convert a *D* to a *T* flip flop.
24. Convert any flip flop to any other flip flop you can think of.
25. Think of any count sequence. Design the network and then analyze it.
26. Make up a control pulse requirement. Design it, implement it on paper, and analyze it; then draw timing diagrams.

chapter 12

D/A and A/D converters

12–1 INTRODUCTION

Digital and analog circuits must sometimes work together in the same system. Measurements taken in the real world are usually analog in nature. If a digital system is to do any processing on such measurements, an *A/D* conversion is needed. The continuous analog values must be changed to digital numbers. Conversely, it is often necessary to look at a stream of changing digital values in analog form or control analog circuits with digital inputs. This requires a *D/A* converter. Furthermore, *D/A* converters are needed in the feedback loop of counting and successive approximation *A/D* converters. It is therefore appropriate to study the *D/A* converter first.

12–2 BITS AND WEIGHTS

In its simplest form a *D/A* converter is a weighted conversion of each bit of a word into a corresponding current. The sum of all the "bit currents" is an analog equivalent of the digital value.

Each bit in a word has only one of two binary states, either 1 or 0. The weight of each bit depends on its position within the word. An eight-bit byte, for example, has weights ranging from 2^7 to 2^0. The value of each bit is either equal to the weight or equal to zero, depending on whether the particular bit is 1 or 0.

EXAMPLE 12–1

Suppose we have an eight-bit byte = 11010011. To find the analog equivalent in decimal form, the byte under consideration is rewritten vertically together with the analog value:

D/A and A/D Converters

Bit 7: $1 = 2^7 \times 1 = 128$
Bit 6: $1 = 2^6 \times 1 = 64$
Bit 5: $0 = 2^5 \times 0 = 0$
Bit 4: $1 = 2^4 \times 1 = 16$
Bit 3: $0 = 2^3 \times 0 = 0$
Bit 2: $0 = 2^2 \times 0 = 0$
Bit 1: $1 = 2^1 \times 1 = 2$
Bit 0: $1 = 2^0 \times 1 = \underline{1}$

211

This is the familiar binary-to-decimal conversion method. ∎

12-3 A SIMPLE D/A CONVERTER

It is easy to obtain currents from each bit position that are equal (or correspond) to these eight values. Assume, for simplicity's sake, that a 1 is 1.28 V and a 0 is 0 V. A 1K resistor from A_7 to ground from the most significant bit position in figure 12-1 will draw 1.28 mA current. A 2K resistor in the next lower significant position will draw 0.64 mA. A 4K resistor from the next lower position draws 0.0 mA because the input is zero. The resistors in a binary resistance ladder increase in binary fashion to 128K. The 1K resistor draws the largest current and represents the most significant bit. The 128K resistor draws the least current and represents the least significant bit. The total current from all resistors in figure 12-1 flows into ground and is not too useful.

Instead of returning the resistors to a common ground, we can choose the virtual ground of an operational amplifier as in figure 12-2. The virtual ground input of the operational amplifier becomes the summing point of all eight currents. The output of the operational amplifier is -2.11 V if a 1K resistor is used in the feedback loop.

The virtual ground of the operational amplifier is for most practical purposes at 0 V. The current collected at the virtual ground summing node must flow through the feedback resistor to the op-amp output. The polarity of the output is reversed. The op-amp output can be written as

$$V_{out} = -[I(sum) * R_{feedback}]$$

For the input chosen in figure 12-2 the output = -2.11 V is indeed an analog representation of 11010011.

The analog output need not be a decimal equivalent of the binary word. In

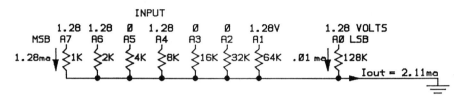

FIGURE 12-1 Binary current summing

most cases it need only be proportional. A different set of resistors can be used, provided that the eight resistors to the summing node retain the binary ratio. The feedback resistor can scale the output to any desired level.

A brief discussion of operational amplifier properties is given in the appendix.

EXAMPLE 12-2

Let us follow the output of the circuit in figure 12–2 if the digital input is incremented from 0000 0000 to 0000 0111, that is, from 0 to 7, in steps of 1.

- An all 0 input results in 0 current and 0 output voltage.
- A single 1 in the least significant position results in 0.01 mA current and 10 mV output. This is arrived at by

$$\frac{1.28 \text{ V}}{128\text{K}} * 1\text{K} = 0.01 \text{ V}$$

- A single 1 in the next higher position results in 0.02 mA current and 20 mV output.
- A 1 in the two least significant positions results in the sum of 0.01 mA + 0.02 mA = 0.03 mA current and 30 mV output.

As the binary word increases in steps of 1, the output increases in steps of 10 mV, generating a staircase, as shown in table 12–1. An eight-bit word will generate 255 analog steps. The output is quantized. A true analog ramp cannot be generated,

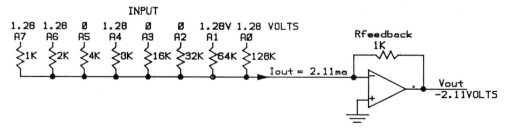

FIGURE 12-2 A simple *D/A* converter

D/A and A/D Converters

TABLE 12-1 Analog staircase generation

Bits 76543210	Current, I_{out} (mA)	V_{out}, $-I_{out} \times R_{feedback}$ (V)	Stepladder
00000000	0.0	0.0	
00000001	0.01	−0.01	
00000010	0.02	−0.02	
00000011	0.03	−0.03	
00000100	0.04	−0.04	
00000101	0.05	−0.05	
00000110	0.06	−0.06	
00000111	0.07	−0.07	
11111111	2.55	−2.55	

but one can come arbitrarily close, within practical limitations, by D/A converting more than eight bits.

The 1.28 V for a 1 level was chosen for ease of explanation. In a real TTL circuit the high level is approximately 3.5 V, and the feedback resistor can be chosen for any desired gain. ∎

The circuit of figure 12-2 has three drawbacks:

1. The binary ladder of different resistances is more difficult to manufacture with the same accuracy and temperature stability on a wafer than a set of identical resistors.
2. Each resistor is a different load to the driving voltage.
3. The output accuracy depends on the TTL (or CMOS) driving voltage.

12-4 THE ITERATIVE BINARY LADDER

The drawbacks listed at the end of the preceding section are overcome by

1. using an iterative ladder network,
2. using a reference source to drive the ladder, and
3. using the bit levels (of 1 or 0) to control switches.

It is possible to construct a resistive network as in figure 12-3 with iterative —that is, repeating—sections such that all resistor values are R and $2R$ and each successive section effects a current division by 2.

The ladder of figure 12-3 has some interesting properties.

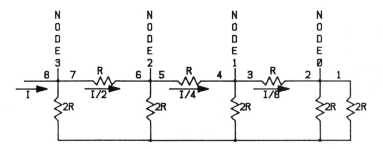

FIGURE 12–3 Iterative binary ladder

- When you look to the right at point 1, you see a resistance of 2R.
- When you look to the right at point 2, you see 2R in parallel with $2R = R$.
- When you look to the right at point 3, you see another R in series; the total is again 2R.
- This repeats to the very input point 8, where you again see R.

The input resistance remains independent of the number of added sections.

Assume now that a precision reference voltage is applied at node 3. The source sees an input resistance $= R$ and will deliver a current $I = V_{ref}/R$. At node 3 there is a 2R resistor to ground and a 2R resistor when you look into the series resistor to the right. The current I will divide evenly into the two branches. $I/2$ flows toward node 2, where the same condition of 2R in each direction is encountered. $I/2$ will again divide by 2. $I/4$ flows toward the next node. The current will divide by 2 as many times as there are sections.

Figure 12–4 is a copy of figure 12–3 except that a set of switches is added,

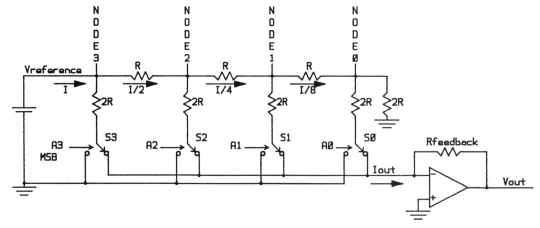

FIGURE 12–4 Binary ladder with controlled bit switches

D/A and A/D Converters

one to each shunt branch, and a virtual ground of an op-amp replaces the real common ground. The switches are controlled by the bits of interest. If a particular bit is a 1, the switch connects to the current summing virtual ground, which is the I_{out} line. If it is 0, the switch connects to real ground and diverts the current from the summing point.

Suppose $R = 1\text{K}\,\Omega$ (and $2R = 2\text{K}\,\Omega$) and $V_{ref} = 16$ V. The current I into node 3 = 16 V/1000 Ω = 16 mA. Let the bit pattern = 1001. Thus switches 3 and 0 are connected to the I_{out} line, and switches 2 and 1 are connected to ground. The I_{out} line receives $I/2 = 8$ mA from node 3 and $I/16 = 1$ mA from node 0. The total current flowing out is the sum of $I/2 + I/16 = 9$ mA, a direct analog representation of the number 1001 = 9 fed into the system. The feedback resistor is conveniently chosen to give the desired output voltage. V_{ref} need not have the value chosen for the example. Output calibration is done with the op-amp feedback resistor.

It is clear that the output voltage is not dependent on the high or low level accuracy of the bits in question but only on the accuracy of the reference voltage V_{ref} and the resistor ladder. The bits of the digital value to be converted serve only to control the switch positions.

Most commercially available D/A converters do not include the op-amp as part of the D/A chip. The I_{out} line is the output.

12–5 BINARY LADDERS WITH SWITCH ISOLATION

A modified version of figure 12–4 in which the switches are isolated by a collector of a drive transistor is shown in figure 12–5. All bases are held at a precise reference voltage; the bit currents, however, are sent out through the collector current source. The output current is therefore independent of switch resistance.

Another version of figure 12–5, shown in figure 12–6, has the current divide ladder in the collector circuit and uses diodes instead of switches to control the bit currents. When a particular bit is high, the diode is reverse biased, and the current

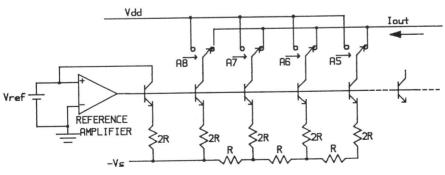

FIGURE 12–5 *D/A converter with current source drive (Courtesy of Datel Corporation)*

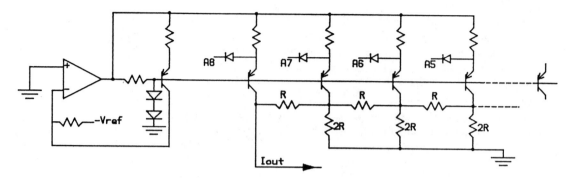

FIGURE 12–6 *D/A converter with current source drive and diode switching (Courtesy of Datel Corporation)*

to the respective transistor is delivered by the reference voltage. If the same bit is 0, the diode is ON, which turns the transistor OFF.

12–6 VOLTAGE DIVISION

If precision is not essential, the ladder can be driven directly from the data inputs as shown in figure 12–7(e). This arrangement is analyzed with Thevenin's equivalent circuit. A_3 to A_0 are four independent voltage inputs, which may be called V_{A0}, V_{A1}, V_{A2}, and V_{A3}.

Looking to the right at node 1 in figure 12–7(e), we see a resistance R and

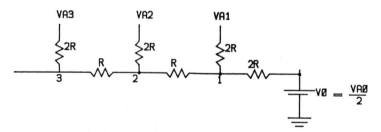

FIGURE 12–7(a)

FIGURE 12–7(b)

D/A and A/D Converters

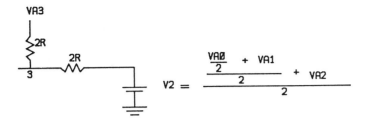

$$V2 = \frac{\frac{V A0}{2} + VA1}{2} + VA2$$

FIGURE 12-7(c)

$$V3 = \frac{\frac{\frac{VA0}{2} + VA1}{2} + VA2}{2} + VA3$$

V3 CAN BE REDUCED TO:

$$V3 = \frac{1}{2}(VA3 + \frac{VA2}{2} + \frac{VA1}{4} + \frac{VA0}{8})$$

FIGURE 12-7(d)

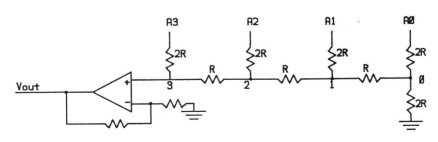

FIGURE 12-7(e)

two 2R resistors in parallel and a voltage source V_{A0}. These can be combined with a voltage source of $V_{A0}/2$ and an output resistance of 2R as shown in figure 12–7(a).

Two voltage sources, both with output resistance 2R, enter node 1, one with a voltage source $V_{A0}/2$ and the other with a voltage source V_{A1}. These, together with the resistor R in series, can be combined with a single voltage source with an output resistance of 2R and an equivalent source as shown in figure 12–7(b).

Looking to the right at node 3, we see a resistance R in series with two resistors of 2R in parallel leading to V_{A2} and V_1. These can be combined with a single voltage source with an output resistance of 2R and an equivalent voltage source as shown in figure 12–7(c).

Finally, figure 12–7(c) is "theveninized" to a single voltage source with an output resistance of R and an equivalent voltage source as shown in figure 12–7(d).

An op-amp isolates the ladder from a load and gives the desired gain as shown in figure 12–7(e).

12–7 A BCD D/A CONVERTER

The resistor ladder of figure 12–2 can easily be modified to a BCD converter, as in figure 12–8. The resistors are looked at in groups of four. The first four remain the same as for binary division. All subsequent sets of four resistors have resistor values stepped up by factors of 10.

Binary ladders of the R, 2R type can also be implemented as shown in figure 12–9. The input resistance to the ladder of BCD digit 1 is R, and a 9R resistor to the second section allows only 1/10 of the current I to flow to the BCD digit 0 ladder. The output current of BCD digit 0 is the second lesser significant decimal digit. A third digit network follows the same reasoning.

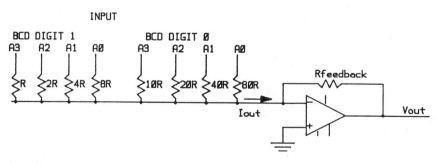

FIGURE 12–8 BCD D/A converter

D/A and *A/D* **Converters**

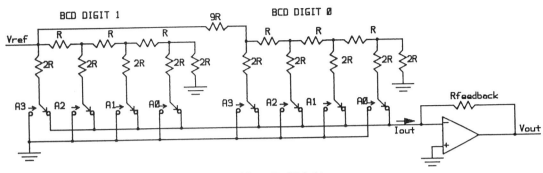

FIGURE 12-9 Two-digit BCD *D/A* converter with an *R*, 2*R* ladder

12-8 A MULTIPLYING *D/A* CONVERTER

The reference voltage input can be used as a multiplying input. The range of the output voltage is proportional to the reference voltage, which in turn can be controlled from another *D/A* converter.

In figure 12–10 the detailed *D/A* converter circuit is replaced by the commonly used symbol shown. The output op-amp provides the reference for the second *D/A*, and the feedback resistor for the op-amp is often included in the *D/A* chip for convenience.

---EXAMPLE 12-3

Data input 1 serves as a multiplier to data input 2 in figure 12–10. Suppose the full output range of both *D/A* converters = 5 V. Set the input to *D/A* converter 1 to 1000 0000 (i.e., midrange). Output polarity reversal can be ignored for this example.

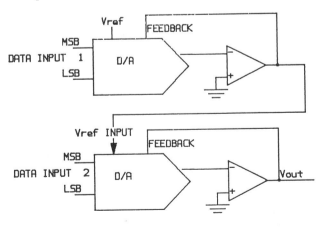

FIGURE 12-10 Multiplying *D/A* converter

For the given input to D/A converter 1, the reference voltage to D/A converter 2 is only 2.5 V. Set the input to D/A converter 2 to 0101 0011 = 83. For a 2.5 V reference the output would be 2.5 V for a full-scale input. For the given input the output is

$$\frac{83}{255} \times 2.5 = 0.8137 \text{ V}$$

Now change the input to D/A converter 1 from 1000 000 to 1001 0000, that is, from 128 to 144. The multiplying factor is 1.125. The reference voltage to D/A converter 2 changes to $2.5 \times 1.125 = 2.8125$. For the same input to D/A converter 2 the output changes to $0.8137 \times 1.125 = 0.9154$ V:

$$\frac{83}{255} \times 2.8125 = 0.9154 \text{ V}$$

Another way to look at D/A converter 1 in figure 12–10 is that the output is really the reference voltage multiplied by the digital input D_1. (D_1 is data input 1.)

$$V_{out1} = -V_{ref1} \times D_1 = V_{ref2} \quad \text{(A proportionality constant is omitted.}$$
$$\text{See the derivation below.)}$$

The output of D/A converter 2 is

$$V_{out2} = -V_{ref2} \times D_2$$

Substituting for V_{ref2}, we have

$$V_{out2} = -(-V_{ref1} \times D_1) \times D_2 = V_{ref} \times D_1 \times D_2$$

It follows that the D/A converter 2 output will be the square of the D/A converter 1 output if both share the same digital input.

If N D/A converters are thus strung in series with all inputs $= D$, the last output will be $V_{ref} \times D^N$. (See figure 12–11.)

The statement that $V_{out} = -V_{ref} \times D$ requires a few comments. The input current to the op-amp in figure 12–4 is derived from the reference voltage divided by the *effective* ladder resistance for a given set of closed switches. When all switches are closed, corresponding to a 1111 data pattern, the current to virtual ground = $15/16 I$. Thus the effective ladder resistance as seen from the op-amp input terminal = $16/15 R$. When only switch 3 is closed, corresponding to a 1000 data pattern, the current to virtual ground = $½ I = 8/16 I$. Thus the effective ladder resistance = $16/8 R$, where R is the ladder input resistance as seen by the reference source.

The numerator for the fractional coefficient of R is always 2^n, where n is the

D/A and A/D Converters

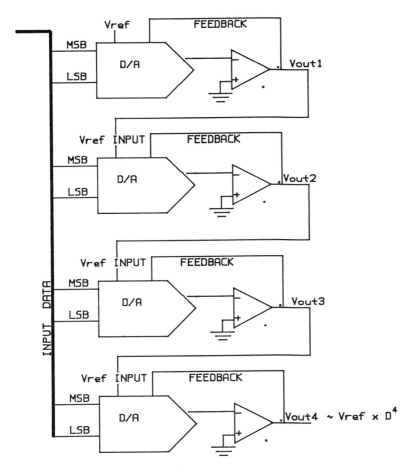

FIGURE 12-11 Power string

number of bits in a word. The denominator corresponds to the input data D. The effective resistance is expressed as

$$R_{\text{eff}} = \frac{2^n}{D} \times R$$

As we know, the op-amp output $V_{\text{out}} = V_{\text{in}} \times (R_f/R_i)$. In the case of the D/A converter, V_{in} is the reference voltage V_{ref}, R_f is the feedback resistance, and R_{in} is the effective ladder resistance. Thus

$$V_{\text{out}} = V_{\text{ref}} \times \frac{R_f}{\frac{2^n}{D} \times R} = V_{\text{ref}} \times \frac{R_f}{2^n \times R} \times D$$

The term

$$\frac{R_f}{2^n \times R}$$

is a constant K. Therefore V_{out} becomes

$$V_{out} = K \times V_{ref} \times D$$

EXAMPLE 12–4

Using the circuit of figure 12–11, we want to find the required feedback resistance R_f for a four-stage D^4 multiplying D/A converter. The given parameters are

- $V_{in} = 1$ V (used as reference)
- n (the number of bits per word) $= 4$
- N (the number of D/A converters in the string) $= 4$
- The ladder resistance $R = 1K$
- The digital input $D = 1011 = 11$ (decimal)
- Full-scale output $= 5$ V
- K must be chosen such that a full-scale input will not saturate the last stage. (A 1111 input will be 5 V.)

Since $V_{out4} = V_{in} \times K^4 \times D^4$,

$$K = \frac{\sqrt[4]{5}}{15} = 0.09968$$

This means that $R_f = K \times 2^n \times R = 1590 \ \Omega$.

$$V_{out1} = 1 \times 0.09968 \times 11 = 1.09648$$
$$V_{out2} = 1.09648 \times 0.09968 \times 11 = 1.20226 = V_{out1}^2$$
$$V_{out3} = 1.20226 \times 0.09968 \times 11 = 1.31826 = V_{out1}^3$$
$$V_{out4} = 1.31826 \times 0.09968 \times 11 = 1.44544 = V_{out1}^4$$

Indeed, $1.09648^4 = 1.44544$.

12–9 A DIVIDING D/A CONVERTER

A D/A converter with an output that is inversely proportional to D can be obtained by placing the effective ladder resistance in the feedback loop of the op-amp. The placement of the feedback resistor and that of the binary ladder are reversed. The

D/A and A/D Converters

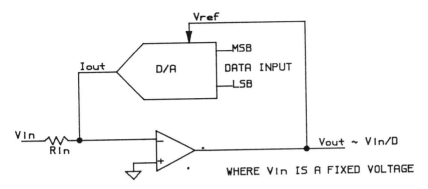

FIGURE 12-12 Reciprocal D/A converter

op-amp output becomes the voltage source for the binary ladder, and the input is driven from the reference voltage in series with an input resistor R_{in} as in figure 12-12. We get

$$V_{out} = V_{in} \times \frac{R_{eff}}{R_{in}} = V_{in} \times \frac{\frac{2^n}{D} \times R}{R_{in}} = V_{in} \frac{2^n \times R}{R_{in}} \times \frac{1}{D}$$

Here $\frac{(2^n \times R)}{R_{in}}$ is again a constant K. Of course, the op-amp output will saturate for an all zero input.

--- EXAMPLE 12-5

An output is needed that is inversely proportional to D.

- $V_{in} = 5$ V
- n (the number of bits per word) $= 4$
- The ladder resistance $R = 1K$
- The digital input $D = 1100 = 12$ (decimal)
- Full-scale output $= 5$ V
- K must be chosen such that a minimum input $= 0001$ will not saturate the op-amp

Since $V_{out} = V_{in} \times K \times 1/D$ (where minimum $D = 0001$),

$$K = \frac{5}{5 \times 1/1} = 1$$

This means that

$$R_{in} = \frac{2^n \times R}{K} = \frac{16 \times 1000}{1} = 16{,}000 \ \Omega$$

$$V_{out} = 5 \times 1 \times \frac{1}{12} = 0.4166 \ V$$

Notice that the minimum output for $D = 1111 = 15$ is 0.333 V and the maximum output for $D = 0001 = 1$ is 5 V. ∎

12–10 SPECIFICATIONS

There is a wide variety of *D/A* converters to choose from. Their specifications may range as follows:

    Accuracy: 0.05–1%

    Nonlinearity: 0.1–1%

    Resolution: 6–14 bits

    Settling time: 0.125–5 μs

    Output: current or voltage

    Input coding: binary, BCD, sign and magnitude, two's complement, companding (compressed)

Accuracy is a measure of the fidelity with which the analog output represents the digital input. Nonlinearity is the output deviation from a straight line when a linearly rising input count is applied. Resolution is a measure of the smallest analog increment that can be detected. 14-bit resolution, for example, detects one increment in 16384 or 0.006% of the full-scale output. For a 5 V full-scale output this represents 0.305 mV for each step. Settling time refers to the speed of the device, the time required from the application of the input to a stable output.

Some *D/A* converters include an op-amp with voltage output. Most *D/A* converters have a current output, and it is the designer's choice to add an op-amp. Some *D/A* converters include special features for input code acceptance or output range compression. The detailed specifications are found in data manuals such as Datel-Intersil, PMI, Analog Devices, National Semiconductor, and Burr-Brown.

12–11 THE *A/D* CONVERTER

There are several methods whereby an analog value can be converted into a digital number. A few are discussed here:

1. the integrating A/D converter,
2. the counter-type A/D converter,
3. the successive approximation A/D converter, and
4. the flash converter.

The choice of an A/D converter depends on the needed resolution, speed of conversion, circuit complexity, and cost.

12–12 THE DUAL-SLOPE INTEGRATING A/D CONVERTER

The dual-slope integrating A/D converter works on the following principle. The input voltage is converted to a time period, which is then measured with a counter. The integrator is switched between a reference and an input voltage. Integration of the input signal begins as soon as the switch is in the input position. If the input is at a given dc level, the output rises linearly with time and a slope proportional to the input voltage V_{in}:

$$V_{out} = -V_{in} * T/(RC)$$

(See figures 12–13(a) and 12–13(b).) After a fixed time period the integration is turned off. The magnitude of the analog output is directly proportional to the magnitude of the input.

The integrator is next switched to a positive reference, and the integrator output discharges until the output crosses zero. The larger the initial charge, the longer it takes to reach zero. The number of clock cycles to reach zero must be directly proportional to V_{in}. This is known as the dual-slope method, as is evident from figure 12–13(b).

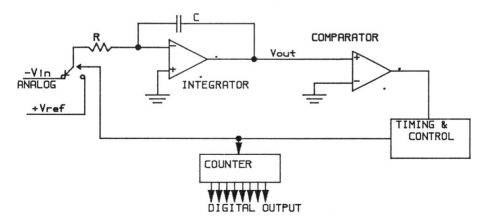

FIGURE 12–13(a)   Dual-slope integrating A/D converter

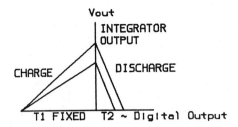

FIGURE 12–13(b) Dual-slope charge-discharge

The accuracy of the output depends on the linearity of the integrator and the precision of the reference voltage. The resolution depends on the clock frequency and the corresponding counter (the higher the frequency, the higher the expected count).

The relationships between V_{in}, V_{out}, T_1, and T_2 are found from the geometry of figure 12–13(b):

$$V_{out} = T_1 * V_{in} * 1/RC \quad \text{when switched to } V_{in} \tag{1}$$

$$V_{out} = T_2 * V_{ref} * 1/RC \quad \text{when switched to } V_{ref} \tag{2}$$

Setting equation (1) = equation (2), we get

$$T_1 * V_{in} = T_2 * V_{ref} \quad (1/RC \text{ cancels})$$

The output count = (Number of fixed clock cycles during T_1) * (V_{in}/V_{ref}).

12–13 THE COUNTER-TYPE A/D CONVERTER

The counter-type *A/D* converter is relatively simple. Let us explain the operation with a few steps:

- Step 1. A counter is cleared and starts counting from 0.
- Step 2. The output of the counter is fed to a *D/A* converter.
- Step 3. The analog output of the *D/A* converter is fed to a comparator.
- Step 4. The comparator compares the fed-back analog signal to the analog input signal.
- Step 5. When the two match (are equal in magnitude), the comparator output turns OFF and stops the counter.
- Step 6. The counter output, after the count was stopped, must be proportional to the analog input level. (See figure 12–14.)

D/A and A/D Converters

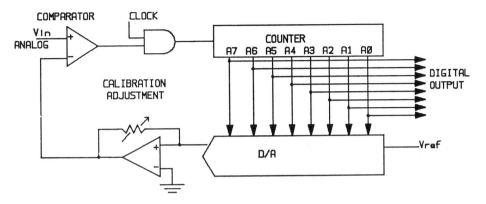

FIGURE 12-14 Up-down counter-type A/D converter

The proportionality factor is set by the D/A converter and op-amp gain.

The counter A/D converter can be used as a tracking or peak-holding A/D converter. The input voltage may change with time. If the input increases, the comparator will turn ON again and continue the count until the feedback voltage has caught up with the increased input. The last value read must correspond to the largest past input. In this mode the A/D converter is peak holding.

If the counter is reset as soon as the comparator switches and the last counter reading is strobed into a holding register, the comparator will reenable the system, and the counter will start again from 0. The next value read will be an update, either higher or lower. In this mode the A/D converter is tracking.

The tracking speed is slow, since the counter always restarts from 0. If an up-down counter is used, as in figure 12–15, the needed counts to find the new input value increases or decreases correspond only to the change. In this case the comparator does not stop the counter but only reverses the mode from up to down. When the input is steady, the output will oscillate around the least significant bit.

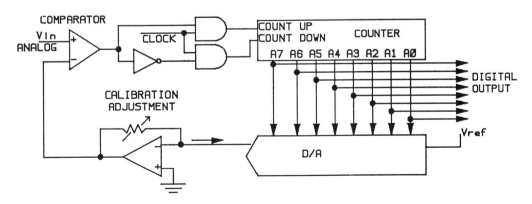

FIGURE 12-15 Counter-type A/D converter

12-14 THE SUCCESSIVE APPROXIMATION A/D CONVERTER

The concept of successive approximation and the successive approximation register were explained in detail in the chapter on synchronous and sequential logic (Chapter 10). The architecture of this A/D converter is similar to the counter type. The main difference is speed. The counter-type converter needs a maximum of 2^N counts (where N is the number of output bits) to reach the final value, and the number of counts (or clock cycles) is a function of the input. The successive approximation type needs only N clock cycles, and the number of clock cycles is always the same. The system clock is stopped after N clock cycles. This can be done with a down counter that is preset to N and stops the system when the count reaches 0 with the borrow output. The borrow also serves as the end-of-conversion flag. (See figure 12-16.)

A start conversion pulse always presets the successive approximation register to half the maximum possible output; that is, the MSB is set to 1, and all other bits to 0.

Comparison of input to feedback voltage in figure 12-16 is done in the same way as for figure 12-14.

The successive approximation converter is very popular. Commercial versions such as the Analog Devices MAH-0801 is as fast as 750 nsec for a complete eight-bit conversion and ¼ LSB differential and integral linearity (more about performance characteristics later). Slower versions are available for microprocessor work such as DATEL's eight-bit ADC 830 with tristate outputs.

For the several A/D converters discussed, we assumed that the input is stable while the A/D converter is performing the conversion. This is usually not the case.

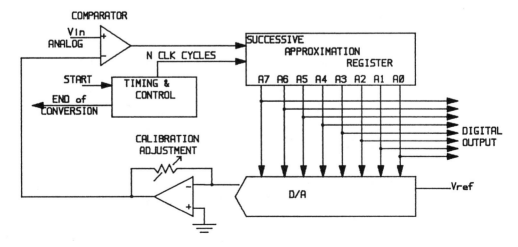

FIGURE 12-16 Successive approximation A/D converter

D/A and A/D Converters

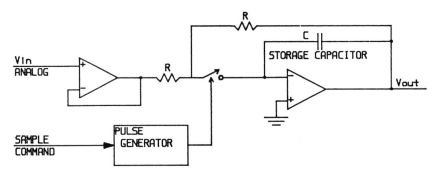

FIGURE 12–17 Sample and hold inverting feedback storage

It is therefore necessary to sample the input for a short period of time and hold it for the duration of the conversion. For that purpose, sample and hold circuits are used.

12–15 SAMPLE AND HOLD CIRCUITS

The basic idea of a sample and hold circuit is to briefly sample a small section of a waveform and store the charge in a capacitor for the needed conversion time. This can be done in a variety of ways, three of which are illustrated in figures 12–17, 12–18, and 12–19.

The input stage is a voltage follower with high input impedance and low output impedance. A sampling command closes the switch, and since the dc gain of the second stage is also unity, the capacitor will charge to the sampled voltage. At the end of the sampling period the switch opens. The capacitor has no discharge path, and the output will remain at $-V$(sample). Finite op-amp input impedance, capacitor, and circuit board leakage will eventually discharge the capacitor. (An example is the Analog Devices HTS-025.)

The second version places the capacitor at the input terminal of the second voltage follower. The switch closes for the sampling time and charges the capacitor. The switch opens and holds the charge. The two back-to-back diodes provide a

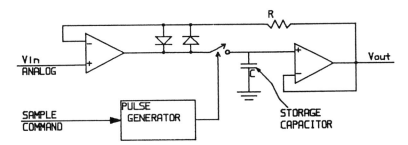

FIGURE 12–18 Noninverting sample and hold (Courtesy of Datel Corporation)

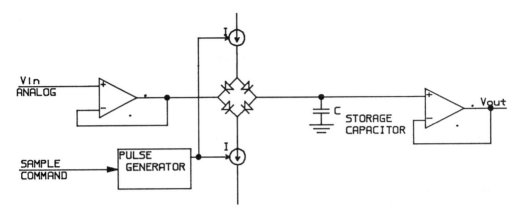

FIGURE 12-19 Noninverting bridge-driven sample and hold (Courtesy of Precision Monolithics Inc.)

feedback path for the first op-amp when the switch is open. (An example is the National Semiconductor LF198.)

The third version is similar to the second except that the switch is replaced with a diode bridge. The sample command allows the input signal to charge the capacitor while the bridge is in a low impedance state. This type is capable of 30 nsec sampling time. (Examples are the PMI SMP10 and OEI 5021.)

12-16 FLASH CONVERTERS

The fastest converter is, as the name implies, the flash converter. The counter-type converter took a variable number of cycles with a maximum equal to the highest count. The successive approximation type takes as many cycles as the resolution requires. The flash converter takes only one clock cycle. A comparison of the four types shows the following:

Integrating type: RC dependent

Counting type: 2^n cycles

Successive approximation: n cycles

Flash converter: one cycle

Of course, there is a price to be paid for such speed. In previously described versions (counter and successive approximation) there was always one comparator in the feedback loop. The flash converter has no feedback loop but 2^{n-1} comparators (where n, as usual, is the number of bits per output word).

For an eight-bit conversion, 255 comparators are needed. Each comparator (analog comparator) is set to an analog switching level that is one resolution level higher than the previous one. If, for example, the maximum input is 5 V, then

$5/255 = 19.6$ mV is the switching level separation between adjacent comparators. Precise switching levels are usually obtained from a serial tapped resistor string and preferably derived with a precision voltage source. (See figure 12–20.)

Imagine, therefore, that you have such a string of resistors and comparators. Each reference input is connected to a tap on the resistor string, and all second comparator inputs are tied together and receive the input to be converted.

For example, let the input = 2.13 V out of a possible 5 V maximum, so $2.13/0.0196 = 108.67$. This means that all lower 108 comparators are turned ON with an output of 1, while all upper 147 comparators are OFF with an output of 0. It is always true with this arrangement that a set of lower comparators is ON while a set of upper comparators is OFF (except for all 0's or all 1's). There remains the task of encoding the desired eight-bit binary number for 256 possible comparator outputs. This is a priority encoder. There are 2^{255} possible combinations. (There are 256 valid outputs and 2^{247} irrelevant states.) Karnaugh mapping for 256 variables is certainly not practical, but inspection of a simple truth table reveals a recognizable pattern from which the needed logic equations are easily derived.

The equation can be derived from table 12–2 with only 16 of the 256 states, and we can see whether an extendable equation pattern emerges. For 16 input states we have only a four-bit output pattern *DCBA*. As the comparators turn on one by one, with an increasing analog input, the output binary number also increases in steps of one.

Table 12–2 is a truth table that shows the 16 possible comparator output states, P_{15} to P_1, and the corresponding desired binary output, *D* to *A*. Fifteen comparator outputs could have 32,768 states, which are mostly irrelevant ("don't care" states) except for the 16 states listed.

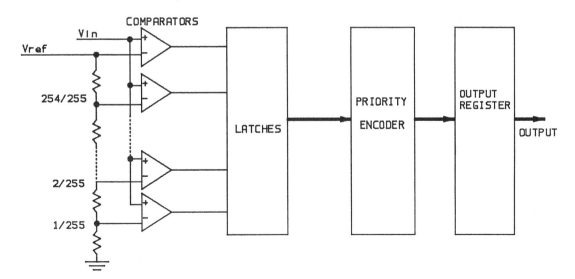

FIGURE 12–20 Flash converter

TABLE 12–2 Priority encoder truth table

Comparator Output															Digital Output			
P_{15}	P_{14}	P_{13}	P_{12}	P_{11}	P_{10}	P_9	P_8	P_7	P_6	P_5	P_4	P_3	P_2	P_1	D	C	B	A
0	0	0	0	0	0	0	0	0	0	0	0	0	0	0	0	0	0	0
0	0	0	0	0	0	0	0	0	0	0	0	0	0	1	0	0	0	1
0	0	0	0	0	0	0	0	0	0	0	0	0	1	1	0	0	1	0
0	0	0	0	0	0	0	0	0	0	0	0	1	1	1	0	0	1	1
0	0	0	0	0	0	0	0	0	0	0	1	1	1	1	0	1	0	0
0	0	0	0	0	0	0	0	0	0	1	1	1	1	1	0	1	0	1
0	0	0	0	0	0	0	0	0	1	1	1	1	1	1	0	1	1	0
0	0	0	0	0	0	0	0	1	1	1	1	1	1	1	0	1	1	1
0	0	0	0	0	0	0	1	1	1	1	1	1	1	1	1	0	0	0
0	0	0	0	0	0	1	1	1	1	1	1	1	1	1	1	0	0	1
0	0	0	0	0	1	1	1	1	1	1	1	1	1	1	1	0	1	0
0	0	0	0	1	1	1	1	1	1	1	1	1	1	1	1	0	1	1
0	0	0	1	1	1	1	1	1	1	1	1	1	1	1	1	1	0	0
0	0	1	1	1	1	1	1	1	1	1	1	1	1	1	1	1	0	1
0	1	1	1	1	1	1	1	1	1	1	1	1	1	1	1	1	1	0
1	1	1	1	1	1	1	1	1	1	1	1	1	1	1	1	1	1	1

Let us look at the output variables one at a time, starting with the least significant bit A.

A is true

- when P_1 is true = 1 and P_2 to P_{15} are false = 0, or
- when P_1, P_2, and P_3 = 1 and all others 0, or
- when P_1, P_2, P_3, P_4, and P_5 = 1 and all others = 0, etc.

There is a boundary for any given row between the 1's and 0's. The only relevant condition is at the boundary; all others are redundant. For A this is true when

$$A = P_1\overline{P_2} + P_3\overline{P_4} + P_5\overline{P_6} + P_6\overline{P_7} + \ldots + P_{13}\overline{P_{14}} + P_{15}$$

At P_{15} there is no boundary; thus P_{15} stands alone.

For B, we must look at two consecutive rows and consider two comparator outputs above the boundary. B is true when

$$B = (P_2\overline{P_3}\overline{P_4} + P_2P_3\overline{P_4}) + (P_6\overline{P_7}\overline{P_8} + P_6P_7\overline{P_8}) + \ldots$$

Groups of two are in parentheses for clarity. For the first term in parentheses, P_3 is redundant. For the second term in parentheses, P_7 is redundant. B can therefore be reduced to

D/A and A/D Converters

$$B = P_2\overline{P_4} + P_6\overline{P_8} + P_{10}\overline{P_{12}} + P_{14} \quad (P_{16} \text{ does not exist})$$

For C we must look at four consecutive rows and consider four comparator outputs above the boundary. C is true when

$$C = (P_4\overline{P_5}\overline{P_6}\overline{P_7}\overline{P_8} + P_4 P_5\overline{P_6}\overline{P_7}\overline{P_8} + P_4 P_5 P_6\overline{P_7}\overline{P_8} + P_4 P_5 P_6 P_7\overline{P_8}) + \ldots$$

P_5, P_6, and P_7 are redundant, and C reduces to

$$C = P_4\overline{P_8} + P_{12}$$

By inspection, D is true when P_8 is true:

$$D = P_8$$

The emerging pattern for more than four output bits is obvious and can extend to n bits:

$$A = P_1\overline{P_2} + P_3\overline{P_4} + P_5\overline{P_6} + P_7\overline{P_8} + \ldots + P(2^n - 2^0)$$
$$B = P_2\overline{P_4} + P_6\overline{P_8} + P_{10}\overline{P_{12}} + P_{14}\overline{P_{16}} + \ldots + P(2^n - 2^1)$$
$$C = P_4\overline{P_8} + P_{12}\overline{P_{16}} + P_{20}\overline{P_{24}} + \ldots + P(2^n - 2^2)$$
$$D = P_8\overline{P_{16}} + P_{24}\overline{P_{32}} + P_{40}\overline{P_{48}} + \ldots + P(2^n - 2^3)$$

The network is partially sketched in figure 12–21.

Figure 12–22 is an application for A/D conversion using the RCA 45051. Because of the high frequency applied to the chip, some clock noise will be seen on the analog input. This can be partially eliminated with an RC filter network at the input as shown. The cutoff frequency should be low enough to attenuate the

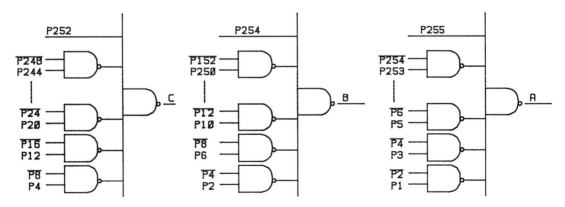

FIGURE 12–21 Partial priority encoder

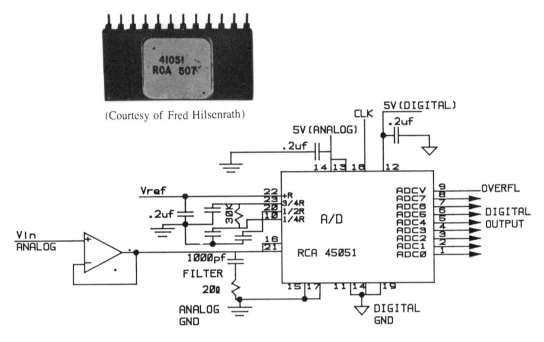

FIGURE 12-22 Flash converter application (Courtesy of GE Solid State)

clock noise but high enough not to attenuate the signal. The input should be driven with an op-amp that has a sufficiently low output impedance at the clock frequency.

To isolate digital noise from the analog signal, it is customary to use separate analog power supplies and ground planes. The ground plane of the digital and analog sections should be tied together close to the conversion chip.

Flash converters are excellent for high-speed data acquisition. They were not on the market until recently because 255 comparators are just too much circuitry. Moreover, it is very difficult to hold that many discrete comparators at the exact switching level. Nowadays they are all on one monolithic structure, and matching properties are inherent. The encode logic is rather extensive, but large-scale integration puts it all on one chip.

The RCA 45051 flash converter further improves on comparator performance by autobalancing each comparator input between cycles. This technique improves comparator sensitivity and temperature tracking. (For a detailed explanation, read the RCA 45051 data sheet.)

Flash converters are particularly useful for video applications. If a video image consists of 512 × 512 pixels (a pixel is one dot of the image) and is displayed 30 times per second, we have $512 * 512 * 30 = 7,864,320$ pixels per second. Flash converters can convert $15 * 10^6$ samples per second. Such real time conversion is not possible with the other converters.

12-17 SOME PERFORMANCE CONSIDERATIONS

Sparkling

When used at a high clock rate, the converter may exhibit what is known as *sparkling*. The output will go to all 1 for some input codes. This problem might be solved in the future.

Differential Linearity

Even if the comparators are reasonably well matched, they are not perfect. Steps from one level to the next are not exactly equal. A converted ramp input might look like figure 12–23 when reconverted to an analog voltage with a high precision D/A converter.

Notice that the step width is not equal, reflecting the uneven switching characteristic of the comparators. Differences in step width (or differences in voltage steps when comparator switching occurs) are a measure of differential linearity. The problem might even be bad enough to skip some levels altogether. This is known as missing codes because no output code for the skipped level exists. The manufacturers try to hold the differential nonlinearity within +/− ½ LSB. What does this mean in the performance of an actual system? To answer that question, we can display the effect of the differential nonlinearity on a linear distribution spectrum.

A good way to measure the differential linearity is to sample a linear ramp randomly and repeatedly. The output for each sample becomes the address to a memory. The content of that memory address is increased by one for every sample height that matches that address. If all step widths are identical, then the probability of accessing all 2^n addresses (n = number of output bits) is equal and $2/n$ for all addresses. The result will be a perfectly flat distribution for an infinite number of samples.

If the differential linearity is indeed +/− ½ LSB, it means that some step widths are only ½, while others are 1½ times as wide as expected. If two adjacent steps have that characteristic, then one address will get 50% of the expected count, and the next will get 150%. This is a 3 to 1 ratio. A plot of such a curve will show substantial *scatter*. This is shown in figure 12–24.

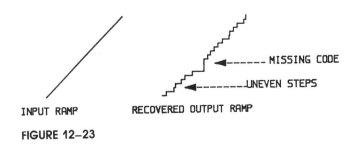

FIGURE 12-23

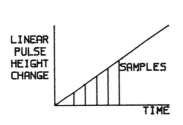
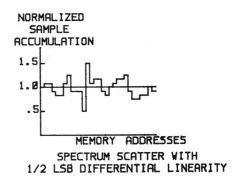

FIGURE 12-24 Flash converter differential linearity test

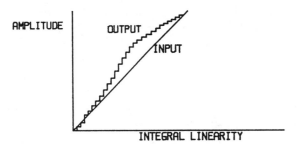

FIGURE 12-25

The flash converter, more than other types of converter, is subject to that problem because of multiple comparators.

Integral Linearity

If the gain of the system is not uniform throughout the allowed range, a distorted output is obtained. A linear input ramp might be a bowed output line. The deviation from the straight line is a measure of the integral linearity. This is also expressed in fractions of the LSB; $+/-$ 1/2 LSB nonlinearity for an eight-bit converter is $+/- = 2/512 * 100\%$, or approximately 0.2%. (See figure 12-25.)

REVIEW QUESTIONS

1. What is a *D/A* converter?
2. What is meant by a six-bit *D/A* converter or an eight-bit *D/A* converter?
3. What is meant by the weight of a bit?
4. Show that the current in each branch of figure 12-1 differs from the adjacent branch by a factor of 2.

D/A and A/D **Converters** 407

5. Why is an *R, 2R* ladder in figure 12–3 preferable to the ladder in figure 12–2?
6. Explain how the input resistance to the ladder in figure 12–3 = *R*. Why is it independent of the number of sections?
7. Why is it necessary to terminate the *R, 2R* ladder with *2R?*
8. Explain the binary current relationship of the ladder branches in figure 12–3.
9. How does a multiplying *D/A* converter work?
10. Discuss the difference between the converter in figure 12–5 and that in figure 12–6.
11. Review the circuit in figure 12–7(e) and show that the voltage contribution of each bit source is related in binary fashion to the adjacent source.
12. Review the circuit in figure 12–8 and show the BCD current relationship.
13. Explain the decade resistance ladder of figure 12–9.
14. Explain how the counter-type *A/D* converter works.
15. Discuss the peak-holding and tracking *A/D* converter.
16. What is meant by successive approximation?
17. How many clock cycles does a successive approximation *A/D* converter need to give the desired output?
18. How is a successive approximation register initialized, that is, what happens immediately after the start command?
19. Describe the dual slope method of *A/D* conversion.
20. What is a sample and hold circuit? Why is it needed?
21. Discuss the three sample and hold circuits of figures 12–17, 12–18, and 12–19.
22. What is the advantage of the flash converter?
23. Review the flash converter encoder truth table (table 12–2).
24. Explain the meaning of accuracy.
25. Explain the meaning of resolution.
26. Discuss differential linearity.
27. Discuss integral linearity.

PROBLEMS

1. In figure 12–2, assign a voltage level of 4 V for bit 0 to bit 7 when high and 0 V when low. Compute a resistance value for the feedback resistor *R* such that the maximum output = 5 V. What is the step size?
2. In figure 12–4, let the reference voltage V_{ref} = 5 V, *R* = 1K. Compute the feedback resistor $R_{feedback}$ for V_{out} max = 5 V.
3. What is the output current I_{out} in figure 12–4 for an input of 1100?

4. Design an eight-bit D/A converter using the voltage division method of figure 12–7(e) for a full-scale output = 5 V.
5. Design a three-digit BCD D/A converter as in figure 12–8.
6. Design a three-digit BCD D/A converter as in figure 12–9.
7. What is the resolution of the multiplying D/A in figure 12–10 if both data inputs are eight bits?
8. Design a D/A converter for V_{out} proportional to D^3.
9. Determine the input resistance R_{in} for the reciprocal D/A converter in figure 12–12 for eight input bits. Full scale output = 5 V, V_{ref} = 5 V, and the ladder resistance is 1K.
10. Design a counter-type A/D converter that includes both features of peak holding and tracking. Select either mode with a switch.
11. Design an eight-bit successive approximation A/D converter (including the timing and control).
12. Design a dual-slope A/D converter:

   Integration time = 10 μs
   Resolution = 8 bits
   V_{in} = −5 V max
   V_{ref} = 5 V

13. Design a four-bit flash A/D converter, including the priority encoder.
14. Determine the maximum spectrum scatter for a flash converter
    (a) with 0.1 LSB differential nonlinearity.
    (b) with 0.5 LSB differential nonlinearity.
    (c) with 0.9 LSB differential nonlinearity.

chapter 13

an introduction to computers

13-1 INTRODUCTION

Until not too long ago, electronics was limited to radio engineering, related home entertainment, and servo controls. The vacuum tube—big, clumsy, fragile, hot and power consuming—was the workhorse of electronics, yet it worked. The flip flop was known as the multivibrator, but digital logic was not much talked about. Electronics was analog. There was talk about a new, rather mysterious device, a transistor. It was doubtful whether the transistor was a new promise or another curious but useless device.

The first point contact transistors were rather fragile structures, but they were soon improved into the junction transistor. In the late 1950s the first commercially available transistors transformed the electronic world. For the first time, engineers could design more complex and versatile electronic systems without the space, power, and fragility restrictions of the large tube. The era of modern electronics was born.

The idea of digital computing is not new. The famous mathematician Gottfried Wilhelm von Leibniz constructed an adding and multiplying machine in the seventeenth century. Punched cards were developed in the eighteenth century to control weaving machines, and Charles Babbage designed an analytical engine in the nineteenth century that did not work.

When George Boole published a paper on theories of logic in 1854, he certainly never imagined them to become the mathematical foundation of the modern-day computer. As late as 1938, C. E. Shannon of MIT published a paper about the mathematical expression of switching circuits that related directly to the work of Boole.

The University of Pennsylvania made an early try at building a computer with tubes, the ENIAC. It took 19,000 tubes and associated passive components to complete it and consumed 200,000 watts of power. Statistical calculations predicted an unacceptable breakdown rate because of tube failures. The ENIAC did not turn out as badly as predicted, but it certainly was not a pocket calculator. It could perform unprecedented thousands of calculations per second but had no memory for stored programs.

It was John von Neumann in 1944 who developed the concept of the stored-program computer using programs that can be changed, not programs that are hard-wired into the system. The UNIVAC I and the IBM 610 and 701 were the first-generation stored program computers built with relays and tubes. The practical foundation was laid with the advent of the transistor, the device that made the modern-day computer a reality.

Second-generation computers followed quickly, such as the IBM 7090 and 7094. These machines were faster and more reliable. Integrated circuits made their appearance, with as many as 20 transistors with associated resistors on a single chip (for example, the 7400 has four dual NAND gates). It was no conceptual breakthrough, but it reflected a maturing fabrication and miniaturization capability and further extended the practical expansion of computer power. It did not stop there. The industry tackled medium integrated circuits with as many as 100 gates per chip such as counters, binary full adders, and encoders.

The third-generation computers followed in the mid-1960s with the IBM 360 and the UNIVAC 1108. These are large computer systems, and access to large computers was inconvenient. There was a need for small-size systems with direct interactive access. Miniaturization made it possible to respond to that need. The minicomputer appeared; the Data General Nova and Eclipse and the Digital Equipment Corp PDP 11 found wide application. However, the industry was not satisfied. The world of high technology had its eyes on "smaller" goals.

Why not put 100,000 or more transistors, bipolar or CMOS, on a single chip? Why not design an entire system on a chip? Manufacturing yields were good enough in the late 1970s to attempt such grand undertakings. The microprocessor was born, featuring the central processing unit of a computer on a single chip such as the INTEL 8080, 8085, 8086, Motorola's 6800, RCA's 1800, Zilog's Z80, and National's NSC800.

Similar developments in solid state memories with 256,000 bits of storage per chip, convenient I/O ports on one chip, timing and control circuits, and bus-structured architecture made the microcomputer so small, compact, and inexpensive as to literally make it available to every household.

Parallel with the hardware development of the computer another discipline evolved—the software. As wonderful as the computer is, it is useless without the programs that run it.

Where is the computer leading us? It certainly opens the way for access to vast information, incredible speeds of data handling, and solutions to mathematical and cosmological problems that simply could not be solved with pencil and paper. Computers can handle and keep track of the intricate web of our financial and management systems. Computers can control robots, which in turn can tend to and repair our machines and automate our factories.

All this is done with a computer programmed by humans, a computer that cannot think on its own. What about computers endowed with learning capabilities? Can electronic brains develop their own sets of experiences and thus their own individual behavior and responses? What can such machines (or shall we still call them machines) do for us? We have indeed embarked on the second phase of the industrial revolution, the "information revolution."

13-2 ESSENTIAL COMPUTER BLOCKS

Up to this point we have studied some components needed to build a computer. They were the arithmetic logic unit (ALU), the memory, and timing and control circuits. In this chapter we will

1. show how to tie subsystems together with a bus structure,
2. describe a keyboard and a display as simple I/O devices, and
3. discuss the basic architecture of a computing system.

13-3 THE BUS

How does data travel between subsystems? In other words, how do basic building blocks communicate with each other? Furthermore, there must be some means to communicate with the outside world, the user. It is not necessary for every block in the system to be uniquely connected to all other blocks in the system. This is somewhat similar to a human community in which everybody needs access to everybody else. The solution is not a private road from each home to each other home but a shared roadway. Such a shared roadway in a computer (and many other digital systems) is called a *bus*.

By itself the bus is nothing more than a set of wires. Data from many sources are tied to these wires; conversely, the data on the bus is connected to many destinations. One obvious question arises. How do we avoid one set of data interfering with another set of data when both want to occupy the common bus simultaneously? Timing and control logic are arranged in such a way as to never allow this situation to occur. When data is allowed to enter the bus from one source, all other sources must be electrically isolated. Two gate types are available for this application, the open collector gate and the tristate gate.

13-4 THE OPEN COLLECTOR GATE AS A BUS DRIVER

The open collector gate is nothing more than a NAND gate in which the collector load resistor has been left out. Rather, a load resistor is added to the circuit externally. This feature allows the tying together of several collectors as shown in figure 13-1.

Either input *A*, *B*, or *C* can pull the bus wire voltage to ground, provided that

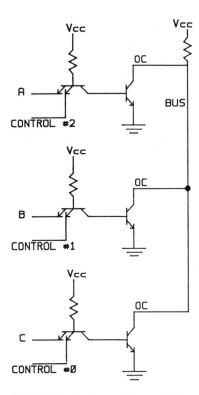

FIGURE 13-1 Open collector (OC) gates, three signals on one bus line

all control inputs are high. If input A is high, current will flow through T_1. If input B is high, current will flow through T_2. If input C is high, current will flow through T_3. If all inputs are high, the load current is shared by T_1, T_2, and T_3. This is called a wired OR connection. In a computer system, input A may be one bit of a longer word A and input B one bit of a longer word B. Both should not try to occupy the bus line at the same time. What is needed is an OR connection by which either A, B, or C, but not all, can put data on the bus line. The addition of a control input with each data input allows the computer to select which of the data is to occupy the bus. Only one control input is high at any one time in figure 13–2. A failure of proper control sends two data words to the bus simultaneously. This is called a *bus contention.*

There is no limit to the number of data inputs that can be connected to that common collector point (except for the increased capacitance hanging on that line). Note that the bus is *high* when all control inputs are disabled. The standby state, when the bus is idle, is a high state. When a control input allows data to enter, the bus is the complement of the input data. The logic on the bus wire is therefore reversed. This is negative logic. The negative logic is preferred on this bus because it means that no current flows and no power is dissipated in the standby state.

The bus is always terminated with a set of terminating resistors. These provide

An Introduction to Computers

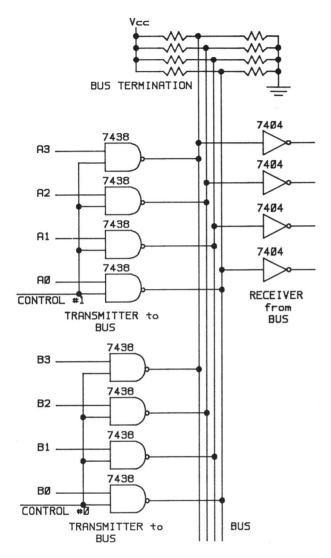

FIGURE 13-2 Two four-bit words on a bus

the load resistance to V_{cc} required by the open collector drivers, the proper TTL output level, as well as a matching termination for the characteristic impedance of the bus. Since the approximate line impedance is 180 Ω, a 330 Ω, 390 Ω combination meets that requirement. (A discussion on cables and transmission lines is given in the appendix.)

Data bits are received by some other subsystem via ordinary inverters, thus reestablishing the positive logic pattern. Open collector bus control works just fine but does require the external load resistor. Figure 13-2 shows two four-bit words to be transmitted to or received from a bus.

Open collector circuits have other applications such as driver circuits for some load that connects directly to V_{cc}, lamp or LED display drivers, and cable drivers terminated at the receiving end.

For the open collector bus driver there is the 7438 (TTL) quad dual-input NAND. The receiver may be the ordinary 7404 (TTL) hex inverter. (See figure 13–2.)

13–5 THE TRISTATE GATE

Over the last few years the tristate gate has replaced the open collector gate. The load resistor is part of the circuit, but isolation from the bus is achieved in a different manner. (See figure 13–3.) When the control input enables the gate (note that the control signal is low), data bit A appears on the bus wire, either inverted or not inverted (depending on the designer's choice).

When the control signal in figure 13–3 is low—that is, the gate is enabled—both diodes are reverse-biased, and the states of T_1 and T_2 depend on the input signal. If the input is high, T_2 is OFF; T_1 is ON and provides a load current path to ground. If the input is low, T_1 is OFF; T_2 is ON and provides a current path through the load resistor to V_{cc}. In either case the output resistance is low.

When the control input disables the gate, that is, the control signal is high, both transistors T_1 and T_2 are open, and the bus wire floats. If we look back at the gate from the bus wire to the data input, we see a high impedance. No data can be transmitted in this state.

Thus there are three states, two states when the control gate is enabled and the bus is in a low impedance state and the output level is either high or low, depending on the data bit, and the third or high impedance state when the control gate is disabled. The bus floats at approximately 2.5 V.

The CMOS version, in figure 13–4 is explained in a similar fashion with the exception that T_2 returns the output high level directly to V_{dd}.

Since each tristate gate has its own built-in load resistor, the bus sees an active load resistor when any one of the tristate gates in figure 13–3 is enabled. When all bus drivers are disabled, the bus floats, a condition that the receiver gates do not tolerate. Thus a resistor is still needed to tie the standby level to either ground or the power supply. The bus is then assured to be at a known level when it is not busy. TTL gates at the receiving end see a well-defined level, and CMOS gates are prevented from drawing excess current. (CMOS gates draw current when inputs are left open.)

Of course, this is needed only when the system bus is expected to be idle part of the time. The load resistor can be a large resistor such as 47 KΩ. This works well as long as there are not too many drivers in parallel on the bus. For example, 50 pF capacitance and 47 KΩ result in a time constant of 2.35 μs or 5 μs rise time. This is too slow for many applications.

Another, more elegant solution to the problem of bus loading is to use an active terminator rather than a passive resistor. The active terminator is nothing more than

An Introduction to Computers

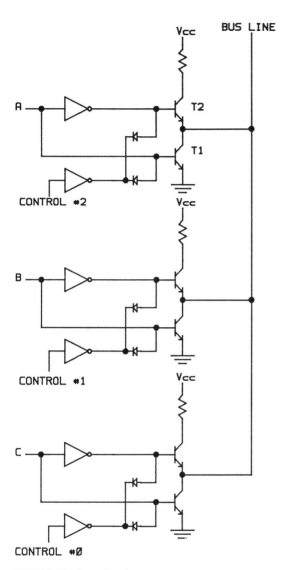

FIGURE 13-3 TTL tristate gates

another tristate gate with its data input pins tied low (or high) that is enabled when no other device demands access to the bus. The active terminator (74HC126, for example) will improve the rise or fall times to 30 nsec. (See figure 13–5.)

The 74HC245 and 74LS245 are CMOS or TTL octal bus transceivers that combine eight drivers and eight receivers in one chip. Most systems send and receive data on the same line, but not at the same time. Transceivers are controlled with a data direction control to send or receive data. These are certainly more compact than the open collector version, which requires four chips for the same

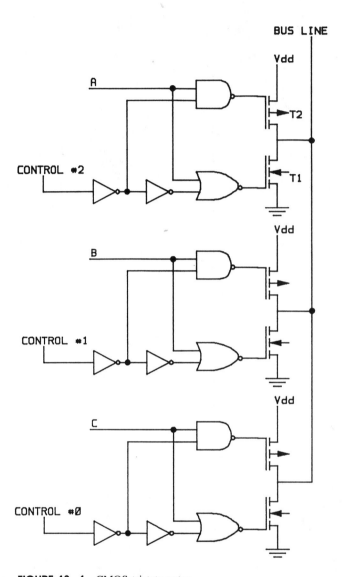

FIGURE 13-4 CMOS tristate gates

purpose. The chip has two control inputs; one enables the chip with an active low level, and the other determines the direction of the data flow. (See figure 13-6.)

74HC245
Function Table

$\overline{G}$	Dir	Operation
0	0	B to A
0	1	A to B
1	X	ISOLATION

An Introduction to Computers

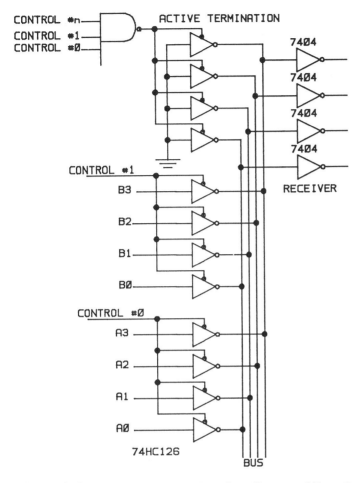

FIGURE 13-5 CMOS bus driver and receiver (Courtesy of Texas Instruments Incorporated)

13-6 I/O PORTS

It is often necessary to hold output data in a latch until the bus is ready to receive data or latch the data from the bus before receiving the data into the system. Such a device is called an *input-output port* or *I/O port*. The INTEL 8212 is an example and is reproduced in figure 13-7. In addition to the eight data lines it has two device select lines, which in conjunction with the mode or strobe line enable the output. The mode input is used to control the state of the output buffer and to determine the source of the clock input (C) to the data latch.

- Output mode: *MD* is high, and the output buffers are enabled with $\overline{DS_1}$ and DS_2 true.

- Input mode: *MD* is low, and the output buffers are enabled with the strobe *STB* as well as device select $\overline{DS_1}$ and DS_2.

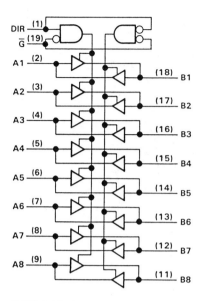

FIGURE 13-6 Octal bus transceiver (74HC245) (Courtesy of Texas Instruments Incorporated)

The chip also has a service request flip flop and an interrupt output line ($\overline{INT}$) to a microprocessor. It alerts the microprocessor that an external source needs attention.

Data from or to a system is often transmitted serially (interface to a CRT terminal, modem, etc.); that is, data bytes are transmitted one bit at a time. This requires parallel-to-serial conversion on the way out and serial-to-parallel conversion on the way in. Such conversion is easily done with a shift register. The data is parallel strobed into the register, then serially shifted out. The reverse happens in the input mode. A standard chip used for this purpose is the universal asynchronous receiver transmitter (UART) and the universal synchronous/asynchronous receiver transmitter (USART). Examples are the INTEL 8251 and Motorola MC6850. A block diagram of the latter is shown in figure 13-8.

13-7 STANDARD BUS STRUCTURES

The bus, of course, consists of as many wires (or conductors) as the system requires. Typically, the bus handles three distinct sets of bits:

1. Data bits are the information that we wish to operate on.
2. Address bits tell the system where to send or retrieve data.
3. Control bits handle the timing and execution of information flow.

(Power lines can be considered a fourth set of bus lines.)

An Introduction to Computers

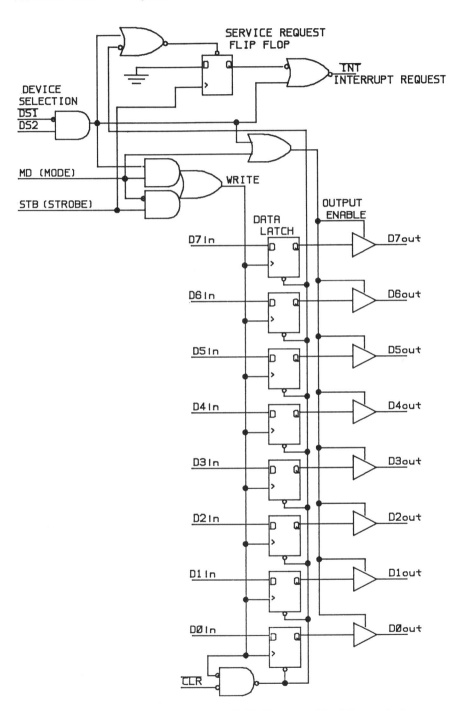

FIGURE 13–7 Eight-bit input/output port (8212) (Courtesy of Intel Corporation)

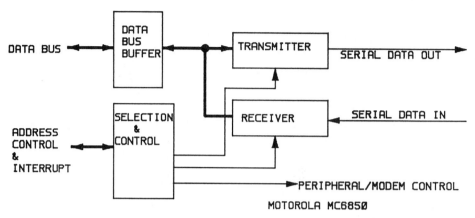

FIGURE 13-8 Asynchronous receiver transmitter (MC6850)

The first standardized bus system, developed by MITS Inc. for the 8080 (INTEL microprocessor) based Altair kit in 1975, was called the S-100 bus. As the name implies, it has 100 conductors dedicated to

- 16 data lines
- 16 address lines
- 62 control lines
- 6 power lines

The bus structure as a main road for all data traffic allows a modular system architecture. Every block in the system ties directly (through appropriate bus drivers and receivers) to the bus. If you imagine each block, or subsystem, to be on a pluggable card, then each can be plugged into a backpanel board. The backpanel is called a *motherboard,* and each of the subsystem cards is called a *daughterboard.* Such an architecture is expandable without rewiring. It is like adding more houses to a street.

The standard for the above described S-100 bus is flexible to some extent. Some of the control bits have multiple designations. The address space can be extended by using four of the control bits as additional address bits. Other control bits may be used as additional power and ground lines.

In spite of the allowed flexibility, such standardization makes for compatibility of products from various manufacturers. Since the introduction of the S-100 bus, more sophisticated computers and microprocessors have stimulated the development of other bus standards. To list but a few microcomputer bus developments:

1. The Benton Harbor Bus developed by the Heath Company for the 8080-based H8 system. It is a more modest bus of 50 conductors.
2. The SBC Multibus by INTEL for the SBC (single-board computer) series of boards with 86 conductors.

3. The TRS 80 Bus for Radio Shack's microcomputer with 40 conductors.
4. The Digital Group Bus, designed for compatibility with many microprocessors such as INTEL's 8080, Zilog's Z80, or Motorola's 6800.
5. The Apple II Bus for the microcomputer of the same name.
6. The SS 50 Bus by the South West Products Corp for Motorola's 6800 microcomputer.
7. The PET Bus for Commodore's PET computer.

13-8 THE KEYBOARD

The simplest input system is a set of toggle switches. Some minicomputers (like the Data General Eclipse) have a row of 16 switches on the front panel. To enter data into the accumulator, the switches are set to the desired 1 0 pattern and a "deposit" switch is activated. A set of neon bulbs above the switches reflects the contents of the accumulator. An entire program is thus entered and executed with a start switch.

When an operation is done and the computer halts, the final contents of the accumulator are displayed in the set of neon bulbs. This is a tedious process, particularly since all instructions have to be entered in binary mode. It is used at best to do some system troubleshooting.

An easier way to enter data is to use a keypad consisting of 16 hexadecimal keys together with some command keys. This is done on some single-board microcomputer trainers such as the INTEL SDK 85 and 86. Such keypads are certainly not the last word in operating convenience but are excellent as microprocessor training boards. They put the student in direct contact with the computer, since instructions are entered in the only language the computer understands—binary patterns.

In a hexadecimal system we combine four bits into one hex digit. The computer still sees four bits, but the user has fewer keys to press and to read, a feature that makes data entry easier and less error-prone. Since our decimal system has only ten digits—0–9—the other six digits for the hex system are

$$A = 10, \quad B = 11, \quad C = 12, \quad D = 13, \quad E = 14, \quad F = 15.$$

The hex system is convenient because two hex digits form one eight-bit byte. (Note how inconvenient a three-bit octal system would be.)

One way to recognize a switch closure is to use a 16-line to 4-line decoder. Another, more typical arrangement, which requires an 8-line to 4-line decoder, is to use a matrix coincident scheme (somewhat similar to address selection in a memory). The 16 switches are placed over the 16 intersections of a wire matrix as in figure 13-9.

When the switch is depressed, it short-circuits the column to the rows, which are scanned by successively pulling the rows to ground. At that time the node is identified by reading the row and column output. When none of the keys are depressed, all four rows and columns are high.

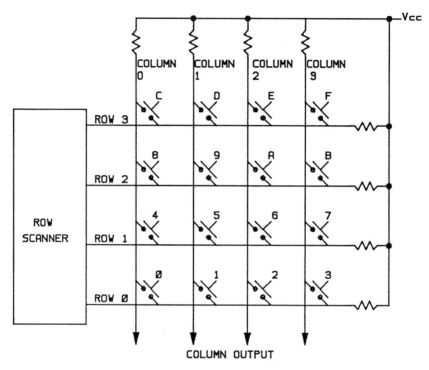

FIGURE 13-9 Keyboard matrix with row scanner

The eight-bit pattern completely identifies the switch position. That pattern can then be decoded into the appropriate hex value with either a simple gating network or a table look-up memory.

The matrix of figure 13-9 can also be arranged into eight columns and three rows for a 24-position keypad for 16 hex digits and eight control inputs as for the SDK 85 or SDK 86 single-board computer. (See figure 13-13, page 426.)

Another method replaces the decoder with two two-bit counters, one for column identification and the other for row identification. The rows are set to 0 one at a time as determined by the row counter. For every row setting the columns are scanned. As long as no 0 is found, the process continues. When a zero is found in the column scan, the search stops.

For example, suppose key 9 is depressed. Row 2 = 10 and column 1 = 01 are low. The combination output = 10 01 = 9. If key E is depressed, row 3 = 11 and column 2 = 10 are low. The combination output = 11 10 = E.

Keyboard encoders must also recognize faulty keyboard use, such as two keys accidentally depressed at the same time. Two 0's are expected any time a key is depressed, one for the rows and one for the columns. Any deviation is easily detected, and the encoder will respond by either not accepting the input, accepting the first pushed key only, known as *lockout,* or accepting the second pushed key for subsequent entry, known as *rollover.*

An Introduction to Computers

TABLE 13-1 Keyboard truth table

Key	Keyboard Response Column 3210	Row 3210	Decoded Binary Output
0	1110	1110	0000
1	1110	1101	0001
2	1110	1011	0010
3	1110	0111	0011
4	1101	1110	0100
5	1101	1101	0101
6	1101	1011	0110
7	1101	0111	0111
8	1011	1110	1000
9	1011	1101	1001
A	1011	1011	1010
B	1011	0111	1011
C	0111	1110	1100
D	0111	1101	1101
E	0111	1011	1110
F	0111	0111	1111

Notice that for Key 9, row 2 and column 1 are pulled to ground; 0's are in the corresponding bit positions

13-9 THE SINGLE-BOARD COMPUTER LED DISPLAY

The LED display serves two functions:

1. It displays the address and data currently being entered.
2. It enables the user to examine the contents of a register or memory location.

While a series of keys are depressed in the process of entering addresses and data, it is convenient to see these displayed. We also wish to examine the contents of various memory locations and read the result of an operation just carried out.

Each display chip need only be capable of showing the hex digits (alphanumeric displays are not needed for this simple display). Two hex digits represent an eight-bit byte. INTEL's SDK 85 display, for example, has six hex digits. Four digits (16 binary bits) display one of 64K addresses, and two digits display the eight-bit content of a register or memory location. The SDK 86, which is a 16-bit microprocessor, has four digits to display the 16-bit content of a register or two adjacent eight-bit memory locations.

A numeric display chip has seven segments. Each segment is separately accessible. When activated, it lights up. Ten numbers and a limited set of letters can

0 1 2 3 4 5 6 7 8 9 A b C d E F

FIGURE 13-10 Hex display

thus be formed with the properly encoded input, which activates the desired combination of segments as in figure 13-10. The LED segments are typically labeled *a* through *g*.

Four-line to seven-segment decoders (7447, 7448) are available for BCD systems, in which only the ten numerics are needed. The BCD four-line to seven-segment decoder is discussed in the laboratory manual. Hex decoding in microcomputer systems can be generated in software. Software decoding methods are usually discussed in a microprocessor course.

There are two kinds of seven segment display chips: the common cathode type and the common anode type. As the name implies, the common anode means that all diode anodes are tied together (the word common does not imply that the common point is ground) and connect to V_{cc}. The seven inputs labeled *a* to *g* are activated with a low output from a 7447 BCD (four-line) to seven-segment decoder as in figure 13-11(a) or from a driver that in turn receives the proper code from a programmable keyboard-display interface such as the INTEL 8279. In either case a series resistor controls the magnitude of the current (See figure 13-13).

The common cathode implies that all cathodes are tied together and connected to ground. In this case the active input to any segment is high. The 7448 BCD to seven-segment decoder provides the active high output. A built-in 2K limiting resistor avoids the need for an external current-limiting resistor. An external shunt resistor may be used if more current is needed. (See figure 13-11(b).)

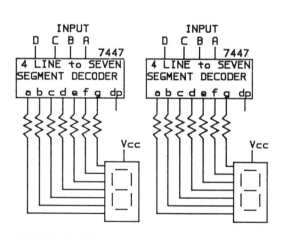

FIGURE 13-11(a) Common anode display

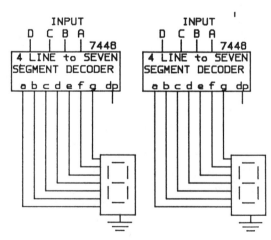

FIGURE 13-11(b) Common cathode display

An Introduction to Computers

The preceding discussion implies the use of one four-line to seven-segment decoder per displayed digit. The circuit of figure 13–11 can be changed to figure 13–12, which uses only one seven-segment decoder but requires a multiplexer. An input selector points toward the digit to be read. The corresponding select transistor grounds the designated display chip. As the pointer scans all inputs, the corresponding select transistors are activated. The scanning rate must be sufficiently high to avoid flicker.

The input selector is not needed if the input source is a register, as in a microprocessor system, whose content is changed at the scan rate. INTEL's 8279 programmable keyboard-display interface provides the functions required to interface the 8085 microprocessor to a keyboard and display. (See figure 13–13.)

The keyboard section of the interface scans the keyboard rows and provides automatic switch debounce as well as N-key rollover. The display section of the interface has an eight-byte FIFO to store keyboard information, a 16-byte internal RAM for display refresh, and the scanner for the multiplexed digits. The CPU is thus relieved from carrying out these functions.

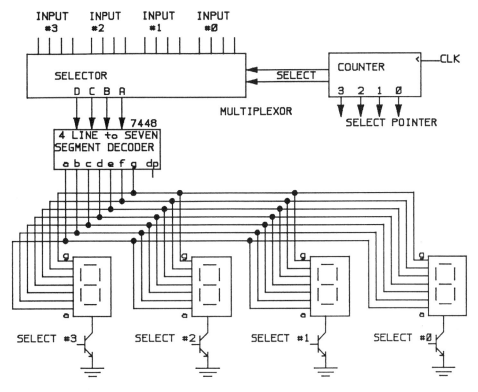

FIGURE 13–12 Multiplexed display

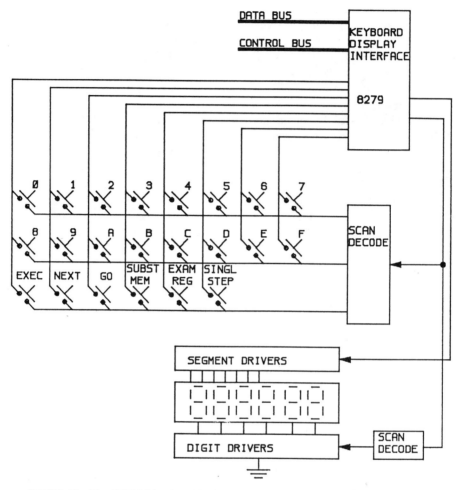

FIGURE 13-13 SDK 85 keyboard/display subsystem (Courtesy of Intel Corporation)

13-10 A SINGLE-BOARD MICROCOMPUTER

We have looked at most of the required bits and pieces to draw a computer system. To review, we have

- The arithmetic logic unit (part of the central processing unit)
- Timing and control circuits (part of the central processing unit)
- The memory
- The bus (the main roadway over which communication flows)
- I/O ports (which allow entering and leaving the bus)

An Introduction to Computers

- The keyboard (to enter hex data into the computer)
- The display (to examine the contents of registers and memory locations)

Let us put all these blocks into a single workable system in figure 13–14 and generally discuss the function of the blocks in the central processing unit (CPU).

13–11 THE CPU

The Arithmetic Logic Unit (ALU)

The ALU performs

    additions of two words

    subtractions of two words

    increment of a word

    decrement of a word

    comparison of two words

    shifting of a word left or right

    logic operations like AND, OR, and Exclusive OR

The Accumulator

The accumulator is a register. It is an integral part of the ALU. The ALU operates on the contents of the accumulator, and the result of the operation is placed in the accumulator.

The Temporary Register

The temporary register holds data from the bus to be entered into the ALU.

The Flag Register

Another register, the flag register, is also part of the ALU. It has an important function: It stores status information. Designated bits in the flag register are set or reset depending on the outcome of an operation. These designated bits typically consist of a carry flag, a zero flag, a sign flag, a parity flag, and an auxilliary carry flag.

    The instruction set of any microprocessor lists the conditions that set the flags. The flags are used to make decisions. What the decisions are, of course, is a function of a particular program. We can make a few comments to illustrate the possibilities.

    The carry flag, for example, is set as a result of an addition overflow. This can be used to carry out additions to more bits than the accumulator holds. A

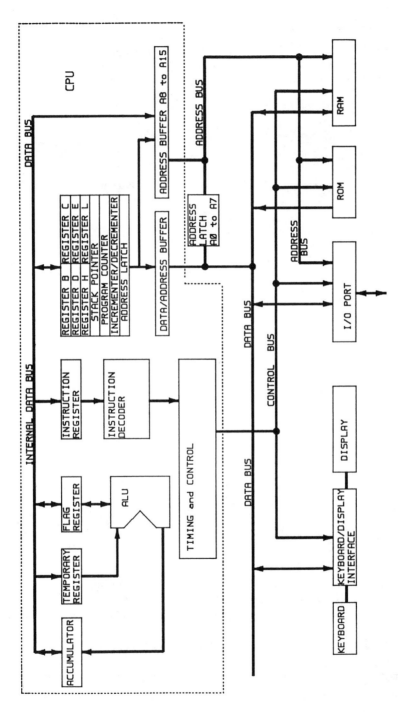

FIGURE 13–14 SDK 85 single-board computer (Courtesy of Intel Corporation)

compare will set the carry flag when one word is larger than the other. This information can be used to branch a program. A shift through the carry flag will set the flag if the shifted bit is a 1. This is a way to test a particular bit within a word in order to make a decision.

The zero flag states that all bits in a tested word are 0. This may indicate the end of an operation or the exit or branch point for a loop.

The sign flag gives information about the most significant bit in a word. In a complement add operation a 1 indicates a negative result or will indicate the end of a shift left operation if a word is to be left justified.

The parity flag is 1 when the sum of all 1's in a word is even. This is useful for error detection.

The auxilliary carry flag is set to 1 in decimal addition to detect overflows past decimal 9. That carry can then be added to the next BCD digit.

Additional Registers

A number of registers are needed to hold and manipulate data that are to be used in current operations. These registers can also be incremented and decremented, thus setting flags. This is most useful for loop counting. One of the registers, the HL register, will hold the program memory address, which is used by the processor as a pointer to send or fetch data.

The Stack Pointer It is often necessary to exit from a main program to a subroutine or service routine. When that happens, the processor must save the last address from which it left in order to be able to return to the main program and continue where it left off. Some interim results and register contents must also be stored in a safe place to make room in the registers for use by the other routines. The stack pointer points to a selected memory space where the last program count and register contents are stored.

The Program Counter The program counter keeps track of the program's progress. It holds the present memory address, the contents of which are being executed by the processor. After each execution the counter is incremented.

The Instruction Register and Decoder An instruction arriving on the data bus is loaded into the instruction register. The instruction decoder receives the information from the instruction register and proceeds to decode the message. Every instruction demands that the processor carry out a set of predetermined steps. The instruction decoder guides the control logic through the sequence of steps. At the end of the sequence the next instruction is fetched.

The Address Register The address register latches the outgoing address for the required execution time.

13-12 COMPLETING THE SYSTEM

The blocks inside the dotted line in figure 13-14 are part of the 8085 processor chip. The blocks outside the dotted line are needed for a workable computer system.

A_0 to A_7 Address Latch

Some address and data lines on a microprocessor are typically shared. INTEL's 8085 has 16 address lines, which are capable of addressing 64K addresses. The eight lower address lines A_0 to A_7 are shared with the eight data lines. A_0 to A_7 are sent out through a buffer and latched again in order to free the data bus for data.

Keyboard and Display

Instruction and commands from the keyboard are sent to the CPU data bus through an interface, as was discussed earlier.

Memory

Memory consists of RAM and ROM. The ROM holds the monitor program. It is a set of instructions that allow the user to execute the keyboard commands to load data and address, to examine contents of memory and registers, and to execute a complete program or step the program one instruction at a time. In computer systems that work with CRT terminal, disk, and printer the operating system is stored on disk. Hence the name disk operating system or DOS.

The RAM in the SDK 85 is the user space. Any program written by the user is stored in RAM, ready for execution. Data input sets and results obtained are also stored in RAM. In the single-board training computer the data is lost when power is shut off. If the system has a disk, programs are loaded into RAM from the disk instead of the keyboard, and results are sent back to the disk.

I/O Port

The I/O port in the SDK 85 allows the sending and receiving of data bytes, in this case from a set of eight switches or to a set of eight neon bulbs.

13–13 DATA FORMAT AND INSTRUCTION SET

There are two sets of information the computer must have in order to carry out an instruction:

1. The *op code,* or operation code, which tells the processor what to do.
2. The *operand,* the data, register, or memory it is to operate on.

The op code tells the CPU whether the operand is data or an address where the data is found. If it is actual data, it is called immediate data. If it is a register, the operand specifies which one. If it is memory, the operand specifies only memory but not the address. The address will have been previously stored in one of the registers of the CPU. The CPU will use that address to carry out the instruction.

Together the op code and the operand form an *instruction*. An instruction may be one byte or several bytes long. The instruction set in the user manual describes each instruction in detail and ascribes a mnemonic to each.

For example, we wish to move the contents of one register to another. The mnemonic in the 8085 instruction set is MOV r1,r2. One of the registers is the source, the other the destination. The instruction byte is

$$0\ 1\ D\ D\ D\ S\ S\ S \qquad \text{D: Destination,} \quad \text{S: Source}$$

01 is the op code, and DDD SSS are the operand.

If the accumulator is the destination and has a code = 111 and the *B* register is the source and has a code = 000, then by substituting for DDD and SSS, the complete instruction is 01111000 = 78_{Hex}. The 78 is a code the machine understands. It is *machine language*. MOV r1,r2 is *assembly language;* it must be translated into machine language for execution.

The machine code for this instruction is the same for the National Semiconductor NSC800 or Zilog's Z80. The mnemonics differ. The mnemonic is LD rd,rs and LD r,r', respectively. Not all instructions are identical because later-built microprocessors include improvements. The idea is that if you learn to program one processor, it is easy to switch to the next.

Assembly language is much easier to use than machine language, and if your system has a disk, the *assembler* program will translate the assembly language into machine language. If all you have is a keyboard, you must hand-assemble the program with the help of the instruction set description. For learning purposes this method does have the advantage of putting you in direct contact with the microprocessor.

The instructions fall into several categories:

- Data transfer
- Input/output (I/O)
- Arithmetic
- Logic
- Control transfer
- Processor control

Data transfer handles moves from any register to any other, data fetching from memory to any register, or data storing from any register to memory. I/O sends or receives data to or from the outside world. Arithmetic, as the word implies, handles addition, subtraction, incrementing, and decrementing. Logic instructions perform the Boolean functions of AND, OR, Exclusive OR, complementation, and compares. The control transfer is done with the *jump* instructions. They test flags and conditionally branch the program. The power of the computer is due to the ability to make decisions. It is this property that makes the computer "smart."

Finally, the computer must be able to respond to external demands. For that purpose, microprocessors have one or several interrupt inputs. The processor control instructions specifically handle program interruptions without destroying the ongoing program.

13–14 A SIMPLE PROGRAM

We want to store 6 in memory and 11 in the accumulator, add the two numbers, and store the result in a second location in memory. We must choose a location in memory where the program starts. That can be address 2000. Using the INTEL 8085 instruction set, the program will be as shown in table 13–2.

A few words of explanation are in order. First, notice that the program is located in location 2000 to 200A. The data is located in address in 2060, away from the main program. The computer does not know where to send the data unless we inform it that 2060 is where we want it to be.

LXI HL 2060 is the mnemonic to store the address in the HL register, which the processor will look up when needed. MVI A, 11 is the mnemonic for moving the number 11 into the accumulator. MOV M,A is the mnemonic to send the contents of the accumulator to memory. MVI A,06 is the mnemonic for moving 6 into the accumulator.

Now we are ready to add the two numbers. ADD M is the mnemonic for adding the contents of the memory, whose address is still in HL, to the contents of the accumulator. The result will be in the accumulator. INX HL is the mnemonic to increment the contents of the HL register. The HL register now points to the address where the result will be stored. MOV M,A is the mnemonic for moving the contents of the accumulator to the memory.

Finally, the CPU must be told to stop and return to the monitor. RST1 is a reset instruction that sends the program back to the monitor.

A detailed study of assembly language and more complex programs is left for a course in microprocessors.

TABLE 13–2

Memory Address	Assembly Code	Machine code	Comments
2000	LXI HL 2060	21 60 20	Initialize the address
2003	MVI A, 11	3E 11	Move 11 to memory
2005	MOV M,A	77	Move ACC to memory
2006	MVI A,06	3E 06	Move 6 to accumulator
2008	ADD M	86	Add 11 to 6
2009	INX HL	23	Increment the address
200A	MOV M,A	77	Send sum to memory
200B	RST1	CF	Stop

REVIEW QUESTIONS

1. List the main components of a computer.
2. What is a bus?
3. Why do we need open collector or tristate circuits to put data on a bus?
4. What are the essential differences between open collector and tristate circuits?
5. How is an open collector–driven bus terminated?
6. How is a tristate-driven bus terminated?
7. Why is the tristate-driven bus response slow when terminated with a pull-up or pull-down resistor?
8. What is the purpose of an active terminator?
9. What is an I/O port?
10. How does one prevent bus contention (more than one set of data entering the bus simultaneously)?
11. How is a keyboard decoded?
12. Discuss keyboard scan decoding.
13. Why does a BCD or hex display chip need seven segments?
14. Why is a decoder driver needed to drive a display chip?
15. Could we drive a display segment with a 7404 hex inverter?
16. Are limiting resistors needed between the decoder driver and the display chip?
17. What is a computer instruction?
18. Why does the instruction contain an op code and an operand?
19. What is the fundamental difference between a computer and a calculator?
20. Why is an instruction decoder needed?
21. Are memories, ALUs, and control circuits used in computers only?
22. What is the purpose of extra registers in the CPU?
23. Explain the purpose of the stack pointer.
24. Explain the purpose of the program counter.
25. Explain the function of the flag register.
26. Explain how flags enable the computer to make decisions.
27. Why are some data and address lines shared?
28. How does one separate the address and data lines for the external bus?
29. Discuss the function of ROM in the SDK 85 system.
30. Discuss the function of RAM in the SDK 85 system.

chapter 14

a survey of computer peripherals

14–1 INTRODUCTION

There must be some means to enter programs and data into the computer and retrieve the results. We are talking about *input/output* (I/O) devices. The most immediate, direct interface between user and computer is the cathode-ray tube (CRT) terminal. The computer and CRT terminal form a workable if impractical system, rather like trying to solve complex problems without pencil, paper, and books. The computer system needs a means to store interim as well as permanent information. The internal memory is usually not large enough to accommodate this requirement. A disk is the appropriate medium for the purpose. (See figure 14–1.)

The disk also serves another most important function: It stores the *operating system*. When the computer is turned on, the internal memory is empty. The computer cannot understand any of the commands. After the operating system is loaded from disk into the internal memory, the computer will respond to and carry out a given set of input commands.

Another storage medium, slower but with more storage capacity than the disk, is the magnetic tape. The tape recorder is typically used for long-term storage of massive amounts of data. A computer user will generate an amazing amount of data in a short time period. Much of the information might be useless and can be discarded, but many results and programs must be kept for future use; these are conveniently stored on tape. It is also a good idea to *back up* the disk periodically onto tape. Disks enjoy reasonable reliability, but disk "crashes" are not unheard of, and when they happen, much precious effort, programs, and accumulated data disappear.

Finally, it is frequently necessary and convenient to get a permanent, hard copy printout of a finished product on a printer.

There are, of course, more peripherals than this basic set. An external memory

A Survey of Computer Peripherals

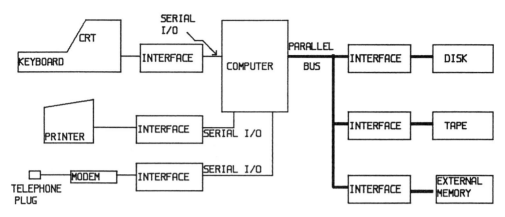

FIGURE 14-1 Computer and peripherals

can be added to the system as a peripheral for applications when the internal memory has insufficient capacity. The graphics display and the graphics printer take a data set and display it on the screen or as hard copy, respectively, in graph or pictorial form. The modem allows your terminal and/or computer to communicate with an off-site computer system over the telephone lines.

The card punch has gone out of use with multiuser interactive computers, and so has the teletypewriter (TTY) with the introduction of the CRT terminal.

Each peripheral communicates with the computer via an *interface,* or control logic. Typically, all interfaces connect to the data, address, and control bus with tristate gates which are high impedance isolated from the bus when inactive and low impedance connected when active. Thus all interfaces are in parallel, and only one can "talk" to the computer at a given time. The computer arbitrates which of the peripherals is to be serviced. The others have to wait their turns. The control or interface logic is designed to work on a *handshake* basis; that is, Busy and Ready status is passed back and forth between the computer and the peripheral interfaces. Commands and requests are carried out at the appropriate time. The interface logic also stores the set of commands that are to be executed and alerts the computer with an interrupt signal when done. Since each peripheral works with its own internal clock, all signals transferring to and from the bus are resynchronized.

Data transfer is either done in byte or word parallel for high-speed operation or in bit serial for slower applications. Disks and tapes are parallel; modems and terminals are serial.

Interface control logic can be quite complex, particularly for cases in which mechanical motion is involved. For example, an interface board for a Pertec tape recorder to a Data General Nova computer requires 142 TTL chips. Much of the control logic for standard components has, in the last few years, been miniaturized into single chips. The detailed description and design considerations for such applications are beyond the scope of this book.

14-2 THE CRT TERMINAL

In some respects the CRT terminal is similar to the home television set. A cathode-ray tube displays a given set of information by horizontally scanning one line at a time. The data, stored in a RAM, provides a 1 and 0 pattern to the scanning beam. A 1 intensifies the beam, producing a bright spot, and a 0 leaves the spot dark. The number of lines scanned is a function of how many character lines are to be displayed and how many dots represent a character.

The horizontal sweep moves the beam to the edge of the screen, then quickly retraces as a response to the horizontal synchronization (sync) signal. The vertical sweep shifts the beam down one line, and the horizontal sweep traces the next line. When the beam reaches the bottom of the screen, the vertical sweep retraces to the top, responding to the vertical sync signal, and the process repeats. (See figures 14–2 and 14–3.)

The television set repeats 30 complete frames per second, and the CRT terminal repeats 60 frames.

When a computer key is depressed, an ASCII character is generated and stored in a RAM location that is indicated by the cursor position. The cursor is an intensified symbol that corresponds to the RAM address where data will be entered when a key is depressed. The RAM is continuously scanned; and when the particular address is encountered, the ASCII character is encoded into a dot pattern and displayed. This is done with a character generator. The standard code for 128 characters is reproduced for convenience in table 14–1. It has a unique seven-bit code for each keyboard character. There are a number of dot patterns that can represent the desired characters. A 7 × 9 pattern with descenders nicely displays all capital and lowercase letters as well as numbers and symbols. For example, look at the display in figure 14–4 of 1A2p&w7y.

The dot pattern is permanently loaded into a ROM. A 1 shows where the beam is to be intensified and a 0 where the beam is to be blanked, as in the example in figure 14–4. The ASCII code addresses the particular location, and eight clock pulses scan the content of the first row. Notice that the first row is empty, loaded with 0's, for line separation. The dots are shifted out serially. The entire row (eight dots for each of 80 characters) is scanned, and the beam shifts vertically down one

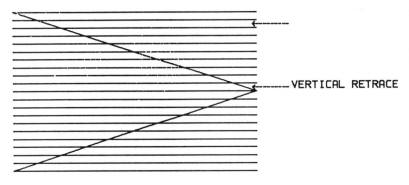

FIGURE 14–2 CRT raster

A Survey of Computer Peripherals

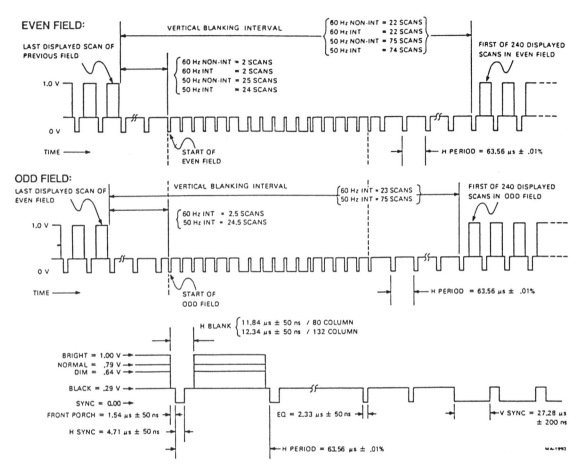

FIGURE 14-3 CRT timing and control waveforms (Copyright © 1983, Digital Equipment Corporation. All rights reserved. Reprinted with permission.)

row to scan the second row of dot patterns. After the beam scans an additional eight rows (10 total), one line of characters appears on the screen.

This typically continues for 24 lines. The beam therefore scans $24 \times 10 = 240$ rows, displaying 80 characters per line (including an end of line marker). Horizontally, each character is seven dots and one space = 8 spaces. Each row is $80 \times 8 = 640$ dots wide. The total number of dots on the screen is $640 \times 240 = 153{,}600$ dots. The screen is scanned 60 times per second. Thus each second, 9,216,000 dots are displayed. The dots are shifted out with a 12.375 MHz dot clock. Each dot takes 80.08125 nsec. Each horizontal scan takes $80.08125 \times 10^{-9} \times 640$ dots = 51.72 μs + 11.84 μs for horizontal retrace = 63.56 μs. The horizontal sweep oscillator is therefore 15.733 KHz. The 240 scans take 15.254 ms + 1.412 ms for vertical retrace = 16.6 ms per frame, or 60 frames per second.

The total number of characters is only $24 \times 80 = 1920$. The ten-row shift

TABLE 14–1 ASCII code

Character	Seven-Bit Octal Code	Character	Seven-Bit Octal Code	Character	Seven-Bit Octal Code	Character	Seven-Bit Octal Code
A	101	a	141	0	060	;	073
B	102	b	142	1	061	<	074
C	103	c	143	2	062	=	075
D	104	d	144	3	063	>	076
E	105	e	145	4	064	?	077
F	106	f	146	5	065	@	100
G	107	g	147	6	066	[	133
H	110	h	150	7	067	\	134
I	111	i	151	8	070	]	135
J	112	j	152	9	071	^	136
K	113	k	153	!	041	_	137
L	114	l	154	"	042	`	140
M	115	m	155	#	043	{	173
N	116	n	156	$	044	|	174
O	117	o	157	%	045	}	175
P	120	p	160	&	046	~	176
Q	121	q	161	'	047		
R	122	r	162	(	050		
S	123	s	163	)	051		
T	124	t	164	*	052	Line feed	012
U	125	u	165	+	053	Return	015
V	126	v	166	,	054	Escape	033
W	127	w	167	–	055	Space	040
X	130	x	170	.	056	Delete	177
Y	131	y	171	/	057	Backspace	010
Z	132	z	172	:	072	Tab	011

```
            1       A       2       p       &       w       7       y
    12345678 COLUMN 8 IS FOR CHARACTER SPACING                          LINE
    .......  .......  .......  .......  .......  .......  .......  .......  0 SPACE
    ...1...  ...1...  .1111..  .......  .1111..  1.....1 1111111  .......  1
    ..11...  ..1.1..  1.....1  .......  1.....1  1.....1  .....1  .......  2
    .1.1...  .1...1.  ......1 11111..  1.....1  1.....1  .....1. 1.....1. 3
    ...1...  1111111  ...11..  1.....1  .1111..  1..1..1 .....1.  1.....1. 4
    ...1...  1.....1  1......  11111..  1....1.  .11111.  ..1....  .11111.  5
    .11111.  1.....1 1111111  1......  .111111  ..1.1..  ..1....  ......1. 6
    .......  .......  .......  1......  .......  .......  .......  1.....1. 7
    .......  .......  .......  1......  .......  .......  .......  .1111..  9
      061     101     062     160     046     127     067     171    OCTAL
    0110001 1000001 0110010 1110000 0100110 1010111 0110111 1111001  BINARY
```

FIGURE 14–4 Characters generated with a dot pattern

A Survey of Computer Peripherals

for each set of characters is part of the character generator scan algorithm. A 2K byte memory is sufficient to store all the information for one screen (or one frame) of display.

The memory is updated for new inputs. An additional 80 character buffer is loaded from an external input, such as keyboard or processor, and transferred to the character-generating logic during the vertical retrace time. (See figure 14–5.)

Let us step through the block diagram of figure 14–5.

- Step 1. All counters are reset and start from 0.
- Step 2. At the beginning of the raster, after a vertical and a horizontal sync, the active line starts. The RAM receives at the start 00000 (five bits for count to 24) from the mod 24 counter for the vertical address and 0000000 (seven bits for count to 80) from the mod 80 counter for the horizontal address.
- Step 3. The ROM receives the ASCII code as an address. The content of that address is the desired eight-bit dot pattern, which is sent to the parallel-to-serial shift register. These are shifted out with the dot clock to the video amplifier.
- Step 4. After eight clock cycles the mod 80 counter steps to 0000001. The next eight bits are read to video. This continues until the mod 80 counter is at 1001111 = 79; 640 dots were sent to video. The mod 80 counter resets to 0000000.
- Step 5. The horizontal sync pulse steps the decade counter to 0001. All other addresses remain unchanged. The next row of dot patterns is sent. This continues until the decade counter is at 1001. Ten rows of dots have been sent, and a complete line of characters is printed. The next count produces a decade counter carry.

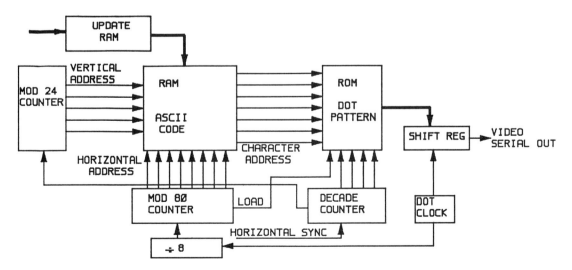

FIGURE 14–5 Character display system

- Step 6. The carry of the decade counter steps the vertical RAM address to 00001. Steps 1–5 repeat until the vertical RAM address = 10111 for 23 additional lines of characters.

The character generator is obviously not limited to ASCII code pattern generation. Any shape can be generated with the proper encoded pattern. Thus the CRT terminal can be designed to produce graphics displays. However, this requires much more memory, particularly if color graphics are desired.

14–3 THE DISK DRIVE

The magnetic disk is the secondary memory of a computer system. In some respects it reminds us of an audio record player. The information is stored in circular fashion serially along a track and is accessed by rotating the disk. But that is where the similarity ends. Whereas the audio recorder rotates at speeds of 33 ⅓ rpm, the digital disk rotates at speeds up to 3600 rpm. The audio recorder stores analog information in a continuous spiraling track in groove variations, but the digital disk stores discrete bits in a concentric track by magnetizing the disk surface. The audio recorder interfaces to an amplifier and speaker; the digital disk interfaces to a computer.

In principle the digital disk is easy to understand. A magnetic head is located over a track on the rotating disk as in figure 14–6. Current in the head coil is switched in accordance with a given bit pattern. This causes corresponding flux reversals in the head core. The flux links with the disk medium and leaves corresponding magnetized spots. The read operation uses the same head. As the head passes over magnetized spots and links with the head core, current is induced in the coil. The signal is amplified and decoded into a corresponding bit pattern.

In practice, many details must be considered. We discuss them in the following sections.

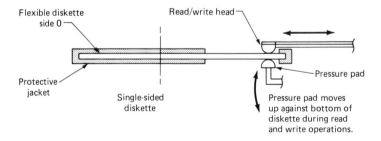

FIGURE 14–6 Head and disk (Redrawn courtesy of Xidex Corporation)

14-4 THE DISK

A hard aluminum disk, 8–14 in. in diameter, is coated with iron oxide, which consists of needlelike particles about 100 μm long and 1 μm thick. Each particle acts as a bar magnet and tends to align itself in the direction of a magnetic field. The total magnetic field of a spot is the sum of all the contributions of the particles. During manufacture the particles are aligned in the direction of the track motion.

Flexible or floppy disks, 3½ to 8 in. in diameter, are made of Mylar plastic coated with iron oxide. Flexible disks have less storage capacity than hard disks but are less costly.

Data on a disk is spatially arranged in a serial bit stream along a given track. The circular track is divided into sectors. The sector can be of any length. Typically, sectors are 128, 256, 512, or 1024 bytes long. The sectors are needed to address a given set of information or blocks. One or more blocks make up a file. Clearly, we want to have as many files as possible of varying length on the disk. A new file starts at the beginning of a sector but might not entirely fill the last sector. It is therefore desirable to make the sectors short in order not to waste data space. Each sector, however, needs a number of overhead bits for formatting, a feature that leads us to the opposite conclusion, that sectors should be long to minimize the overhead. Compromises are reached in terms of application.

Sector Formatting

To recognize the beginning and end of a sector and synchronize the data, the format of a sector is arranged as in figure 14-7.

A *preamble* is a burst of 0 bits that is required for sector tolerancing. The preamble, 20–40 bytes in length and preceding the data, is written long enough to ensure that the read electronics acquires data.

A *sync* bit is a single 1 bit written at the end of the preamble. It is used as a flag to signify the beginning of useful data.

A *header* consists of two bytes that contain the cylinder, head, and sector address of the current sector. This information is compared to the expected address before data transfer occurs.

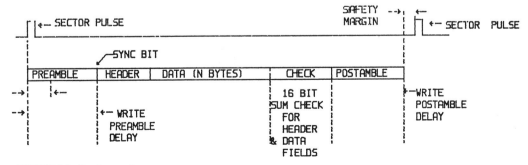

FIGURE 14-7 Sector format

The *data field* portion of the format is made up of *n* bytes of data with the least significant bit of each byte written first. The value of *n* depends on the sector size.

The *check* portion of the format (for error correction) is made up of two bytes of *sum check* information for the header and data fields. In reading from the disk drive these bytes are compared to a recomputed sum check.

The *postamble* is a burst of bits, 2–8 bytes long, that allows time for the erase head to complete the erasing of the check field before the write current is switched off.

Another gap or safety margin of 30–300 bytes is added to the format. Disk warping and speed variations require the added spatial tolerance for head locating.

The total number of bits for formatting is not standardized. It may nearly double the sector length. The formatted disk therefore has less data capacity than an unformatted disk.

There are two ways to mark the sector locations. One method, called *hard sectoring,* has punched holes along the inner or outer rim of the disk. As the holes pass over optical indicators, the holes are counted until the desired sector is reached. *Soft-sectored* disks have only one indicator, and the sectors are magnetically coded. The latter allows flexibility in sector size determination by the user. (See figures 14–8 and 14–9.)

The disk is encased in a jacket for protection against physical damage and dirt. A slot in the jacket allows head access for reading and writing. The disk rotates inside the jacket. A notch at the edge of the jacket is used to flag the disk drive logic to permit or inhibit writing onto the disk. If the notch is uncovered on an 8 in. diskette, writing is inhibited. The disk is *write-protected*. The convention is reversed for a 5¼ in. diskette.

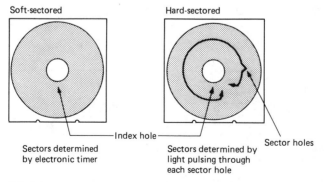

FIGURE 14–8 Disk sector indications (Redrawn courtesy of Xidex Corporation)

A Survey of Computer Peripherals

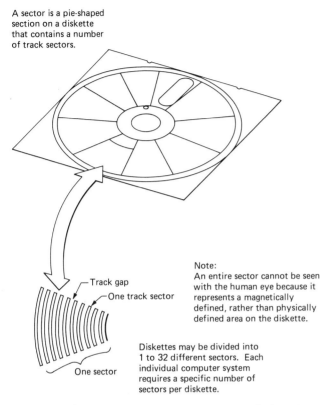

FIGURE 14–9 Track sectors (Redrawn courtesy of Xidex Corporation)

Disk Capacity

The disk capacity is a function of the bit density measured in bits per inch (BPI), the disk diameter, and the number of tracks per disk. Disk system storage capacity is increased when both sides of the disk are used with a dual head and several such disks are packed into a disk pack.

Applied Information Memories produces a disk with 18,534 BPI and 4000 tracks on a 5¼ in. disk with an average track length of 13.92 in.

The disk capacity is easily calculated:

$$\text{Capacity} = \frac{18534 \text{ bits/in.} * 13.92 \text{ in./track} * 4000 \text{ tracks}}{8 \text{ bits/byte}}$$

$$= 129 \text{ megabytes}$$

This is indeed a very respectable number. Table 14–2 shows a few other representative disk characteristics.

TABLE 14-2 A few representative disk parameters

	Applied Information Hard	Dysan Hard	Dysan Flexible	Pertec Hard	Apple McIntosh Flexible
Storage capacity (megabytes)	129	16	0.720	50.75	0.400
Recording density (bits/inch)	18534	6038	8900	2200	8900
Track density (tracks/inch)	1000	384	100	200	135
Data tracks	3550	808	80	812	80
Operating speed (rpm)	3600	3600	300	1500	390 to 600
Transfer rate (megabits/sec)	9.677 max	9.677 max	0.250	1.5625	0.250
Disk diameter (inches)	5¼	14	8	14	3½

14-5 THE HEAD

To transform an electronic pulse, which represents a 1 or a 0, into a magnetic flux that links with the magnetic medium on the disk, an iron oxide core is used. A current pulse induces a magnetic flux in the core. A narrow gap in the core forces the flux to jump across the gap. Magnetic flux lines in the fringe field of the gap curve away from the core edges and, in doing so, link with the disk surface and leave behind a magnetized spot. (See figures 14–10(a) and 14–10(b).)

An actual head is more complex than the head in figure 14–10(a). Head design conforms to the application. Heads for hard disks are designed not to be in contact with the disk surface when the disk is spinning. When the disk is rotating at 3600

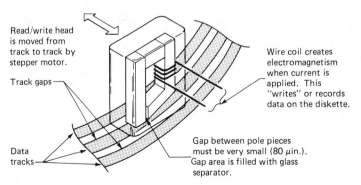

FIGURE 14-10(a) Head and track (Redrawn courtesy of Xidex Corporation)

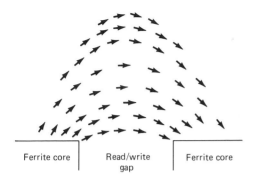

FIGURE 14–10(b) Fringe field linking head and track

rpm, the velocity of the disk surface under the head is about 100 mph. This creates an air current that lifts the head away from the surface. The distance is indeed very small, only ½ μm. This is sufficient to eliminate frictional head and disk surface wear, yet hold the fringe field close to the magnetic medium. (See figure 14–11.) The head must be very light, about 10 g, and be made of a material that yields large magnetization for small currents. Flying heads were first developed by IBM for the Winchester disk.

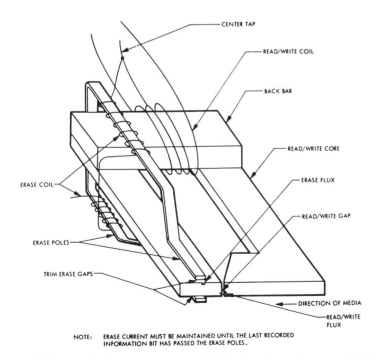

FIGURE 14–11 Head for the Pertec D3000 disk drive (Courtesy of DDC PERTEC, Pertec Peripherals Corporation)

With such small head-to-disk distances it is extremely important to keep the disk surface free of any particles. A human hair is about 100 μm in diameter, and dust and smoke particles or fingerprints by far exceed the 0.5 microinch spacing. Not only can errors occur when there is debris on the head, but the head can bounce and then crash onto the disk, causing irreparable damage.

Flying head technology is not suitable for floppy disks. Here, the head, an iron oxide–coated Mylar sheet, is in contact with the disk. The disk turns more slowly to reduce frictional wear, and the bit transfer rate is correspondingly smaller.

In either the hard or floppy disk system the head is moved from track to track by means of a stepping motor.

14–6 ACCESS TIME

When a particular address is to be found, the head must step to the desired track, then wait until the addressed sector passes under the head. The time to find the track, the *head-positioning time,* varies from a minimum of 0 if the head is already on the chosen track to some maximum if the head must step from position 0 to the last track (or vice versa). Stopping the head causes it to vibrate. An additional *settling time* is required to stabilize the head before proceeding. The time to find the addressed sector also varies. The sector sought might be the first one encountered as the disk rotates, or a complete revolution might pass before it is found. This is called the *rotational latency* and is a function of disk speed.

The sum of the positioning and settling time and rotational latency is the *access time* (also known as the *seek time*). For hard disk drives the maximum access time is typically 25 ms; for floppy disk drives it is as high as 630 ms.

If the system recognizes a *seek error,* the head retracts to the zero position and tries again. Errors can also occur when the head is not properly positioned over the track. This is a *misalignment.* It can happen if data is written on one drive and read on another. Some disk drives incorporate a means for head-positioning correction. It can be activated to read the desired data but should be deactivated for continued normal operation.

The above discussion centered on removable disks. Performance is improved if the disk is fixed (nonremovable). The disk is completely enclosed for dirt protection. Fixed disks also have fixed heads, that is, one head per track, thus reducing alignment problems and eliminating positioning and settling times.

14–7 DATA RECORDING

Logic data is in a form that is not particularly suited for data recording. Assume a data pattern of 1's and 0's as in figure 14–12. A string of several 1's or 0's is void of any changes until the opposite bit appears. Surface magnetization without changes is subject to errors. The system loses track of which bit belongs to which clock

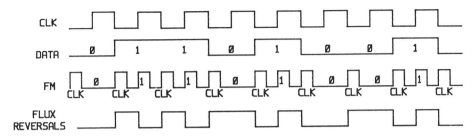

FIGURE 14–12 FM encoding

cycle. It is desirable to have at least one flux change per cycle. Nonencoded data is known as *NRZ data* (nonreturn to zero). To overcome this difficulty, NRZ data can be encoded in several ways.

FM Code

For each clock cycle there is a pulse at the beginning of the clock cycle (or beginning of the *bit cell*). If the data bit is a 1, another pulse exists in the middle of the bit cell. If the data bit is a 0, the bit cell is left blank. For an all 1 pattern, the maximum recording frequency is twice the clock frequency. The minimum recording frequency for an all 0 pattern is the clock frequency. It is a form of frequency modulation called *FM encoding*. The clock is embedded in the data, ensuring a flux reversal for each bit. Notice a flux reversal of either polarity in the middle of the bit cell if the bit is a 1. (See figure 14–12.)

MFM Code

A modified version of FM, called MFM, reduces the number of required flux reversals, thereby improving the data bit density. This is known as *double-density* recording. What FM tried to avoid is the lack of any flux changes for strings of 1's or 0's. But FM produced two flux changes per cycle for each recorded 1. There is more synchronization data than is necessary. MFM eliminates the bit cell edge clock pulse when recording a 1 and inserts a clock pulse only at the bit cell for the second 0 in a string of 0's. The maximum recording frequency is equal to the clock frequency. An all 1 pattern produces pulses in the middle of the bit cell. (See figure 14–13.) An all 0 pattern produces pulses at the edge of the bit cell. The pattern is shifted 180 degrees. The minimum frequency of half the clock frequency occurs for an alternating 1/0 pattern. Figure 14–13 shows the results of that change. The encoding and decoding electronics for the MFM are more complex than for FM. However, MFM does improve recording density and finds use in double-density disk drives.

A modified version of MFM called M^2FM goes a step further. It inserts clock pulses only for every other 0 in a string of 0's.

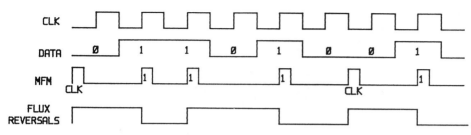

FIGURE 14–13 MFM encoding

14–8 DISK ERROR DETECTION AND CORRECTION

Machines, being what they are, make errors. Storage systems make two kinds of errors. The system recognizes an error and can recover from it by either correcting the error with logic circuits or rereading the sector up to seven times. If the error was corrected, it is called a *soft error;* if not, the error is called a *hard error*. The system can designate a particular bad sector as unusable and avoid ever accessing it again or flag the user to remove the disk.

To a limited extent, errors can be found and corrected. If a single bit is affected, a special code that is part of the format identifies the exact error location and the associated bit and simply inverts the bit in question. If more than one error has occurred, the system usually cannot correct all the errors. There are methods for correcting bursts of errors, but they are complex. Single-error correction and detection is relatively simple.

Single-Error Detection

Any data pattern consists of a string of 1's and 0's. The sum of 1's in a string must be either odd or even. A simple flip flop can reverse for every arriving 1 and will be left in a high state if the number of 1's is odd or in a low state if the number of 1's is even. This information is added as an additional bit to the string as a *parity bit*. Adding this parity bit to the string will result in an even number of 1's, thus making it even parity. Adding the complement of the above bit will result in a string of odd numbers of 1's, making it odd parity. (In this context, *parity* refers to the sum of 1's in a string of bits. The sum can be odd or even.)

Upon reading the data the parity is reestablished and compared to the parity bit just read. If they agree, all is well; if not, there must be an error. (The error could also be in the parity position.) This method detects only a single error. If two or any even number of errors occurs, the parity bit cannot flag the error. The parity check does not tell where the error is, only that there is one.

Single-Error Correction

To correct the error, we must also know the faulty bit position. One must therefore add a sufficient number of bits to identify the address within the string where the fault occurred. The address space must also include the just-added check bits, as well as one nonexisting location, such as address 0, to inform the system that no error has occurred. The following is a description of the well-known Hamming code. The total number of bits transmitted are the data bits plus the check bits.

Let the number of check bits $= k$, and let the number of data bits $= b$. The first question arises: How many check bits are needed? There must be enough check bits such that all combinations of check bits are equal to or larger than the transmitted bit string, so that a check bit pattern can point to the fault position within the string (which includes the check bit positions); k check bits can point to 2^k positions. Thus

$$2^k \rightarrow b + k + 1$$

$d =$ data bits
$k =$ check bits
$1 =$ the extra code for no error

We are now in a position to determine the number of required added check bits for a string of a given length. Table 14–3 gives a few values of k for corresponding strings of b.

We have determined how many k bits are needed. The next task is to find what each of the k bits has to be in a given string of data.

Each check bit is required to check parity of an address column. Assume that we have a byte (or an eight-bit string) to which we attach, according to table

TABLE 14–3 Required check bits for a string of data bits

b	k
4	3
8	4
16	5
32	6
64	7
128	8
256	9
512	10
1024	11
2048	12
4096	13

14–3, four check bits. Twelve bit positions must be checked. The 16 possible combinations are shown in table 14–4.

From table 14–4 we see that

k_0 is true for positions 1, 3, 5, 7, 9, and 11

k_1 is true for positions 2, 3, 6, 7, 10, and 11

k_2 is true for positions 4, 5, 6, 7, and 12

k_3 is true for positions 8, 9, 10, 11, and 12.

Check bits k_0, k_1, k_2, and k_3 are encoded by looking at the even parity of the corresponding positions. Upon reading the encoded word the k bits are read separately and directly yield either a no-error answer or the location of the error.

Let us encode a few of the possible strings of the byte. The data bits d_0 to d_7 are in positions 3, 5, 6, 7, 9, 10, 11, and 12. The check bits are placed in positions 1, 2, 4, and 8. (Any other placement makes the check bits dependent on each other.)

For example, let us encode the data byte 00001100:

12	11	10	9	8	7	6	5	4	3	2	1
d_7	d_6	d_5	d_4	k_3	d_3	d_2	d_1	k_2	d_0	k_1	k_0
0	0	0	0	X	1	1	0	X	0	X	X

X = unknown check bits

TABLE 14–4 Check bit addresses

k_3	k_2	k_1	k_0	
0	0	0	0	No error
0	0	0	1	Error in position 1
0	0	1	0	" " " 2
0	0	1	1	" " " 3
0	1	0	0	" " " 4
0	1	0	1	" " " 5
0	1	1	0	" " " 6
0	1	1	1	" " " 7
1	0	0	0	" " " 8
1	0	0	1	" " " 9
1	0	1	0	" " " 10
1	0	1	1	" " " 11
1	1	0	0	" " " 12
1	1	0	1	
1	1	1	0	
1	1	1	1	

- k_0 = even parity of positions 1, 3, 5, 7, 9, and 11 (see table 14–5). Note: Position 1 is the k_0 that we are trying to determine. We look at positions 3, 5, 7, 9, and 11. There is a 1 in position 7 only; thus $k_0 = 1$.
- k_1 = even parity of positions 2, 3, 6, 7, and 12. We look at positions 3, 6, 7, and 12. There are 1's in positions 6 and 7; thus $k_1 = 0$.
- k_2 = even parity of positions 4, 5, 6, 7, and 12. We look at positions 5, 6, 7, and 12. There are 1's in positions 6 and 7; thus $k_2 = 0$.
- k_3 = even parity of positions 8, 9, 10, 11, and 12. We look at positions 9, 10, 11, and 12. There are no 1's anywhere; thus $k_3 = 0$.

The encoded word for byte = 12 is as shown in table 14–5.

Suppose an error-free encoded byte 126 arrives. Let us see how the check bits check out. Even parity check = 0, meaning no error in that check.

					Position						
12	11	10	9	8	7	6	5	4	3	2	1
0	1	1	1	1	1	1	1	1	0	0	0

k_3: check parity of 12, 11, 10, 9, 8 = 0
k_2: check parity of 12, 7, 6, 5, 4 = 0
k_1: check parity of 11, 10, 7, 6, 3, 2 = 0
k_0: check parity of 11, 9, 7, 5, 3, 1 = 0

No error

TABLE 14–5 A few encoded bytes

	Position											
	12	11	10	9	8	7	6	5	4	3	2	1
	d_7	d_6	d_5	d_4	k_3	d_3	d_2	d_1	k_2	d_0	k_1	k_0
0	0	0	0	0	0	0	0	0	0	0	0	0
1	0	0	0	0	0	0	0	0	0	1	1	1
2	0	0	0	0	0	0	0	1	1	0	0	1
12	0	0	0	0	0	1	1	0	0	0	0	1
13	0	0	0	0	0	1	1	0	0	1	1	0
126	0	1	1	1	1	1	1	1	1	0	0	0
127	0	1	1	1	1	1	1	1	1	1	1	1
254	1	1	1	1	0	1	1	1	0	0	0	0
255	1	1	1	1	0	1	1	1	0	1	1	1

Suppose the same bit string arrives with an error and we want to determine where the error is:

						Position					
12	11	10	9	8	7	6	5	4	3	2	1
							Error ↓				
0	1	1	1	1	1	1	0	1	0	0	0

k_3: check parity of 12, 11, 10, 9, 8 = 0
k_2: check parity of 12, 7, 6, 5, 4 = 1
k_1: check parity of 11, 10, 7, 6, 3, 2 = 0
k_0: check parity of 11, 9, 7, 5, 3, 1 = 1

Error in position 5

The error location was indeed found. It is a simple matter to complement the "bad" bit to correct the error.

A 512 byte sector has 4096 bits and needs 13 check bits. If one more parity bit is added, two errors can be detected, but only one can be corrected.

14-9 THE TAPE TRANSPORT

Tapes are used when access time is not of prime importance but inexpensive mass storage is. By inexpensive we mean the cost per bit, not necessarily the cost of the equipment. Tapes are much too slow to be used interactively with the computer. In some cases, cassette tapes are used to download programs, but the reel-to-reel drive finds more widespread application. There are obvious similarities between tape and disk storage. In both systems, information is stored on an iron oxide surface, and the magnetic medium is moved past stationary read-write heads. The difference lies in the fact that tape is a continuous band and offers an order of magnitude more recording surface. However, this is offset by the fact that disks have higher bit density than tapes, such that disk and tape systems are nearly equal in storage capacity. The major reasons for the use of tape are still cost and ease of tape reel handling.

While disks rotate at constant speed, tapes advance one record at a time. The speed is controlled by the *capstan*. The tape is pressed against the capstan, which rotates at a well-controlled speed. The speed varies for different models from 25 to 125 in./s. The take-up reel and supply reels rotate just fast enough to allow the tape to advance while keeping the tape under controlled tension. This control is achieved either with spring tension arms or with a vacuum chamber mechanism. Along the path the tape is held in place by several tape guides, passes under the tape write, read, and erase heads, and passes the capstan to the take-up reel. A typical configuration is shown in figure 14-14. For fast forward and rewind, the tape-to-capstan pressure is released, allowing high-speed motion of 200-400 in./s.

A Survey of Computer Peripherals 453

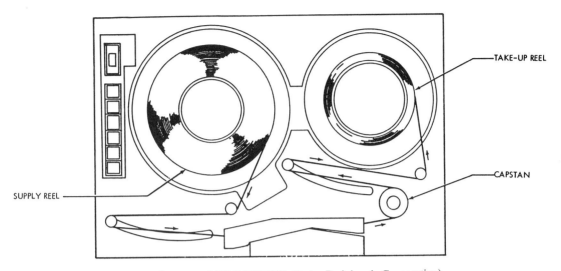

FIGURE 14–14 Tape path (Courtesy of DDC PERTEC, Pertec Peripherals Corporation)

There are three separate heads, each reading or writing all eight tracks and a ninth track for parity bits. The first head the tape encounters is the erase head. In the write mode the tape is erased to be subsequently overwritten by the write head. The read head always reads, whether in read or write mode. In the write mode the function of the read head is to immediately verify the entered information.

14–10 TAPE FORMAT

Tapes are typically ½ inch wide with nine tracks: eight tracks for one byte of data in parallel and one track for lateral parity bits. The beginning of the tape is identified by a BOT tab (there is a similar EOT tab at the end of the tape) and an identification burst of alternate 1's and 0's 1.7 in. long. This is followed by an initial 3 in. gap. The first data block starts here.

Each data block consists of a preamble of 40 bytes of all 0's and one byte of all 1's. In reading in the forward direction the preamble is seen at the beginning of each data block for the purpose of establishing synchronization. When locking on or synchronization is achieved, the all 1's byte defines the end of the preamble and the beginning of the data field. Each data block is followed by a postamble, which is the mirror image of the preamble. This symmetry is useful when reverse read is attempted.

The data field varies in length. A typical record length is 512 bytes. At the end of the data field the formatter computes and adds the appropriate longitudinal parity bits or a sum check to the bits of the data stream. These bits are used to check errors and do some limited correction. Records are separated by a 0.6 in. interrecord (or interblock) gap (IBG). (See figure 14–15.)

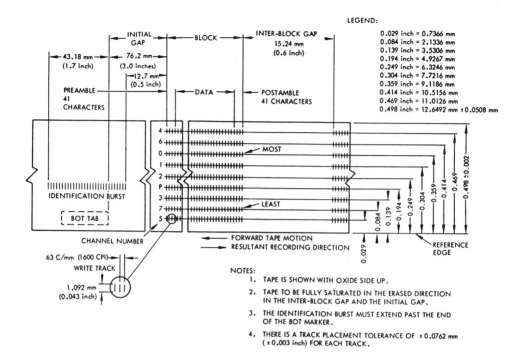

FIGURE 14–15 PE recording format (Courtesy of DDC PERTEC, Pertec Peripherals Corporation)

A user's file may be many records long. Files are separated by file marks. The format for a file mark is written as 40 0 bits (for PE encoding) in tracks P, 0, 2, 5, 6, and 7. Tracks 1, 3, and 4 are dc erased. Each file mark is preceded by a gap of 3.5 in. and followed by a normal IBG. There is no address identification on the tape. This software is controlled by counting file marks from the beginning of tape.

14–11 DATA RECORDING

Codes used for tape transports are nonreturn to zero inverse (NRZI) for 800 bits/in. density, phase encoding (PE) for 1600 bpi, and group code recording (GCR) for 6250 bpi.

NRZI Code

NRZI produces a pulse for each 1 and no pulses for 0's. A long string of 0's threatens to lose synchronization. (See figure 14–16.) An all 1 pattern has a maximum frequency equal to the clock frequency. The minimum frequency is zero.

A Survey of Computer Peripherals

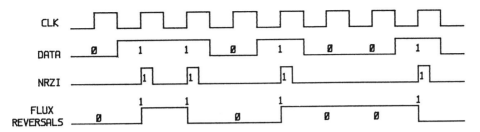

FIGURE 14–16 NRZI encoding

PE Code

Phase encoding requires that a 1 be represented by a positive transition and a 0 by a negative transition. This results in the waveforms of figure 14–17.

The flux reversal pattern is the same as the encoded data pattern. The frequency varies from clock frequency to twice the clock frequency. The bandwidth, however, is the same as for NRZI, and there is an ensured flux reversal for every clock cycle. An all 0 pattern is shifted 180° from an all 1 pattern, simply because an all 1 pattern must have positive transitions at the lagging edge of the clock (or in the middle of the bit cell) and an all 0 pattern must have a negative transition in the middle of the bit cell; hence the name phase encoding. Transitions at the edge of the bit cell serve only to establish the correct polarity of the transition to follow. Note that the inserted level changes at the bit cell edge occur when a 1 follows a 1 or a 0 follows a 0.

GCR Code

Group code encoding first encodes the electronic data into a code that ensures that for any group of four bits there are never more than two consecutive 0's. This can be done as shown in table 14–6 by encoding four bits into five bits. The five-bit code can then be NRZI encoded without its fundamental drawback.

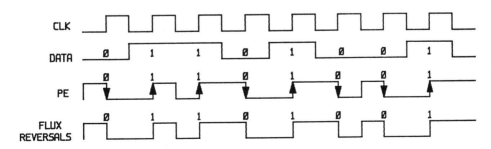

FIGURE 14–17 PE encoding

TABLE 14-6 GCR encoding

Nibble	Encoded Nibble	Nibble	Encoded Nibble
0000	11001	1000	11010
0001	11011	1001	01001
0010	10010	1010	01010
0011	10011	1011	01011
0100	11101	1100	11110
0101	10101	1101	01101
0110	10110	1110	01110
0111	10111	1111	01111

14-12 DATA TRANSFER AND TAPE SPEED

When commanded to read or write, the tape must first come up to speed before attempting any data transfer. The tape ramps up to the desired speed, performs the read or write, and immediately starts a ramp down to a stop. If the tape is commanded to read more blocks (or records), the process of slowing down and speeding up again is not noticeable to the eye. At the end of the file the tape stops. At an operating speed of 125 in./s and 1600 bits/in. (or bytes/in. since there are eight tracks in parallel) the transfer frequency of data is $1600 \times 125 = 200K$ bytes/s.

Access time is another story. Since files are stored serially, it might take the entire reel to find a file. The maximum access time is

$$\frac{2400 \text{ ft} \times 12 \text{ in./ft}}{125 \text{ in./s}} \sim 230 \text{ s} \quad \text{or almost 4 min}$$

These numbers are only an example; the actual numbers vary from model to model.

14-13 TAPE ERROR DETECTION AND CORRECTION

The expected error rate of tape recorders is one error in 10^{12} bits. If an error occurs, it is identified and corrected. The intersection of a lateral and a longitudinal parity bit locates the exact error position. Correction implies only a bit reversal. Many tape systems incorporate more sophisticated error correction codes such as CRC, a cyclic redundancy code.

When an error is detected, the tape rewinds to the beginning of the record and retries a read or write five times. If the error is recovered, a soft error indicator warns the user. If the error is not recoverable, a hard error indicator sends a warning, and the particular section of tape is flagged for avoidance.

FIGURE 14–18 Digital communication link

14–14 THE MODEM

The word *modem* is an acronym formed from MOdulator–DEModulator. A modulator converts digital data, normally in serial format, to an analog carrier for transmission over a telephone network. A demodulator receives a modulated analog carrier and converts the analog signal to digital data. The modem connects your digital equipment—your home computer, for example—to some other computer via the telephone lines, as illustrated in figure 14–18.

The bandwidth of the telephone lines is from 300 Hz to 3400 Hz. A digital data stream of 1's and 0's contains frequencies from 0 (dc) to at least the bit rate of the digital data stream. The telephone line cannot accommodate the dc end. Digital data is therefore encoded with various modulation techniques. Low-speed modems, up to 1200 bits per second, are implemented by using frequency shift keying (FSK) modulation. Medium-speed modems (up to 4800 bps) use phase shift keying (PSK), and higher-speed modems (up to 9600 bps) use quadrature amplitude modulation (QAM).

14–15 TRANSMISSION CODES

These modulation techniques differ in the method of encoding data into an analog carrier—frequency, phase, or amplitude. The term used to describe the rate at which the modulated carrier changes is called the *baud rate*. A baud may consist of one or more bits, depending on the modulation method. The number of bits transmitted per second is the baud rate multiplied by the number of bits per baud.

The simplest to implement, and therefore the least expensive, is the frequency shift keying method; FSK encodes one bit per baud. A logic 1 in the bit stream places a mark frequency (f_m) on the phone line. A logic 0 places a space frequency (f_s) on the line. As the bit stream switches between 1 and 0, the analog signal on the line modulates between f_m and f_s. (See figure 14–19.)

The modulation process generates energy over a broad spectrum, not only at the two frequencies f_m and f_s. This spectrum depends on the sequence of bits in the serial data stream. An alternate 1/0 pattern generates a simple line spectrum (figure 14–20(a)). The spectrum for a 511-bit pseudorandom pattern of data bits is shown in figure 14–20(b). (A good model of most data transferred via a modem) approximates a broad band spectrum.

Since FSK encodes only one bit per baud, it uses approximately 1 Hz of bandwidth for each bit per second of data rate. At 1200 bps a substantial portion of the telephone line bandwidth is used, allowing only a single channel to be

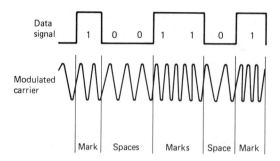

FIGURE 14–19 Modulated FSK signal (Copyright © Advanced Micro Devices, Inc. 1983. Reprinted with permission of copyright owner. All rights reserved.)

transmitted. At 300 bps, two independent channels can be accommodated via frequency division multiplexing (FDM).

A 1200 baud modem provides a reasonably comfortable transmission speed as a terminal to computer interface; 300 baud modems are slow and require much patience on the part of the user.

Phase shift keying (PSK) and quadrature amplitude modulation (QAM) encode more than one bit per baud. These methods are faster but more sensitive to line distortions and require automatic equalizers for compensation.

14–16 PRINTERS

There is a large variety of printers on the market. Initially, ordinary typewriters were modified to include a solenoid with each key that was activated by a computer command. Watching these working was amusing, reminiscent of watching a player piano. Teletypewriters followed soon; instead of keys, a cylinder or ball with embossed characters moved horizontally along the line to be printed, while the paper remained stationary. The cylinder rotated so that the selected character faced

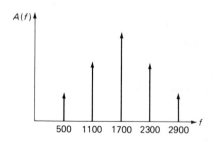

FIGURE 14–20(a) 1/0 pattern (Copyright © Advanced Micro Devices, Inc. 1983. Reprinted with permission of copyright owner. All rights reserved.)

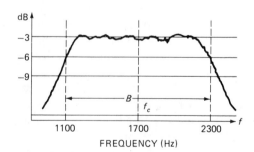

FIGURE 14–20(b) 511-bit pseudorandom pattern (Copyright © Advanced Micro Devices, Inc. 1983. Reprinted with permission of copyright owner. All rights resrved.)

A Survey of Computer Peripherals

the ribbon, and a striking action printed the letter. A typical speed was ten characters per second. The teletypewriter had one nice feature: It served as the computer input terminal as well as the hard copy printer.

Today's printers are output devices only. A number of methods are used to transfer characters to paper.

Thermal printing utilizes heat-sensitive paper that is thermally activated by voltage pulses, leaving behind a dot matrix pattern of the character.

Electrostatic printing is done by placing dot charges on coated paper. Subsequent processing changes the color of the charged spots.

Inkjet printers actually write on paper by spraying ink in well-controlled fashion, either by aiming a nozzle to the desired spot or by electrostatically deflecting the ink beam.

Laser printers are fast (2 seconds per page) and give very good quality output. A laser beam controlled by the computer spot charges a drum, which then transfers the character to paper.

Impact printers write either a line at a time or a character at a time. Line printers are fast, ranging from 200 to 600 lines per minute. They use either a cylinder or a chain horizontally facing the paper; 132 hammers are ready to strike a ribbon when the correct character faces the paper. Line printers are used for high-volume output.

The character-at-a-time printer, slower and less expensive, is popular for use with the personal computer. The print head moves along a rail carriage advancing in one-character column steps. (See figure 14–21.)

The head in figure 14–22 consists of nine needlelike hammers actuated in accordance with the dot pattern of the particular addressed dot column of the character. As the head moves in seven steps through the character pattern, it prints out the entire letter.

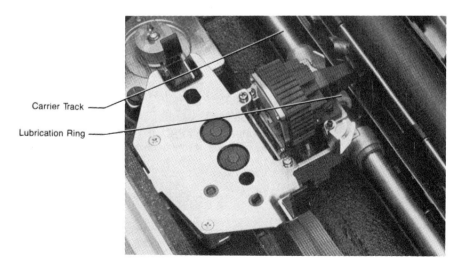

FIGURE 14–21 Head and rail (Courtesy of Apple Computer Inc.)

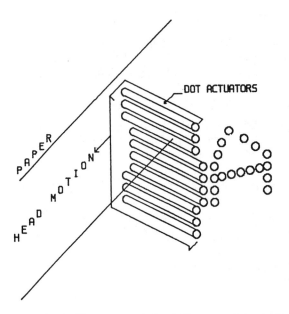

FIGURE 14–22 Dot matrix head (Courtesy of Apple Computer Inc.)

The objection against dot matrix printing lies in the fact that letters are formed by individual dots, which are visible. Full-character printing is preferred for letter-quality output. However, dot matrix printers are capable of filling the "void pots" with an additional set of dots, resulting in a nearly continuous appearance. Such "correspondence-quality" printouts are finding increased acceptance.

The dot matrix method of printing is particularly attractive because it allows the choice of character style, font, and size by simply selecting the proper ROM set for the character generator. (See figure 14–23.)

ELECTRONICS IS FUN

ELECTRONICS IS FUN

FIGURE 14–23 Example of font and style change

REVIEW QUESTIONS

1. Discuss serial and parallel I/O.
2. What is the ASCII code?

A Survey of Computer Peripherals

3. How many dots are there in an alphanumeric display character?
4. Why are descenders needed in a character display?
5. Explain the horizontal and vertical sweep in a CRT monitor.
6. Review the character display system of figure 14–5.
7. Why are there five vertical and seven horizontal addresses in the ASCII code RAM in figure 14–5?
8. Why is the dot clock divided by 8 before driving the mod 80 counter in figure 14–5?
9. What is meant by a flying disk head? Are flexible disk heads different?
10. Compare disk and tape access times.
11. Compare FM and MFM encoding in disk systems.
12. Compare NRZI and PE encoding in tape systems.
13. How is tape motion controlled?
14. Review the Hamming code for error detection and correction. Expand table 14–5 to include the encoded bits for 16, 187, and 220.
15. Suppose an error is in bit position 7 for 187. Find k_0, k_1, k_2, and k_3.
16. How are errors detected and corrected on tape?
17. What is the advantage of GCR encoding? See table 14–6.
18. What is a modem?
19. Why are 1200 baud modems preferable to 300 baud modems?
20. Discuss dot matrix printing.

appendix a

the operational amplifier

A-1 INTRODUCTION

Some of the laboratory experiments as well as the *D/A* converter and *A/D* converters use operational amplifiers (op-amps). It is therefore useful to include a brief treatment of the subject in this digital book. The student without prior knowledge of op-amps will find this discussion sufficient for the needed understanding in this course; for others it is at least a practical review.

What is an operational amplifier? How does it differ from any other amplifier? A single transistor stage, for example, is an amplifier, but the single stage has a number of drawbacks:

1. The gain is limited.
2. The input impedance is too low.
3. The output impedance is too high.
4. The output is dc shifted.
5. Output distortion is a function of transistor parameters.
6. Common mode rejection is poor.

Let us discuss the above statements with the help of the single-stage schematic of figure A–1. For convenience we assume that the internal base resistance $r_b = 0$ and the internal emitter resistance $r_e = 0$. The basic transistor equations are:

$$I_c + I_b = I_e$$

$$I_c = \alpha I_e$$

$$\beta = \frac{\alpha}{1 - \alpha}$$

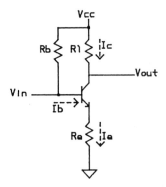

FIGURE A-1 Single transistor amplifier

The voltage gain G is defined as the ratio of output voltage (V_{out}) to the input voltage (V_{in}). V_{out} is the product of the collector current (I_c) and the load resistance (R_1). The collector current (I_c) is a function of the emitter current (I_e), which in turn is a function of the input voltage (V_{in}). The voltage gain is thus derived by the following set of equations:

Gain

$$I_e = \frac{V_{in}}{R_e} \qquad V_{eb} \text{ is assumed to be 0 for ac gain.} \qquad \text{(I)}$$

$$I_c = \alpha \frac{V_{in}}{R_e} \approx \frac{V_{in}}{R_e} \text{ if } \alpha \text{ is close to 1} \qquad \text{(II)}$$

$$V_{out} = -I_c * R_1 \cong -\frac{V_{in}}{R_e} * R_1 \qquad \text{(III)}$$

$$Gain = G = \frac{V_{out}}{V_{in}} \cong -\frac{R_1}{R_e} \qquad \text{(IV)}$$

Input Impedance

The input impedance is defined as the ratio of $\frac{V_{in}}{I_b}$, but

$$I_b = \frac{I_c}{\beta} = \frac{V_{in}}{(\beta + 1) * R_e} \qquad \text{(V)}$$

Thus the impedance seen by the input source

The Operational Amplifier

$$Z_{in} = (\beta + 1) * R_e \qquad (VI)$$

A high input impedance requires a large R_e. That, however, decreases the gain.

Output Impedance

Looking into the output node, we see R_1 in parallel with the collector. The collector is an infinite impedance. The output impedance therefore $= R_1$. To get a low output impedance, R_1 should be small. This again decreases the gain.

DC Offset

For dc biasing purposes, V_{eb} cannot be considered $= 0$. The minimum input voltage V_{in} must exceed V_{eb} so that current may flow. The dc output voltage $= V_{cc} - I_c * R_1$. The minimum output voltage must be more positive than the maximum input voltage ($-$ the V_{bc} drop) in order to allow a reverse bias for the base–collector junction. This results in an inherent dc shift between output and input as shown in figure A–2.

Distortion

The output of equation (IV) depends only on the external resistors. In reality the output also depends on r_e, r_b, and V_{eb}, all of which vary with current and temperature. The current, in turn, is a function of the input voltage. Thus the output is not only a function of the input, as is desirable, but also a function of the parameter variations.

Common Mode Rejection

Finally, any ripple in the power supply adds without attenuation to the output voltage V_{out}, since the collector is an infinite impedance. If a differential stage is used, the ripple will appear on both collectors. The difference voltage between the collectors rejects the ripple as shown in figure A–3.

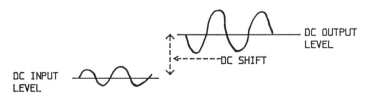

FIGURE A–2 DC offset

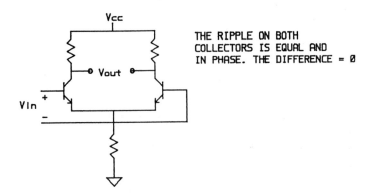

FIGURE A-3 Differential amplifier

Remedies

To cure all the ills described above, the following steps are taken.

- Step 1. Multiple dc-coupled stages are added to get high gain from dc to some cutoff frequency.
- Step 2. Emitter followers or field effect transistors are added as an input stage to get high input impedance.
- Step 3. An emitter follower output stage is added to get low output impedance.
- Step 4. Complementary NPN PNP stages alternate the offset such that the final offset at the output = 0.
- Step 5. Differential stages cancel common mode signals such as power supply ripple or common mode input signals.
- Step 6. As will be shown, a high-gain amplifier makes the amplifier system independent of transistor parameter variations, therefore yielding a low distortion output.

An amplifier with all the above properties is easy and convenient to operate. The name "operational amplifier" comes from early uses in the 1950s in analog computers to perform arithmetic operations. Such an amplifier deserves its own symbol, shown in figure A–4.

The input is differential, and the output is usually single ended. As in any differential amplifier, one input will be inverting and the other not. The inverting input is labeled "−," the noninverting input is labeled "+."

The terms "op-amp" and "amplifier system" were mentioned earlier. The op-amp has more gain than is usually needed, on the order of 10^6. Some external components are needed to tailor the performance to our needs. The op-amp and the external components make up the amplifier system (in short, just called the amplifier).

The Operational Amplifier

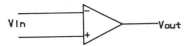

FIGURE A-4 Operational amplifier

A-2 THE INVERTING AMPLIFIER

In its simplest form, which is frequently used, two resistors are needed to make an inverting amplifier with a specific gain. Let us analyze the circuit in figure A-5.

The circuit is much simpler to analyze if three assumptions are made:

The internal gain K of the amplifier $= \infty$

The input impedance $= \infty$

The output impedance $= 0$

These assumptions are close to the truth, and the answers obtained by using these assumptions are quite close to observed performance. Next, we want to make two observations:

1. The gain is by definition $\dfrac{V_{out}}{v} = \infty$ as assumed above. Therefore $v = \dfrac{V_{out}}{K} = 0$.

   This means that the voltage measured across the amplifier input terminals for a given finite output must be 0 or close to it.

2. Any current flowing through R_1 in figure A-5 toward the input terminal must continue to flow through resistor R_f because the input impedance to the op-amp is infinite.

Notice that we differentiate between the applied voltage V_{in} and the voltage v appearing across the input terminals. The differential voltage v, as was stated before, is 0 or close to 0. If the positive terminal is physically grounded, then the measured voltage at the negative terminal is for all practical purposes at ground potential. It always is, no matter how much V_{in} is applied (provided that the maximum allowable input conditions are not exceeded). The negative input terminal is said to be at *virtual ground*.

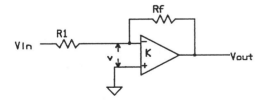

FIGURE A-5 Inverting amplifier

We can now proceed to obtain an expression for the gain of the amplifier with the following steps.

- Step 1. The input current through R_1 must be $I_{in} = \dfrac{V_{in}}{R_1}$ because the current flows from V_{in} to ground potential.
- Step 2. Since all the current flows through R_f toward the output, the voltage across R_f must be $I_{in} * R_f = \dfrac{V_{in}}{R_1} * R_f$. Since the current flows away from the negative input terminal, which is at ground, the output node must be negative and equal to the drop across R_f. Thus

$$V_{out} = -\frac{V_{in}}{R_1} * R_f \tag{VII}$$

or

$$\text{Gain} = G = \frac{V_{out}}{V_{in}} = -\frac{R_f}{R_1} \tag{VIII}$$

The input impedance to the amplifier system, in this case $= R_1$, and the output impedance is low (a function of the output stage and the applied feedback).

The question may arise as to why use an op-amp with almost infinite gain if all we use is a gain of R_f/R_1, which is considerably smaller. Moreover, the internal gain of the op-amp does not even appear in the gain equation (VIII). Does that mean that the op-amp was not needed in the first place?

All the op-amp did was to provide the virtual ground because of the high gain. It is due to that property that we obtained a gain expression independent of the op-amp gain K. The latter is indeed a function of nonlinear internal parameters; but because the final gain does not depend on K, the amplifier system is a function of the external resistors only. The external resistors are passive elements and inherently more linear than active components.

If the gain G of the amplifier approaches the op-amp gain K, the advantages discussed begin to vanish. The ratio of K/G is called the excess gain, and it is good practice to keep it at 100 or higher.

A derivation including the open loop gain K and the input resistance R_{in} terms is shown in Section A–11.

Resistor R_f is a feedback resistor. It feeds some of the output signal back to the input terminal. If, for example, the output rises, it feeds the increase back to the negative terminal, which in turn sends the opposite command to the output. This is called *negative feedback*. The amplifier is in equilibrium in accordance with equation (VIII). The negative feedback tends to control the output from making wild excursions. *Positive feedback* has the opposite effect. A rising output fed back to the positive terminal commands the output to rise some more. The effect snowballs, and the output runs away until it is limited by the supply level. Or worse, it begins to oscillate. Stable systems must have the feedback applied to the negative input terminal.

The op-amp becomes the basic component for a variety of applications. The inverting amplifier is one application and was the vehicle for the initial discussion. Let us now develop several useful op-amp circuits:

A-3 THE NONINVERTING AMPLIFIER

In this case the input signal is fed to the positive input terminal, but the feedback must still be returned to the negative input terminal as in figure A–6. There is no virtual ground in this configuration, but v must be 0. Therefore the potential at the negative terminal $= V_{in}$.

Since the potential at the negative terminal $= V_{in}$, the current through $R_1 = V_{in}/R_1$. That current cannot come from the negative terminal because of the high input impedance. It must come through R_f, the feedback resistor. If we know the current through R_1 and R_f, the output voltage V_{out} must be

$$V_{out} = \frac{V_{in}}{R_1} * (R_1 + R_f) \tag{IX}$$

and the gain is

$$G = \frac{V_{out}}{V_{in}} = \frac{R_1 + R_f}{R_1} = 1 + \frac{R_f}{R_1} \tag{X}$$

The gain expression is similar to the inverting amplifier except that the minus sign is missing, and the gain is incremented by 1.

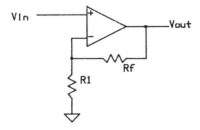

FIGURE A–6 Noninverting amplifier

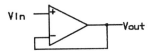

FIGURE A-7 Voltage follower

A-4 THE VOLTAGE FOLLOWER

A special case of the noninverting amplifier is the voltage follower. It is reminiscent of the emitter follower. If we set $R_1 = \infty$ and $R_f = 0$ in equation (X) and figure A-7, the gain = 1:

$$V_{out} = V_{in}$$

This circuit is useful if the source voltage V_{in} has a larger than desirable output impedance, that is, insufficient drive capability. V_{out} is equal to V_{in} with low output impedance and good drive capability.

There is no voltage gain, but there is power gain because the current output capability is improved. It is said that the voltage follower is used for isolation because it also provides high impedance, nonloading input to the driving source. (The emitter follower works in a similar fashion except that the output is slightly attenuated and shifted by V_{eb}.)

A-5 THE DIFFERENCING AMPLIFIER

Taking advantage of the differential input, we can obtain the difference of two inputs shown in figure A-8. The voltage at the positive terminal must be $V_{in2}/2$. V_{in2} is just divided in half by the two equal resistors R. Since $v = 0$, the same voltage must exist at the negative terminal. Thus the input current from V_{in1} is

$$I_{in} = \frac{V_{in1} - V_{in2}/2}{R} \qquad (XI)$$

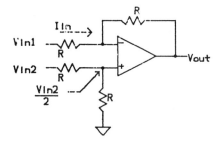

FIGURE A-8 Difference amplifier

I_{in} flows to the output such that V_{out} must equal the voltage drop across the feedback resistor R subtracted from the voltage at the negative terminal:

$$V_{out} = V_{in2}/2 - I_{in} * R \qquad \text{(XII)}$$

$$V_{out} = V_{in2}/2 - \frac{V_{in1} - V_{in2}/2}{R} * R = V_{in2} - V_{in1} \qquad \text{(XIII)}$$

A–6 THE SUMMING AMPLIFIER

Both inputs V_{in1} and V_{in2} can be summed at the negative terminal. If the positive terminal is grounded and the negative terminal is at virtual ground, the sum of the currents from V_{in1} and V_{in2} will flow through the feedback resistor. The output in figure A–9 is therefore

$$V_{out} = -\left(\frac{V_{in1}}{R} + \frac{V_{in2}}{R}\right) * R = -(V_{in1} + V_{in2}) \qquad \text{(XIV)}$$

The output is inverted but could be inverted again with a unity gain inverter.

A–7 A VOLTAGE-TO-CURRENT CONVERTER

If the output in the noninverting amplifier of figure A–6 is taken across the feedback resistor, then the output impedance is infinite. This is the definition for a current source. Figure A–6 is redrawn with a perspective change in figure A–10.

Since $v = 0$, the current through the reference resistor R_{ref} is V_{in}/R_{ref}. That current must flow through the feedback path to the load resistor and the op-amp output. The current is fixed by R_{ref}. The current is not a function of the load resistor. Any source that delivers a constant current independent of the size of the load resistor must be a current source. The impedance looking back from the load resistor

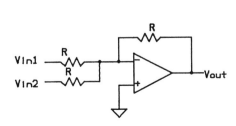

FIGURE A–9 Summing amplifier

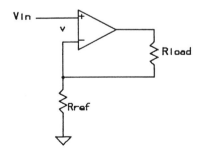

FIGURE A–10 Voltage-to-current converter

terminals into the amplifier is therefore infinite. It is curious, at least, that a device with nearly zero output impedance can be used in such a way as to have infinite output impedance.

The converter finds application in situations in which a signal is to be sent over a long distance (such as a measurement from the field to a control room). Voltage is attenuated over a long line. Current at both ends of the wire must be the same.

A–8 THE INTEGRATOR

Integration is the process of summation of an infinite number of samples along a function over infinitely small increments, or in other words, the continuous accumulation of all points. Consider, for example, a square pulse (figure A–11) of width T. Divide the pulse into n small sections, each with width dt. Sum the amplitudes $A_1 + A_2 + \ldots$ of all n sections. The total area $= (A_1 + A_2 + \ldots) \, dt$. But $dt = T/n$. The area is the integral for any pulse shape (if dt is small and n is large).

Integration is done by placing a capacitor in the feedback loop of an op-amp as in figure A–12. The capacitor receives a current proportional to the instantaneous amplitude of the input. The charge $Q = i \, dt$. Thus the charge builds up in proportion with the instantaneous input amplitude over a time interval T, and the voltage across the capacitor is proportional to the charge. If the input is a square pulse and the current is constant, then

$$V_{out} = -V_{cap} = \frac{-IT}{C} = \frac{-V_{in}T}{RC} \qquad \text{(XV)}$$

The current through R is constant because of the virtual ground at the negative terminal. The voltage across the capacitor builds up linearly with time. V_{out} is proportional to the capacitor charge.

For the student who is familiar with integral calculus,

$$V_{out} = -\int_0^T \frac{1}{C} \frac{V_{in}}{R} \, dt = -\frac{1}{RC} \int_0^T V_{in} \, dt \qquad \text{(XVI)}$$

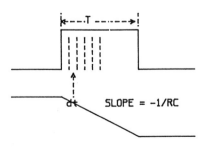

FIGURE A–11 Integrated pulse

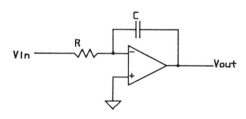

FIGURE A–12 Integrator

The Operational Amplifier

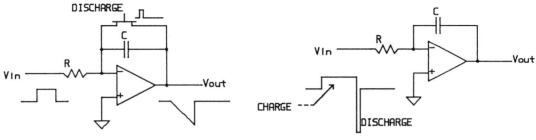

FIGURE A–13(a) FET discharge

FIGURE A–13(b) Opposing pulse discharge

In practice there should also be a means to discharge the capacitor. This is done either with an FET switch across the capacitor that is activated after the integration is done or with an opposite polarity input pulse of equal area. (See figure A–13.)

A–9 THE DIFFERENTIATOR

Differentiation is the process of measuring rate of change. It measures the slope of a function. A capacitor placed across a voltage source will charge to that voltage. If the resistance of the circuit is assumed to be 0, the charging current is directly proportional to the capacitance and the rate of change in voltage:

$$i = C\, dv/dt \qquad \text{(XVII)}$$

Let us place the capacitor in the input branch of the amplifier as in figure A–14. The capacitor current flows through the feedback resistor such that the output voltage is

$$V_{\text{out}} = -RC\, dv/dt \qquad \text{(XVIII)}$$

The input terminal, however, cannot be left floating from dc considerations. A resistor (R_b) must be used in parallel with the capacitor. This renders the dif-

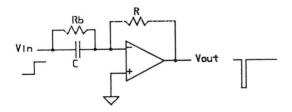

FIGURE A–14(a) *RC* differentiator

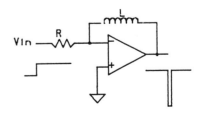

FIGURE A–14(b) *LR* differentiator

ferentiation process imperfect. The source V_{in} feeding the differentiator has a finite output impedance, which further deteriorates the performance. Thus only an approximate differentiation can be obtained. Differentiation is also an inherently noisy operation. The capacitive input does not allow dc to pass but passes all high frequencies.

It is for all these reasons that differentiation is avoided if possible. In analog computer work, for example, differential equations are changed into integral equations before modeling the circuit for a solution.

An alternative way is to use an inductor in the feedback loop. The output becomes

$$V_{out} = -\frac{L}{R} * \frac{dV}{dt} \tag{XIX}$$

A high Q coil should be chosen to minimize the effect of resistance.

A–10 A SIMPLE LOW PASS FILTER

An RC filter using an op-amp has the advantage of isolation and signal gain over a passive RC network, as shown in figure A–15:

$$V_{out} = V_{in} * \frac{1}{RC} * \frac{1}{jw + 1/RC} \tag{XX}$$

The cutoff frequency is defined as the frequency at which $w = 1/RC$ or $f = 1/(2\pi RC)$. At that frequency the phase shift $= 45°$, and the output is 3 dB down. It will continue to decrease at a rate of 6 dB/octave. The cutoff frequency of the passive network is a function of the source impedance as well as the load resistance of any circuit it is feeding.

The output of the active circuit in figure A–16 is

$$V_{out} = V_{in} \frac{Z}{R_1} = V_{in} \frac{1}{R_1 C} \frac{1}{jw + 1/R_f C}$$

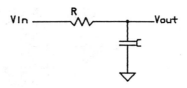

FIGURE A–15 RC low pass filter

The Operational Amplifier

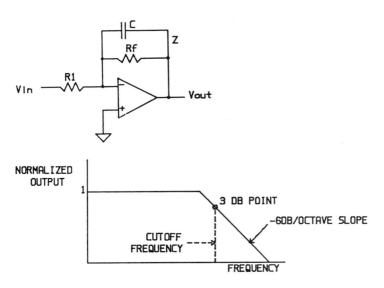

FIGURE A–16 Active low pass filter

The gain can be set with R_1. The cutoff frequency ($1/2\pi R_f C$) is not affected by the source impedance, load, or gain setting.

In choosing an op-amp for a particular application, several characteristics must be considered.

1. Frequency response: 1–100 MHz. There is no need to choose a higher frequency than the application requires. Lower bandwidth op-amps are usually internally compensated, simplifying the external circuit design, and are less prone to oscillate.

2. Slew rate. When an op-amp is driven with a large fast input change, it might internally saturate temporarily. It will slowly recover as internal voltages catch up to the input. The output will not resemble the input during that time but will linearly follow in its own good time. This is called *slewing*. The slew rate for various op-amps will vary from 0.5 V/μs for the often used 741 to 1000 V/μs for ultrafast devices such as Analog Devices ADLH0032.

3. Common mode rejection. This is the ability of the device to discriminate against common mode signals, typically 1000 to 1 (741).

4. Input impedance: 200K Ω for the 741. If very high impedance is needed, choose an FET input device.

5. Output current: typically 10 mA. If more current is needed, choose a power op-amp. Keep in mind that the input, feedback, and load resistor of a particular design cannot draw more current than the device is capable of.

6. Required power supplies. Most op-amps operate off +/− 15 V. This allows signal swings above and below ground. Some op-amps operate off a single 5 V supply. This is convenient, but the design shall not allow signal inversions that require signal swings below ground.

For additional characteristics, consult available data books.

A–11 OP-AMP GAIN DERIVATION WITH *K* AND R_{in}

For the inquisitive reader, let us *not* assume that K and R_{in} are infinite. Equations (VII) and (VIII) are derived by taking a finite K and R_{in} into account. (K = open loop gain, and R_{in} = input resistance.)

If K is not infinite, then v must be finite (other than 0) and is accounted for in the equations. If R_{in} is not infinite, then not all the current flows into R_f, but some current is diverted into the op-amp. (See figure A–17.)

We obtain the following set of equations:

$$\frac{V_{in} - v}{R_{in}} = I_i$$

$$I_f = I_i \frac{R_{in}}{R_{in} + R_f}$$

$$v = \frac{V_{out}}{K}$$

$$V_{out} = v - I_f R_f$$

The simultaneous solution of the four equations gives

$$G = -\frac{R_f}{R_1} * \frac{1}{(1 + R_f/R_{in})(1 - 1/K + R_f/(KR_1(1 + R_f/R_{in})))} \qquad \text{(XXI)}$$

Note that the first term is the same as equation (VIII). The second term is the correction term. If R_{in} alone is considered infinite, then the expression reduces to

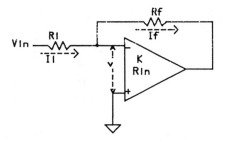

FIGURE A–17 Op-amp with finite gain and input impedance

The Operational Amplifier

$$G = -\frac{R_f}{R_1} * \frac{1}{(1 - 1/K + R_f/KR_1)} \qquad \text{(XXII)}$$

If K is also considered infinite, then equation (XXII) further reduces to $G = -R_f/R_1$, which is exactly equation (VIII).

A–12 THE COMPARATOR

The comparator is not too dissimilar from the op-amp. It is also a high-gain amplifying device. The output is not required to faithfully reproduce the input. Rather, the output is to switch high or low depending on which input terminal exceeds the other in potential. Linearity is not a requirement, but switching speed is. Allowable input voltage differences must be equal to the maximum input signal swing. This is not the case for the op-amp in which the voltage at the input terminal is close to 0. An op-amp should not be used when a comparator is required.

A commonly used device is the quad comparator LM 339 with 1.3 µs switching speed. Operated with a 5 V supply, it can drive TTL logic directly. The output is open collector, and an external 10K load resistor is needed. (See figure A–18(a).)

When hysteresis is needed, it can be included as in figure A–18(b). Here the reference is a function of the output such that when switching occurs, the reference shifts to provide an increased input differential. For reverse switching the input

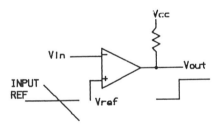

FIGURE A–18(a) Inverting comparator

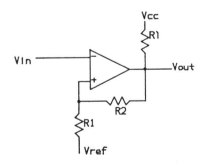

FIGURE A–18(b) Inverting comparator with hysteresis

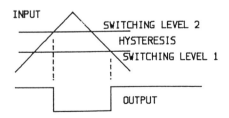

FIGURE A–18(c) Switching occurs at different levels

must reach the newly established reference level. The reference level shift is the magnitude of the hysteresis.

When the input is low, the output is high, and the switching level at the positive input to the comparator is

$$\text{Switching level 2} = V_{ref} + (V_{cc} - V_{ref}) * R_1/(R_1 + R_2)$$

When the input is high, the output is at ground, and the switching level at the positive input to the comparator is

$$\text{Switching level 1} = V_{ref} - V_{ref} * R_1/(R_1 + R_2)$$

appendix b

cabling

A subject that is not directly related to digital logic, yet that cannot be ignored, is cabling. The student who is familiar with the theory and application of transmission lines will find nothing new in this discussion. For those who do not have this familiarity a few qualitative and quantitative points will be explained.

Signals are transmitted from one point to another through a wire. As long as the distance is short, there is no problem. When a signal is to be transmitted over a distance of 30 ft or more, reflection problems arise.

Let us carry out a mental experiment with a battery, a switch, and a long wire as in figure B–1. Suppose the wire is open ended at the far end and at time = 0 we close the switch for 10 nsec and then open it again. The 10 nsec pulse will not get to the other end instantaneously. The pulse travels down the line with a speed of approximately 1 nsec per foot. We expect to see the pulse 100 nsec later, as indeed we will.

The source—in this case, the battery—has no information at $t = 0$ as to the nature of the load at the far end. Therefore the source does not know how much current to send along with the voltage pulse. If the wire is lossless, that is, there is zero resistance, and there is no inductance and no capacitance associated with the wire, the source might assume a short circuit load and send an infinite current. But what happens to that current when it gets to the dead end?

In reality, even if the wire is considered lossless, the wire has a distributed inductance along its length and a distributed capacitance to a nearby ground. The wire can be represented in approximate form by a lumped LC network as shown in figure B–2.

The battery does see the ac impedance of the wire as soon as the switch is closed. That is the only information the battery has upon which to decide how much current to send. It might not be the required current of the load, but we will deal with that problem later.

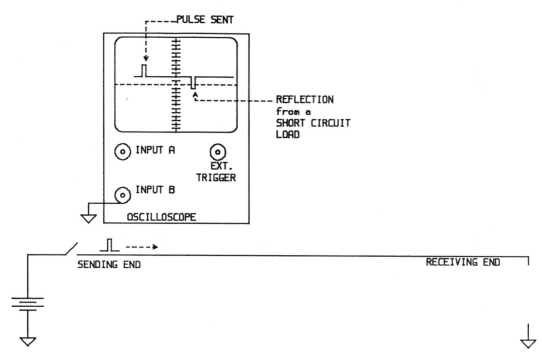

FIGURE B–1 Signal transmission over a long wire

The input impedance to the wire is a function of the inductance and capacitance. The inductance is a function of the wire size, its material, and the medium the wire is placed in. The capacitance is a function of the distance to ground, the size of the wire, and the medium between the wire and ground. The impedance thus formed is characteristic of the particular setup. It is therefore called the *characteristic impedance* of the wire or cable. A proper derivation of an expression of the characteristic impedance requires a solution of a set of differential equations, which we will avoid in this discussion. Instead, we will look at the lumped network of figures B–2 and B–3 to see whether we can arrive at the answer in simpler terms.

Figure B–3 is a modification of figure B–2. Each inductor is split in half, such that each lumped section is now a symmetrical T network instead of an unsymmetrical L network. The T network looks identical in each direction. If all T networks are identical and we assume an infinite number of them, then the input impedance to each T network is the same. Such networks are called *iterative networks*. It is

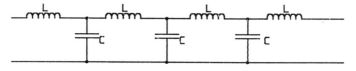

FIGURE B–2 Lumped approximation of a cable

Cabling

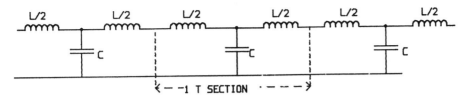

FIGURE B–3 Modified lumped approximation of a cable

therefore necessary to analyze only one T network, loaded or terminated in the same impedance as is seen at the input. (See figure B–4.)

The input impedance Z_{in} is easily calculated:

$$Z_{in} = X_1 + \frac{X_c(X_1 + Z_{in})}{X_c + X_1 + Z_{in}}$$

$$Z_{in} * X_c + Z_{in} * Z_{in} * X_c + Z_{in}^2$$
$$= X_1 * X_c + X_1^2 + X_1 * Z_{in} + X_1 * X_c + X_c * Z_{in}$$

Note that four terms cancel and Z_{in} reduces to

$$Z_{in}^2 = 2 * X_1 * X_c + X_1^2$$

Substituting the impedance expressions for $L/2$ and C, we get

$$Z_{in}^2 = 2 * \frac{jwL/2}{jwC} - (wL/2)^2$$

$$Z_{in} = \sqrt{\frac{L}{C} - (wL/2)^2}$$

The last equation is true for a lumped iterative network. The approximation to a transmission line assumes that each lumped section is an infinitely small section of the line, of which there are an infinite number. The inductance and capacitance of such an infinitely small section is almost zero. Therefore $(wL/2)^2$ is so close to zero that it vanishes, particularly since the term is squared. The ratio of L/C is finite even if both numerator and denominator are small.

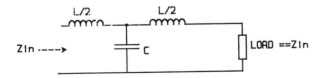

FIGURE B–4 Lumped cable section

Thus Z_{in} reduces to

$$Z_{in} = \sqrt{\frac{L}{C}}$$

Notice that the characteristic impedance is independent of frequency and independent of the cable length.

It was assumed in the above derivation that L and C were uniform and constant throughout the length of the cable. In practice this is achieved by building a coaxial cable in which the distance from the signal wire to ground (or shield) is maintained constant. A reasonably good cable can also be obtained with a twisted pair of wires. Twisted pair cables are commonly used in the computer industry because they are adequate and less bulky than coaxial cables. Coaxial cables are used where signal cleanliness is of utmost importance. Coaxial cables are available with 10–300 Ω characteristic impedance; 50 Ω cables are the most common. Twisted pair cables have a typical impedance of 160 Ω.

Coming back to our experiment of figure B-1, we can now say that the battery has information as to how much current to send initially. Let us resume the experiment and see what happens.

The instant the switch is closed, the battery sees only the cable characteristic impedance. For the experiment, let the battery be 5 V and the characteristic impedance 165 Ω. The initial current must be 5/165 = 30 mA. The 5V, 10 nsec pulse accompanied by 30 mA flows down the cable until the end of the line is reached, where an open termination exists. What must happen?

According to Kirchhoff's law, the sum of currents into a node must be zero. If the end of the cable is considered to be a node, then the arriving 30 mA have nowhere to go when reaching the open-ended terminal node. The only existing possibility is for the current to reverse direction and flow back to the battery. The voltage, of course, accompanies the current on the way back.

An open termination causes a current reversal but no voltage reversal. The *voltage reflection* is in phase.

A scope placed at the sending end of the cable will observe the returning pulse another 100 nsec later. Since the switch had been opened, the sending end is also an open termination, and the current will reflect again. If the line is indeed lossless, the pulse will ping-pong forever between the two open ends. In reality the line is resistive, and the pulse is continuously attenuated until it disappears.

What would happen if we placed a short circuit at the receiving end? We close the switch again for 10 nsec and observe our scope. We cannot possibly see a pulse at the receiving end, since there is a short circuit. The current arriving at the receiving end simply flows through the short circuit and returns via the ground lead. But what happened to the voltage pulse? Just before arriving at the short, the pulse could have been observed anywhere along the line, yet at the termination it has disappeared. The only way to satisfy this terminal condition is to assume that an equal and opposite voltage pulse existed at the shorted termination such that the sum of the positive arriving pulse and the negative reversed pulse equaled zero. This is indeed what happens. The scope at the sending end displays the reflected negative pulse. In this short circuit experiment the current did not reverse, but the voltage did. A short circuit termination causes a voltage reversal but no current reversal. The voltage reflection is out of phase.

There are many direct applications for this phenomenon. To cite one: When a power failure occurs in a power distribution network, a pulse is sent down the failed branch. Some microseconds later, a pulse returns. An in-phase reflection indicates an open wire; an out-of-phase reflection indicates a shorted wire. The time it took for the pulse to return is twice the distance to the failure location.

There is nothing surprising in this phenomenon. Electromagnetic waves reflect, as we notice every morning when we look into a mirror. But it is a phenomenon that must be taken into account when building electronic systems. Reflections might be falsely received as true signals at either the receiving or the sending end. How do we avoid them? (There are cases in which reflections are introduced on purpose. Such applications are discussed in the transmission line literature.)

It is reasonable to assume that if an open termination causes an in-phase reflection and a short circuit termination causes an out-of-phase reflection, there exists a termination somewhere between zero and infinite resistance that balances the two extremes.

Neither an open nor a short circuit termination can absorb the arriving energy. Therefore it must return. If we were to use a resistor at the receiving end of just the right size to absorb the energy, then all the energy would be dissipated and no reflection would exist.

In our experiment the energy is 5 V * 30 mA for 10 nsec:

$$5 \text{ V}/30 \text{ mA} = 165 \text{ }\Omega$$

If we were to use a 165 Ω termination, all the energy would be absorbed. But 165 Ω is exactly the value of the characteristic impedance.

We conclude that a line should be terminated with a resistance that is equal to the characteristic impedance R_o of the line. In other words, the termination matches the line impedance.

If the terminating resistor does not exactly match the line impedance, some of the energy is dissipated and some is reflected. Every time the reflected pulse returns to the termination, some more of the energy is dissipated until it is all used. The ever-decreasing pulses bouncing back and forth are observed on a scope as a ringing waveform. Ideally, these decreasing pulses maintain their square shape. In reality, because of bandwidth limitations, they are deteriorated and appear as nearly sinusoidal ringing. (See figure B–5.)

The characteristic impedance of manufactured cables is specified. We can also

FIGURE B–5 Line response to an improperly terminated line

make our own twisted pair cables. Take two wires of the desired length, fasten one end of the pair to a fixed position, and twist the other end of the pair with a slow-turning drill. Twist as tightly as possible. When released, the wires will partially untwist.

The characteristic impedance can be measured in the following manner:

- Step 1. Feed a string of short pulses down the line.
- Step 2. Observe the pulses at the sending end with a scope.
- Step 3. Adjust a variable resistor at the receiving end until all reflections disappear (or are minimized).

The measured resistance value is equal to the characteristic impedance.

To drive and terminate a cable in practice, the terminating resistor is placed at the receiving end of the cable. From a dc consideration the terminating resistor also becomes the load resistor for the driving circuit. The driving circuit therefore does not need another built-in load resistor. An open collector driver is appropriate. A load resistor should return to V_{cc} (not to ground). From an ac viewpoint, V_{cc} is the same as ground. This means, however, that the signal swing at the receiving end is V_{cc} to ground. (See figure B–6.) For TTL logic it is desirable to swing from 3.5 V to ground.

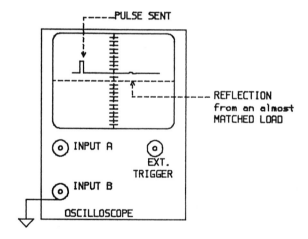

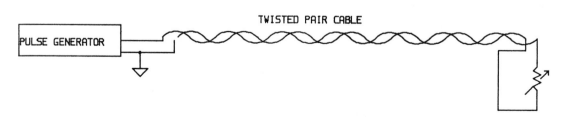

FIGURE B–6 Characteristic impedance measurement

Cabling

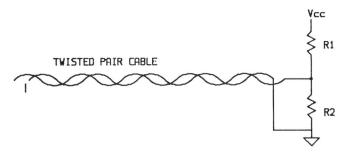

FIGURE B–7 Cable with termination

Two resistors arranged as in figure B–7 will provide the proper terminating impedance as well as the correct dc level. The two resistors are effectively in parallel. The resistance of the parallel combination is equal to the characteristic impedance, and the voltage divider level is equal to the desired TTL level:

$$R_o = \frac{R_1 * R_2}{R_1 + R_2}$$

$$V_{(TTL)} = V_{cc} \frac{R_2}{R_1 + R_2}$$

$V_{(TTL)}$ is the desired output level. Solving the two equations gives

$$R_1 = R_o * \frac{V_{cc}}{V_{(TTL)}} \quad \text{and} \quad R_2 = R_o * \frac{V_{cc}}{V_{cc} - V_{(TTL)}}$$

Terminating resistor pairs are available in dual in-line chips. One side of all R_1 resistors are tied together for the V_{cc} input. One side of all R_2 resistors are tied together for ground. This leaves 14 input pins for cable terminations as in figure B–8.

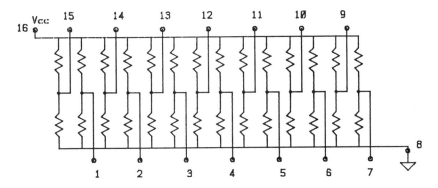

FIGURE B–8 Packaged terminating resistors

index

Access time, 265–266
  with disk, 446
  with tape, 456
Accumulator, 220, 427
Accuracy
  of binary number, 12–14
  of D/A converter, 394
A/D converters, 394–406
  counter-type, 396–397
  dual-slope integrating, 395–396
  flash converters, 400–404
  performance considerations with, 405–406
  sample and hold circuits with, 399–400
  successive approximation, 398–399
  survey of, 201–202
Adders, 191
  accumulating, 220
  fixed point one's complement, 225–227
  fixed point two's complement, 227–230
  full, 216
  half, 214
  serial, 254
Addition, 214–220
  BCD, 246–252
  complement, 221–230, 240–242, 256–258
  serial, 254–256
Addressing, 268–279
  in core memory, 297–299
  in dynamic memory, 295–297
Address latch, 430
Address register, 429
Alpha, 35, 40
Alphanumeric displays, 207
ALU, 242, 243–246, 427
Amplification. *See* Gain.

Amplifier
  differencing, 470–471
  inverting, 467–469
  noninverting, 469
  operational. *See* Operational amplifier (op-amp).
  sample and hold, 202–203, 399–400
  sense, 296
  summing, 471
Analog switches, 203–204
AND gate, 65–66
  part numbers for, 184
  use of, 70–71
Anode, 33
Arithmetic, 214–260
  accumulator for, 220
  addition, 214–220
  circuits for, 190–191
  comparator for, 258–260
  complement addition, 221–230, 240–242, 256–258
  decimal, 246–254
  division, 236–240, 242, 253–254
  floating point operations, 240–246
  for logic functions, 243–246
  multiplication, 230–236 242, 252–253
  serial, 254–258
  subtraction, 221–230
Arithmetic logic unit (ALU), 242, 243–246, 427
ASCII codes, 436, 438
Assembly language, 431
Associative laws, 75
Associative memories, 302–303
Astable multivibrator, 118–119
  555 timer as, 123–125
Asynchronous counter, 141
Atomic number, 29

Index

Atomic structure, 29–33
Auxiliary carry flag, 429
Avalanche injection, 291

Back up, 434
Base, 39
Base current, 35, 39, 40
Base eight system. *See* Octal system.
Base 16 system. *See* Hexadecimal system.
Base ten system. *See* Decimal system.
Base two system. *See* Binary system.
Baud rate, 457
Beta, 35, 40
Binary coded decimal (BCD), 15–16
    arithmetic units, 246–254
    D/A converter, 388
Binary ladder, 382
    iterative, 383–385
    with switch isolation, 385–386
Binary system, 1, 4–5
    accuracy of numbers in, 12–14
    decimal conversion to, 6–8
    fractional conversion to decimal, 10
    fractional decimal conversion to, 11–12
Bistable multivibrator, 100–101
    NAND gate, 103
    NOR gate, 101–103
    Schmitt trigger, 127–129
Bit, 5
    in D/A conversion, 380–381
    least significant, 5
    most significant, 5
Bit cell, 447
Bit plane, 270
Bit slice architecture, 246
Bloch wall, 300
Boolean algebra, 75–78
Bootstrap, 128
Bubble memory, 300–302
Bucket brigade, 304
Buffers, 185
Bus, 411
    open collector gate with, 411–414
    standard structures for, 418–421
    tristate gate with, 414–416
Bus contention, 412
Busy map, 369, 372, 373
Busy table, 370–372
Byte, 216n

Cabling, 479–485
Capacitance, 58–59
Capstan, 452
Card punch, 435
Carry, 217
Carry bypass enable function, 218–219
Carry flag, 427–429
Carry look-ahead, 191, 217–218
Cathode, 33
Cathode-ray tube (CRT) terminal, 434, 436–440
CCD memory, 303–305
Central processing unit (CPU), 427–429
Character generator, 436–440
Characteristic impedance, 480–485
Charge-coupled device (CCD) memory, 303–305
Charge deficit transfer, 304
Charge storage, 291–292
Check, 442
Chip select line, 272–273
Circuit hysteresis, 128–129, 267
Circuit simplification, 83
Clear
    in core memory, 298, 299
    J/K flip flop with, 110–113
Clock
    crystal oscillator as, 127
    555 timer as, 123–125
    R/S flip flop as, 104–105
    timing chips for, 197–198
CMOS, 170–172
    design considerations for, 173–175
    interfacing with ECL, 176
    interfacing with TTL, 175–176
    NAND gate in, 172–173
    speed and power for, 183–184
Coaxial cables, 482
Code(s). *See also* Numbering systems.
    ASCII, 436, 438
    excess 6, 246
    excess 3, 246
    FM, 447
    Gray, 16–20
    GRC, 455
    MFM, 447
    missing, 405
    NRZI, 454
    operation (op), 430–431
    PE, 455
    transmission, 457–458

weighted, 5, 16
Code conversion, 6–9
  decimal to binary, 6–8
  decimal to hexadecimal, 9
  decimal to octal, 8–9
  with fractional components, 9–15, 23–25
  general method for, 20–23
  short form for, 25–26
Coefficients, 240–241, 242
Collector, 39
Collector current, 35, 40
Column address strobe, 296
Common anode display, 424
Common cathode display, 424
Common mode rejection, 465, 466, 475
Commutative laws, 75
Comparator(s), 477–478
  as arithmetic elements, 191, 258–260
  Karnaugh map as, 92
  survey of available, 204–206
Complement adder
  fixed point one's, 225–227
  fixed point two's, 227–230
Complement addition, 221–230
  floating point operations in, 240–242
  serial, 256–258
Complement control flip flop, 256–257
Computer(s), 409–432
  accumulator of, 427
  address latch of, 430
  arithmetic logic unit (ALU) of, 427
  bus in, 411
  CPU of, 427–429
  data format and instruction set for, 430–432
  essential components of, 411
  flag register of, 427–429
  history of, 409–411
  input/output (I/O) ports for, 417–418, 430
  keyboard of, 421–422
  memory of, 430
  open collector gate in, 411–414
  registers of, 427–429
  simple program for, 432
  single-board LED display for, 423–425
  single-board microcomputer, 426–427
  standard bus structure in, 418–421
  temporary register of, 427
  tristate gate in, 414–416
Computer peripherals, 434–460
  CRT terminal, 436–440
  disk, 441–443
  disk access time, 446
  disk data recording, 446–447
  disk drive, 440
  disk error detection and correction, 448–452
  head, 444–446
  modem, 457
  printers, 458–460
  tape access time, 456
  tape data recording, 454–455
  tape data transfer and tape speed, 456
  tape error detection and correction, 456
  tape format, 453–454
  tape transport, 452–453
  transmission codes, 457–458
Conductors, 30
Control pulse generation, 357–374
  for eight clock cycles and stop, 363
  with FIFO, 368–374
  for handshake situation, 363–365
  output pulse with, 359
  for peak detection, 365–368
  with serial two's complement logic, 362–363
  single-pulse, 357–358
  two-pulse, 358–359
Conversion
  A/D. See A/D converters.
  D/A. See D/A converters.
  decimal to binary, 6–8
  decimal to hexadecimal, 9
  decimal to octal, 8–9
  flash, 201, 401–404
  flip flop, 374–376
  fractional, 9–15, 23–25
  parallel-to-serial, 418
  serial-to-parallel, 418
  TTL to ECL, 169
  voltage-to-current, 471–472
Core memories, 267, 297–299
"Correspondence-quality" printers, 460
Count-down device, 138–139, 194
Counter(s), 135–149, 153–160
  asynchronous, 141
  with chosen sequence, 336–338
  decade, 148, 333
  defined, 135

Index

down, 137–139, 194
feedback, for nonstandard moduli, 139–141, 149
Gray, 336
illegal states with, 145–148
Johnson, 144–145
presettable, 154
program, 429
ring, 141–144, 334
ripple, 135–139, 194
74192/74193, 153–160
shift, 144–145, 148, 338–340
survey of available, 194–195
synchronous, 135, 141, 194
two-mode (up-down), 334–336
up, 137–139, 326–333
Counterpotential, 36–37
Counter-type A/D converter, 396–397
Count modulus, 136
nonstandard, 139–141, 149
Count-up device, 137–139, 326–333
Couplers, 207
Covalent bonding, 31, 32
CPU, 427–429
CRT terminal, 434, 436–440
Crystal oscillator, 127, 197
Crystal structure, 31
Current
base, 35, 39, 40
diffusion, 37
emitter, 35, 39–40
inhibit, 297
input, 180, 182
output, 180, 182
specifications for, 180, 182
voltage conversion to, 471–472
Cursor, 436
Cutoff frequency, 474–475

D/A converters, 380–394
basic components of, 381–383
BCD, 388
binary ladders with switch isolation in, 385–386
bits and weights in, 380–381
dividing, 392–394
flash converters, 400–404
iterative binary ladder in, 383–385
multiplying, 389–392
performance considerations with, 405–406
specifications for, 394
survey of, 201–202
voltage division in, 386–388
Data dimension, 270
Data field
of disk, 442
of tape, 453
Data format, of computer, 430–432
Data line interconnections, 273, 275
Data recording
with disk, 446–447
with tape, 454–455
Data sheets, 54–58, 179–184
for memories, 310–322
Data transfer, 435
and tape speed, 456
Daughterboard, 420
dc offset, 465, 466
Decade, 5
Decade counter, 148, 333
Decimal adjust, 252
Decimal arithmetic, 246–254
Decimal system, 1–5, 10
conversion to binary, 6–8
conversion to hexadecimal, 9
conversion to octal, 8–9
fractional binary conversion to, 10
fractional conversion to binary, 11–12
fractional conversion to hexadecimal, 15
fractional conversion to octal, 15, 24–25
fractional hexadecimal conversion to, 14
fractional octal conversion to, 14
short form for conversion of, 25–26
Decoders, 189
instruction, 429
De Morgan's Theorems, 78–80
Destination, 431
Destructive readout, 299
D flip flop, 106–107, 193
up counter, 333
Differencing amplifier, 470–471
Differential linearity, 405–406
Differentiation, 365–366
Differentiator, 473–474
Diffusion, 36–37
Digit. *See also* Bit.
tens, 2, 4
unit, 2, 3–4

Diode, 33
 forward-biased, 38–39
 reverse-biased, 34–35, 38
 light-emitting (LED), 206, 423–425
DIP, 56
Disable time, of tristate output, 181
Disk(s), 265, 434, 441–443
 access time with, 446
 capacity of, 443
 data recording with, 446–447
 error detection and correction with, 448–452
 head for, 444–446
 parameters of, 444
 sector formatting of, 441–442
Disk drive, 440
Disk operating system (DOS), 430
Displays, 206–207, 423–425
 graphics, 435, 440
Distortion, 465, 466
Distributive law, 75–76
Divide-by-sixteen ripple counter, 137–139
Dividend, 236
Dividing D/A converter, 392–394
Division, 236–240
 decimal, 253–254
 floating point operations in, 242
Divisor, 236
D latch, 151, 193
Doping, of silicon, 36
DOS, 430
Dot matrix displays, 207
Dot matrix printing, 460
Double dabble method, 8, 11
Double-density recording, 447
Down counters, 137–139, 194
Drain, 171
Drift, 36
Drivers, 185–186
Dual inline package (DIP), 56
Dual-slope integrating A/D converter, 395–396
Dynamic memory, 295–297

ECL. *See* Emitter-coupled logic (ECL).
Edge triggering, 104–105
Electrically erasable programmable read-only memory (EEPROM), 292
Electric potential, 36
Electrons, 29–30, 31, 32
Electrostatic printing, 459

Emitter, 39
Emitter-coupled logic (ECL), 164–167
 design considerations for, 167–169
 interfacing with CMOS, 176
 TTL logic level conversion to, 169
 two families of, 164, 170
Emitter current, 35, 39–40
Enable time, of tristate output, 182
Encoders, 189–190
End around carry, 247
Erasable programmable read-only memory (EPROM), 291–295
Error detection and correction, 192–193
 disk, 448–452
 tape, 456
Excess gain, 468
Excess of holes, 170
Excess 6 code, 246
Excess 3 code, 246
Excitation tables, 324–325
Exclusive NOR (XNOR) gate, 69–70
Exclusive OR (XOR) gate, 68–69
Exponents, 240–241, 242
External memory, 434–435

Fall time, 59, 180
Fanout, 51–53, 57–58
Fault simulation program, 209–210
FDM, 458
Feedback
 negative, 468
 positive, 468–469
Feedback counters, for nonstandard moduli, 139–141, 149
File marks, 454
First-in-first-out (FIFO) memory, 200, 287
 control pulse generation with, 368–374
555 timer, 119–122, 197
 as astable multivibrator, 123–125
Fixed point numbers, 225
Fixed point one's complement adder, 225–227
Fixed point two's complement adder, 227–230
Flag register, 427–429
Flash converters, 201, 400–404
Flatpak package, 56
Flexible disks, 441
Flip flop(s), 100–129. *See also* Multivibrators.
 clocked, 104–105
 complement control, 256–257

Index

conversion of, 374–376
D, 106–107
defined, 100
excitation tables for, 324–325
hold sign, 256
J/K, 107–115
master/slave, 114–115
octal, 193
with Preset and Clear, 110–113
programming of, 323
quad, 193
race problem with, 113–114
R/S, 101–105
survey of available, 193
T, 109, 193
Floating gate, 291
Floating point operations, 225, 240–246
Floppy disks, 441
Flying head, 445, 446
FM code, 447
Forward-biased diode, 38–39
Four-bit up counter, 330–331
Fractional conversion, 9–15, 23–25
Free-running oscillator, 119
Frequency division multiplexing (FDM), 458
Frequency modulation (FM) code, 447
Frequency multiplier, 198
Frequency response, of operational amplifier, 475
Frequency shift keying (FSK), 457
Full adder, 216

Gain
  derivation of, 476–477
  excess, 468
  with inverting amplifier, 468
  with noninverting amplifier, 469
  with operational amplifier, 464, 466, 476–477
  with transistor, 39–40
Gate(s)
  AND, 65–66, 70–71
  exclusive NOR (XNOR), 69–70
  exclusive OR (XOR), 68–69
  floating, 291
  NAND, 44–50, 67, 172–173
  NOR, 68
  open collector, 411–414
  OR, 66
  part numbers for, 184
  transmission, 212
  tristate, 414–416
  universal NAND, 94
Gate arrays, 179, 210
GCR code, 455–456
Germanium, 31, 33
Glitch, 182
Graphics display, 435, 440
Gray code, 16–20
Gray counter, 336
Group code recording (GCR) code, 455–456

Half adder, 214
Handshake situation, 363–365, 435
Hard error, 448
Hard sectoring, 442
Head, 444–446
Header, 441
Head-positioning time, 446
Hexadecimal system, 1, 3, 4
  decimal conversion to, 9
  fractional conversion to decimal, 14
  fractional decimal conversion to, 15
Histogram generator, 285–287
HL register, 429
Hold sign flip flop, 256
Hold time, 181
Holes, 32
  excess of, 170
Hybrid devices, 201–204
Hysteresis, 128–129, 267

Illegal counts, 145–148
Image memories, 304
Immediate data, 430
Impact printers, 459
Impedance
  characteristic, 480–485
  input, 464–465, 466, 475
  output, 465, 466
Inert elements, 30
Information register, 299
Inhibit current, 297
Inhibit wire, 297
Inkjet printers, 459
Input circuit, TTL, 50–51
Input current, 180, 182
Input impedance, 464–465, 466, 475
Input offset voltage, 205
Input/output (I/O) devices, 434

Input/output (I/O) port, 417–418, 430
Input slew rate, 205, 475
Input voltage, 179–180, 182
Instruction, 431
Instruction decoder, 429
Instruction register, 429
Instruction set, for computer, 430–432
Integral linearity, 406
Integrating A/D converter, 395–396
Integrator, 472–473
Interfacing
   CMOS to TTL, 175
   computer, 435
   ECL to CMOS, 176
   TTL to CMOS, 175–176
Inverter circuit, TTL, 43–44
Inverting amplifier, 467–469
I/O devices, 434
I/O port, 417–418, 430
Iterative binary ladder, 383–385
Iterative networks, 480

J/K flip flop, 107–110
   conversion of R/S to, 374–375
   master/slave, 114–115
   with Preset and Clear, 110–113
   race problem with, 113–114
   survey of available, 193
Johnson counter, 144–145
Jump instructions, 431
Junction, 32
   forward-biased, 38–39
   PN, 33
   reverse-biased, 34–35, 38

Karnaugh map, 83–88
   complementing, 88–89
   five-variable, 90–91
   practical uses for, 92–94
   simplifying functions using, 85
Keyboard, 421–422

Large-scale integrated (LSI) circuit, 179
Laser printers, 459
Last-in-first-out (LIFO) memory, 288
Latch(es), 150–151
   address, 430
   D, 151, 193
   parallel-loaded, 152–153

   serial-loaded, 150–151
   survey of available, 193
Law of Perfect Induction, 76
LCDs, 206
Least significant bit (LSB), 5
LIFO memory, 288
Light-emitting diode (LED), 206
   computer display, 423–425
Linearity
   differential, 405–406
   integral, 406
Line drivers, 185–186
Line printers, 459
Liquid crystal displays (LCDs), 206
Lockout, 422
Logic arrays, 207–209
   components of, 210–212
   design procedure for, 209–210
Logic elements, 65–95
   algebraic analysis of existing circuits, 80–81
   AND circuit, 65–66
   Boolean algebra, 75–78
   circuit simplification, 83
   De Morgan's Theorems, 78–80
   exclusive NOR gate, 69–70
   exclusive OR gate, 68–69
   Karnaugh map, 83–94
   NAND gate, 67–68
   NOR gate, 68
   NOT function, 67
   OR gate, 66
   other theorems, 82
   product of sums, 81–82
   sum of products from truth table, 73–74
   universal NAND gate, 94
   usage of, 70–73
Logic functions, arithmetic, 242, 243–246
Low pass filter, 474–475
LSB, 5
LSI circuit, 179
LS series, 60–61

Machine language, 431
Macrocells, 210
Magnetic core storage, 267, 297–299
Magnetic tape. *See* Tape(s).
Majority carriers, 36
Mask-programmable memory, 291
Master/slave J/K flip flop, 114–115

Index

Medium-scale integrated (MSI) circuit, 179
Memory, 265–322
   access time to, 265–266
   addressing and data line interconnections, 268–279
   application examples, 283–289
   associative, 302–303
   available chips for, 199–200
   bubble, 300–302
   CCD, 303–305
   components of, 266–268
   core, 267, 297–299
   data sheets, 310–322
   dynamic, 295–297
   external, 434–435
   first-in-first-out (FIFO), 200, 287
   in histogram generator, 285–287
   image, 304
   last-in-first-out (LIFO), 288
   mask-programmable, 291
   nondestructive, 266, 268
   ping-pong, 283
   random access, 265–266, 430
   read-only, 289–295, 430
   sectioned, 283–284
   small, 199
   static, 199, 295
   table look-up, 288–289
   timing, 279–282
   volatile, 266, 268
MFM code, 447
Microcomputer, single-board, 426–427
Minority carriers, 36
Misalignment, of head, 446
Missing codes, 405
Modem, 435, 457
Modified frequency modulation (MFM) code, 447
Modulus ten system. *See* Decimal system.
Monitor program, 430
Monostable multivibrator, 115–118
   with output pulse width of 1.0 ms., 122–123
   retriggerable, 126–127
   74121 and 74123, 125–127
   as timer, 197–198
Most significant bit (MSB), 5
Motherboard, 420
M-Q register, 232, 233, 235
MSB, 5
MSI circuit, 179
Multiplexing, frequency division, 458

Multiplication, 230–236
   decimal, 252–253
   floating point operations in, 242
Multipliers, 191
   frequency, 198
Multiplying D/A converter, 389–392
Multivibrators, 100–129. *See also* Flip flop(s).
   astable, 118–119, 123–125
   bistable, 100–101
   crystal oscillator, 127
   555 timer, 119–125
   monostable, 115–118, 122–123, 125–127
   NAND gate bistable, 103
   Schmitt trigger, 127–129

NAND gate, 67
   in bistable multivibrator, 103
   CMOS, 172–173
   part numbers for, 184
   TTL, 44–50
   universal, 94
Negative feedback, 468
Net-list, 209
Nibble, 216n
Noise margin, 42
Nondestructive readout, 266, 268, 299
Noninverting amplifier, 469
Nonlinearity, of D/A converter, 394
Nonreturn to zero inverse (NRZI) code, 454
Nonreturn to zero (NRZ) data, 447
NOR gate, 68
   exclusive (XNOR), 69–70
   part numbers for, 184
   in R/S flip flop, 101–103
NOT gate, 67
Now-next table, 326, 350
NPN transistor, 34
NRZ data, 447
NRZI code, 454
N type material, 32–33, 170–173
Numbering systems, 1–26
   binary (base two), 1, 4–5
   binary coded decimal (BCD), 15–16
   decimal (base ten), 1–5
   decimal to binary conversion, 6–8
   decimal to hexadecimal conversion, 9
   decimal to octal conversion, 8–9
   fractional conversion of, 9–15, 23–25
   general method for conversion of, 20–23

Gray code, 16–20
hexadecimal (base 16), 1, 3, 4
octal (base eight), 1, 3, 4
short form conversion of, 25–26

Octal system, 1, 3, 4
decimal conversion to, 8–9
fractional conversion to decimal, 14
fractional decimal conversion to, 15, 24–25
Odd/even parity generators, 192
Offset, 465, 466
Offset voltage, 205
One's complement adder, 225–227
One shot. *See* Monostable multivibrator.
On the fly reading, 141
Op-amp. *See* Operational amplifier.
Op code, 430–431
Open collector buffers, 185
Open collector gate, with bus, 411–414
Open collector output circut, 53–54
Operand, 430–431
Operating system, 434
Operational amplifier (op-amp), 463–478
choice of, 475–476
common mode rejection in, 465
comparator, 477–478
dc offset in, 465
differencing, 470–471
differentiator, 473–474
distortion in, 465
gain in, 464, 476–477
input impedance in, 464–465
integrator, 472–473
inverting, 467–469
noninverting, 469
output impedance in, 465
simple low pass filter, 474–475
summing, 471
voltage follower, 470
voltage-to-current converter, 471–472
Operation code (op code), 430–431
Optoelectronic components, 206–207
OR gate, 66
exclusive (XOR), 68–69
part numbers for, 184
Oscillators, 197
crystal, 127, 197
free-running, 119
voltage-controlled, 198

Output circuit
open collector, 53–54
TTL, 51–54
Output control map, 372, 373
Output current, 180, 182
Output enable line, 273
Output impedance, 465, 466
Output table, 372
Output voltage, 180, 182

Parallel-loaded latch, 152–153
Parallel-to-serial conversion, 418
Parity bit, 448
Parity flag, 429
Parity generators, 192
Peak detection, 366–368
PE code, 455
Perfect Induction, Law of, 76
Peripherals. *See* Computer peripherals.
Phase encoding (PE) code, 455
Phase-locked loop, 198
Phase shift keying (PSK), 458
Pin assignments, 56
Ping-pong memory, 283
PN junction, 33
PNP transistor, 39
Pop instruction, 288
Positive feedback, 468–469
Postamble, 442
Power consumption, for TTL and CMOS, 183–184
Power dissipation, 35
Preamble, 441
Preset, J/K flip flop with, 110–113
Presettable counters, 154
Printers, 458–460
Priority encoder, 189–190, 401
Product of sums 81–82
Program counter, 429
Programmable read-only memory (PROM), 291
Propagation delay, 59–60
in comparators, 205
on specification sheets, 181, 183
Protons, 29
PSK, 458
P type material, 32–33, 170–173
Pull-up resistor, 53–54
Pulse duration, 181
Push and pop operation, 288

Index

QAM, 458
Q output, 101
$\bar{Q}$ output, 101
Quadrature amplitude modulation (QAM), 458
Quotient, 236

Race problem with J/K flip flop, 113–114
Random access memory (RAM), 265–266, 310–322, 430
Read cycle, 299
Read cycle time, 279
Read-modify-write mode, 282, 285, 299
Read-only memory (ROM), 289–295, 430
Read-write timing control circuit, 282
Receivers, 185–186
Refresh cycle, 295, 296–297
Regenerative circuit, 116
Register(s)
  address, 429
  flag, 427–429
  HL, 429
  information, 299
  instruction, 429
  M-Q, 232, 233, 235
  program counter, 429
  shift, 195–196, 338–340
  stack pointer, 288, 429
  successive approximation, 191–192, 342–347
  temporary, 427
Resolution, with D/A converter, 394
Restore operation, 299
Retriggerable monostable multivibrator, 126–127
Reverse-biased junction, 34–35, 38
Ring counter, 141–144, 334
Ring effect, 141
R input, 101
Ripple counter, 135–137, 194
  divide-by-sixteen, 137–139
Rise time, 58–59, 180
Rollover, 422
ROM, 289–295, 430
Rotational latency, 446
Row address strobe, 296
R/S flip flop, 101–103, 193
  clocked, 104–105
  conversion to J/K flip flop, 374–375

Sample and hold amplifiers, 202–203, 399–400

Saturation, 60
Scatter, 405
Schmitt trigger, 127–129
Schottsky (S) series, 60–61
Secondary assignment, 336
Sectioned memory, 283–284
Sector formatting, of disk, 441–442
Seek error, 446
Seek time, 446
Selectors, 189
Self-correcting pointer, 340–342
Semiconductors, 30–31
Sense amplifier, 296
Sense inhibit wire, 298, 299
Sense wire, 268, 298
Sensitivity, of comparator, 205
Sequential analysis, 353–357
Sequential logic, 323–376
  with chosen-sequence counter, 336–338
  in control pulse generation, 357–374
  with decade counter, 333
  defined, 323
  with D flip flop up counter, 333
  in flip flop conversion, 374–376
  flip flop excitation tables, 324–325
  with four-bit up counter, 330–331
  with self-correcting pointer, 340–342
  sequential analysis, 353–357
  with shift register, 338–340
  state diagrams, 350–353
  with successive approximation register, 342–347
  with synchronous counters, 326–347
  with three-bit up counter, 326–330
  with two-mode (up-down) counter, 334–336
Serial adders, 254
Serial arithmetic, 254–258
Serial-loaded latch, 150–151
Serial-to-parallel conversion, 418
Settling time
  of D/A converter, 394
  for head, 446
Setup time, 181
Seven-segment display, 92–94, 424
Shift counter, 144–145
  decade, 148
Shift register(s)
  design of, 338–340
  survey of available, 195–196
Sign flag, 429

Silicon, 31-32, 33
  doping of, 36
Simulation, 209
Single-board computer LED display, 423-425
Single-board microcomputer, 426-427
Single-error correction, 449-452
Single-error detection, 448
Sinking, 50
S input, 101
Slew rate, 205, 475
Small-scale integrated (SSI) circuit, 179
Soft error, 448
Soft sectoring, 442
Source, 171, 431
Sparkling, 405
Specification data sheets, 54-58, 179-184
  for memories, 310-322
Speed, for TTL and CMOS, 183
S series, 60-61
SSI circuit, 179
Stack, 288
Stack pointer, 288, 429
State diagrams, 350-353
Static memory, 199, 295
Strobe, 194-195
Subtraction, by complement addition, 221-230
Successive approximation, 6-8
  A/D converter, 398-399
  registers, 191-192, 342-347
Sum check, 442
Summing amplifier, 471
Sum of products, 73-74
Switch(es), analog, 203-204
Switch bounce, 197-198, 357
Switch isolation, binary ladders with, 385-386
Sync bit, 441
Synchronous counter, 135, 141, 194
Synchronous sequential logic, 323-347
  with chosen-sequence counter, 336-338
  with decade counter, 333
  with D flip flop up counter, 333
  and flip flop excitation tables, 324-325
  with four-bit up counter, 330-331
  with self-correcting pointer, 340-342
  with shift register, 338-340
  with successive approximation register, 342-347
  with three-bit up counter, 326-330
  with two-mode (up-down) counter, 334-336
Synchronous switching, in flip flops, 104

Table look-up memory, 288-289
Tape(s), 265, 434, 452-453
  data recording with, 454-455
  data transfer and speed of, 456
  error detection and correction with, 456
  format of, 453-454
Teletypewriter (TTY), 435, 458-459
Temporary register, 427
Tens digit, 2, 4
Test program, 210
T flip flop, 109, 193
Thermal printing, 459
Three-bit up counter, 326-330
Threshold level, 180
Timing, memory, 279-282
Timing chips, survey of available, 197-198
Timing conditions, specifications for, 180-181, 182-183
Totem pole output, 49
Transceivers, 186-188
Transfer characteristic tables, 289
Transistor(s), 29-41
  action of, 38-41
  NPN, 34
  PNP, 39
  two-junction device, 33-38
Transition region, 36
Transition table, 326, 350
Transmission code, 457-458
Transmission gate, 212
Transmission lines, 479-485
Tristate gate, with bus, 414-416
Tristate output
  disable time of, 181
  enable time of, 182
Truth table, 66. *See also* Karnaugh map.
  product of sums from, 81-82
  sum of products from, 73-74
TTL, 41-42
  fall time in, 59
  fanout in, 51-53
  input circuit in, 50-51
  interfacing with CMOS, 175-176
  interpreting specification data sheets for, 54-58
  inverter circuit in, 43-44
  logic level conversion to ECL, 169
  NAND gate in, 44-50
  open collector output circuits in, 53-54
  other series in, 60-61

Index

output specifications in, 51–53
propagation delay in, 59–60
rise time in, 58–59
speed and power for, 183–184
voltage specifications for, 41–42
TTY, 435, 458–459
Twisted pair cables, 482
Two-junction device, 33–38
Two-mode counter, 334–336
Two's complement adder, 227–230

UARL, 418
Unit digit, 2, 3–4
Unit load (U.L.), 42
Universal asynchronous receiver transmitter (UART), 418
Universal NAND gate, 94
Universal synchronous/asynchronous receiver transmitter (USART), 418
Up counters, 137–139
  D flip flop, 333
  four-bit, 330–331
  three-bit, 326–330
Up-down counter, 334–336
USART, 418

Valance shell, 30, 31, 32
VCO, 198

Verify program, 209
Very large scale integrated (VLSI) circuit, 179
Virtual ground, 467
Volatile memory, 266, 268
Voltage
  input, 179–180, 182
  input offset, 205
  output, 180, 182
  specifications for, 179–180, 182
Voltage-controlled oscillator (VCO), 198
Voltage division, 386–388
Voltage follower, 470
Voltage gain. *See* Gain.
Voltage reflection, 482
Voltage-to-current converter, 471–472

Weight(s), in D/A conversion, 380
Weighted codes, 5, 16
Word, 217
Write cycle, 298, 299
Write cycle time, 281
Write protection, 442
Write release time, 282

XNOR gate, 69–70
XOR gate, 68–69

Zero flag, 429